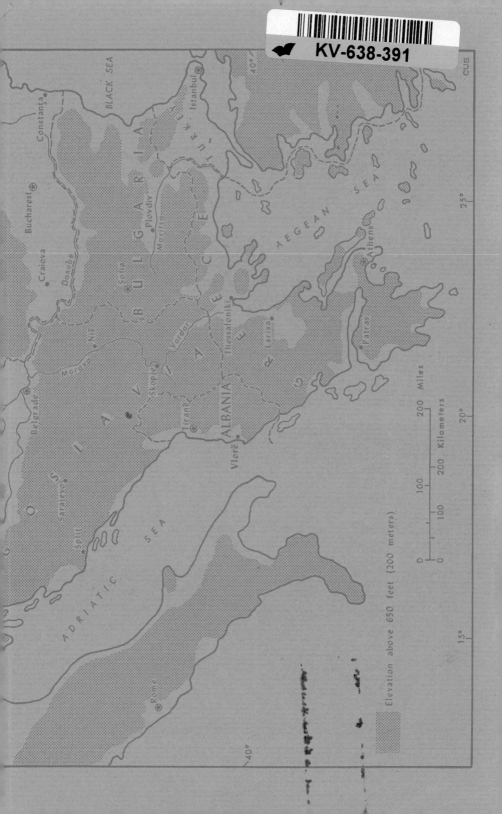

Eastern Europe: Essays in
Geographical Problems

Eastern Europe: Essays in Geographical Problems

EDITED BY GEORGE W. HOFFMAN

METHUEN & CO LTD
11 NEW FETTER LANE EC4

First published in Great Britain in 1971
by Methuen and Co Ltd, 11 New Fetter Lane,
London, EC4
© *1971 Methuen and Co Ltd*
Printed in Great Britain by
Butler & Tanner Ltd, Frome and London

SBN 416 15990 7

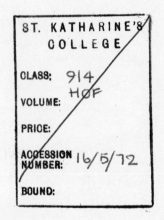

Contents

Preface	vii
	xiii
Contributors	
Biographical Sketches of Commentators	xvii
Participants in the Conference	xxi
List of Tables	xxv
List of Maps	xxvii

1 GEORGE W. HOFFMAN   1
Regional Synthesis: An Introduction

2 NORMAN J. G. POUNDS   45
The Urbanization of East-Central and Southeast Europe: An Historical Perspective
Comments by Ian M. Matley

3 DEAN S. RUGG   83
Aspects of Change in the Landscape of East-Central and Southeast Europe
Comments by Leszek A. Kosiński
Remarks

4 JACEK I. ROMANOWSKI   127
Geographic Research and Methodology on East-Central and Southeast European Agriculture
Comments by Barbara Zakrzewska
         by Paul B. Alexander
Remarks

5 F. E. IAN HAMILTON   173
The Location of Industry in East-Central and Southeast Europe
Comments by S. Earl Brown
         by Karel J. Kansky
Remarks

6 THOMAS M. POULSEN 225
Administration and Regional Structure in East-Central and Southeast Europe
Comments by Joseph Velikonja
by Leszek A. Kosiński
Remarks

7 FRED E. DOHRS 271
Nature versus Ideology in Hungarian Agriculture: Problems of Intensification
Comments by Jerzy F. Karcz
Remarks

8 JACK C. FISHER 301
The Emergence of Regional Spatial Planning in Yugoslavia: The Slovenian Experience
Comments by Ivan Crkvenčić and Joseph Velikonja
Remarks

9 ORME WILSON, JR. 365
The Belgrade–Bar Railroad: An Essay in Economic and Political Geography
Comments by Guido G. Weigend
Remarks

10 HUEY LOUIS KOSTANICK 395
Significant Demographic Trends in Yugoslavia, Greece and Bulgaria
Comments by Leszek A. Kosiński
Remarks

11 GEORGE W. HOFFMAN 431
Regional Development Processes in Southeast Europe: A Comparative Analysis of Bulgaria and Greece
Comments by Roger E. Kasperson
Remarks

Statistical Summary Tables 491
Index 497

Preface

The papers published in this volume, with the exception of the introduction, were presented at a conference at the University of Texas at Austin, April 18–20, 1969, under the title of East-Central and Southeast European Geography.* The purpose of the conference and the gathering of such a distinguished group of 40 scholars and teachers† interested in the various geographic problems of the area was twofold. One, to show the wide range of problems in the area of possible interest to geographers and the variety of research techniques being used, as well as to survey the state of geographic research, its future direction and possibilities; and two, the assessment of research opportunities, educational exchanges and graduate as well as undergraduate training. The participation by area specialists from other social sciences was of special value in the interdisciplinary interpretation of the research studies presented.

The countries included in the area of East-Central and Southeast Europe cover 542,000 sq. miles (roughly twice the size of Texas) with a population of over 136 million and with the exception of Greece are characterized by their socialist type of government. It is this type of government which brings certain common characteristics to the area and is the main justification for grouping the eight countries of the area, The German Democratic Republic (East Germany), Poland, Czechoslovakia, Hungary, Romania, Bulgaria, Yugoslavia and Albania under one umbrella. Greece's inclusion in the region is based on intimate historical ties and cultural contributions

---

* Throughout this volume the term Eastern Europe is used interchangeably with East-Central and Southeast Europe, excluding the Soviet Union, but including Greece and The German Democratic Republic (East Germany).

† The participants came from 27 American Universities, one from Canada, from four European countries, including three from the countries of Eastern Europe, from the United States Department of State and the Foundation of Education and World Affairs. One of the East European participants, Dr Muhamed Hadžić was an Eisenhower Fellow and professor of economics of the University of Sarajevo. He recently was appointed Federal Secretary for Foreign Trade of the Yugoslav Government.

to the countries of Southeast Europe and often way beyond. These countries are a unique laboratory for examining a variety of geographical processes. By studying the various cultures and divergent ideas which come into conflict geographers, just as other social scientists, are offered a unique opportunity to view historical, economic and political processes, often in multinational and multiethnical environments, thus contributing to our knowledge of empirical comparisons of countries often with different ideologies, and testing methodological principle.

The topics selected for formal papers purposely included a wide range of geographic research, focusing on the work and interests of American geographers (the exception was Professor F. E. Ian Hamilton of London). Limitations as to scope and subject were obvious. The authors were asked to present interpretive and provocative papers for the purpose of stimulating discussion and further research. For this reason, each paper was distributed before the conference convened to allow all participants to read the papers and thus spend most of the time during the formal sessions on discussion, with only a brief time allotted to each author for the purpose of summarizing his findings and a formal rebuttal. Each paper was discussed by one or two designated commentators, and these comments appear in the volume following each research paper. With the exception of one evening paper, an open discussion followed the presentation of formal comments. These general discussions also are briefly summarized after each paper. Authors subsequently received transcripts of the comments, a summary of the general discussions, and each participant was free to modify his paper after reviewing comments. Authors also supplied map sketches which were redrawn for this volume.

In addition to the ten papers presented at the conference, together with Comments and Remarks which appear in this volume, the editor was asked to write a broad introduction, a regional synthesis, to serve as background for those readers not intimately acquainted with the region. It is envisaged that this purposely very general introduction to the area may give greater utility to the volume by contributing to a better understanding of the problems and processes presented in the individual contributions and thus serve a somewhat larger reading public. No effort was made to restrict individual contributions to adhere to similar length, though all authors received only 20 minutes for their oral presentation. Obviously, the choice of the topic dictated the size of the contribution.

In spite of the importance of events in the countries of East-Central and Southeast Europe and the unique laboratory this area offers for a great number of geographic research problems, the area, according to a recent survey‡ has been given insufficient attention by geographers. The reason for such a long-term neglect lies in the peculiarities of the area itself, especially the tremendous language problem and the fact that previously a number of senior geographers interested in the Slavic realm had limited themselves to Soviet studies. It is a fact that there were few rewards resulting from scholarly attainment in the region. There also is the difficulty encountered in offering academic course work in the area which was a limiting factor in stimulating interest by American geographers, though in all fairness it must be said that important improvements in this situation are being noticed.

As a result of the findings of the Survey in Geography, the editor of this volume, after consultation with many colleagues, took the necessary initiative to organize the Austin conference, hoping in this way to contribute to increased interest and scholarly work in the area. The Subcommittee on East-Central and Southeast Europe of the Joint Committee on Slavic Studies (JCSEES) voted a basic grant in support of a conference for geographers with the hope of increasing the geographer's interest in the problem of the region. The conference was sponsored by the Department of Geography of the University of Texas at Austin. The Conference itself was made possible through the earlier mentioned financial support of the Sub-

‡ George W. Hoffman, 'Geography', in *Language and Area Studies – East-Central and Southeastern Europe. A Survey*, edited by Charles Jelavich (Chicago and London: The University of Chicago Press, 1969), pp. 199–224. This survey in geography, one of 18 surveys published, is based on a careful perusal of the literature in all western languages, a questionnaire which was mailed to every known geographer in the United States teaching and/or researching in the area (90% of those questioned returned the questionnaire), and the consultation with colleagues active in the field. The purpose of this detailed survey according to Professor Jelavich 'was to indicate strength and weaknesses, the accomplishments and the failures in American training and scholarship, and to indicate the direction in which the participants in this project would like to see research and training in this field move in the future.'

committee on East-Central and Southeast Europe of the JCSEES of the American Council of Learned Societies and the Social Science Research Council, the International Research and Exchange Board (mainly in providing funds which enabled one geographer each from Bratislava and Zagreb to attend the meetings and thus establish contacts with a large group of American colleagues), the Austin Committee on Foreign Relations and the University of Texas at Austin, the latter providing the major share of the funds, including the expense involved in drafting the maps appearing in this volume.

A brief word should be said about the standardization of place names, transliteration and statistical data presented. The spellings of place names generally followed the practice of the United States Board of Geographic Names and if necessary the area maps published by the National Geographic Society (diacritical marks were omitted in the Hungarian place names). Basically the names used locally with the exception of those names for which a well established Western spelling exists were followed. For example, Warsaw was used for Warszawa, Prague for Praha, Belgrade for Beograd and Danube for the various locally used names. This practice has resulted in the use of different names for certain geographical features, especially rivers when crossing international boundaries, i.e., Odra and Oder, Labe and Elbe, Vardar and Axios, Maritsa, Meriç and Evros, Maros and Mureş, Hornad and Hernad respectively. Obviously, transliteration of the Slavic language presented problems in various chapters, in which case the Library of Congress system with minor modifications has been followed. Statistical data are now generally obtainable for the countries of the area; although those for subsystems and comparative data cause problems for the researcher. National yearbooks are now obtainable for 1968 and 1969, and most of them contain data for as late as 1967 or 1968. For some data, only the last censuses are available, some of which go back to 1961.

As editor of this volume and responsible for the convening of the Austin Conference I must first of all extend my deep appreciation to my colleagues in the field of Eastern Europe who have given so generously of their time to this project. Their contributions in the preparatory discussions leading to the basic idea and organization of the conference, in their participation and their contribution in solving the many details associated with the final publication of this volume, have been invaluable. Special appreciation is also extended to the conference participants from economics, history, political science and sociology and the Department of State, Drs Jerzy F. Karcz, Stephen Fisher-Galati, R. John Rath, Ladis Kristof and Irwin T. Sanders, and Mr Guy E. Coriden. The willingness of my University colleagues to serve as session chairmen is also appreciated, Professors Donald D. Brand, Charles E. Butler, Robert C. Mayfield and F. J. Simoons. Appreciation is extended to Dean W. Gordon Whaley of the Graduate School for his challenging opening remarks. A few other individuals must also be mentioned, especially those without whose encouragement and active participation the conference and this volume would never have materialized. Particular appreciation must be accorded Dr Gordon B. Turner, Vice-President, American Council of Learned Societies for his encouragement and assistance, to Professor Charles Jelavich of Indiana University for his leadership and untiring effort on behalf of creating a greater area awareness and increased standards of scholarly performance, to Professor Chauncy D. Harris of the University of Chicago and past chairman of the JCSEES for his encouragement and contributions to the conference and his long-time efforts on behalf of international scientific co-operation, especially among geographers.

Colleagues in the Department of Geography were most helpful in all facets of the Conference. To the graduate students of the Department of Geography in residence in the spring term of 1969 who assisted in handling both the mass of preparatory details called for in such a conference and who assisted in the smooth operation of the various sessions, including the work of many as recording secretaries during the discussions, the editor is especially indebted. Special mention must here be made of the contributions of Claude M. Davidson (now Assistant Professor of Geography at Texas Tech) in charge of the mechanics of the conference sessions, my research assistant Jean T. Hannaford (now Assistant Professor of Geography at Trinity University, San Antonio) in charge of papers, and Major Robert S. Shohan (now Lecturer in Geography at West Point) in charge of arrangements and transportation. Only someone who has organized a conference extending over several days is aware of the many details called for and the need for a dedicated group of assistants. Without the contribution of the graduate students of the Department and their work far beyond the call of duty, this conference never could have been run so smoothly. The editor and all participants give recognition to this dedicated group of students. The tedious job of checking the spellings of names and preparing and checking numerous statistical data, as well as the indexing of the book, fell to my research assistant Jean T. Hannaford, who also had the assistance of Michael Helfert, though final responsibility for all errors and omissions in this volume obviously rests with the editor. The drafting of all maps for this volume was in the experienced hands of Jean T. Hannaford, assisted by Christine Studer, but special appreciation is also extended to my colleague, Professor Robert K. Holz, whose advice and availability were of importance in the final preparation of the maps. As usual, my deepest thanks go to my wife, Viola Hoffman, for her varied assistance which among other things included the typing and editing of my own contributions, for her reading of the manuscript and proof-reading of the entire volume and the writing of the 'Remarks' based on the notes provided by the recording secretaries.

Whatever long-term merits this volume containing the proceedings of the Conference on East-Central and Southeast European Geography may have, and whatever future benefits the profession may derive from this experience, I believe I express the hope of all those who encouraged and actively participated in the Conference that this first conference of geographers interested in Eastern Europe will be followed by others, perhaps concentrating on individual research topics. The increased attention by geographers in recent years to the problems of the region should hopefully encourage greater individual research, additional course offerings in American colleges and universities and increased co-operative work between geographers from North America and Western and Eastern Europe. To the fulfilment of these expectations this volume is dedicated.

Austin, Texas GEORGE W. HOFFMAN
April, 1970

Contributors

FRED E. DOHRS
Author of the chapter 'Nature versus Ideology in Hungarian Agriculture, Problems of Intensification', is Professor of Geography, Wayne State University. He has written 'Incentives in Communist Agriculture: The Hungarian Models', *Slavic Review*, Vol. 27 (March, 1968), pp. 23–38 and contributed some comment on 'Soviet Agricultural Policy', in *Soviet Agriculture: The Permanent Crisis*, edited by Roy D. Laird and Edward L. Crowley (New York: Frederick A. Praeger, 1965), pp. 67–70.

JACK C. FISHER
Author of the chapter 'The Emergence of Regional Spatial Planning in Yugoslavia: The Slovenian Experience', is Associate Director of the Center for Urban Studies at Wayne State University and Director of the American-Yugoslav Project for Regional and Urban Planning Studies, Ljubljana, Yugoslavia. He has published *Yugoslavia – A Multinational State. Regional Difference and Administrative Response* (San Francisco: Chandler Publishing Company, 1966), was editor of *City and Regional Planning in Poland* (Ithaca, N.Y.: Cornell University Press, 1966) and has written 'The Yugoslav Commune', *World Politics*, Vol. 16 (April, 1964), pp. 418–41 and 'Planning the City of Socialist Man', *Journal of the American Institute of Planners*, Vol. 28 (November, 1962), pp. 251–65.

F. E. IAN HAMILTON
Author of the chapter 'The Location of Industry in East-Central and Southeast Europe', is Lecturer in Social Studies, University of London, England, a position held jointly in the Department of Geography, London School of Economics and Political Science, and in the Social Science Division, School of Slavonic and East European Studies. During 1968–70 he was Visiting Associate Professor, Department of Geography,

Northwestern University and Portland State University. He has published *Yugoslavia: Patterns of Economic Activity* (London and New York: G. Bell & Sons, Ltd, and F. A. Praeger, Inc., 1968) and written 'Models of Industrial Location', in *Models in Geography*, edited by R. J. Chorley and P. Haggett (London: Methuen & Co. Ltd, 1967), pp. 361–424, and 'Some Aspects of Spatial Behavior in Planned Economies', *Papers, Regional Science Association*, Vol. XXIV (1970).

GEORGE W. HOFFMAN

Author of the chapters 'Regional Synthesis: An Introduction', and 'Regional Development Processes in Southeast Europe: A Comparative Analysis of Bulgaria and Greece', is Professor of Geography, The University of Texas at Austin. He has published *Balkans in Transition*, Searchlight book No. 20 (Princeton, N.J.: D. Van Nostrand Co., 1963), is co-author (with F. W. Neal) of *Yugoslavia and the New Communism* (New York: The Twentieth Century Fund, 1962), editor and contributor of *A Geography of Europe* (New York: The Ronald Press Co., 3rd edition, 1969) and has written 'The Problem of the Underdeveloped Regions in Southeast Europe: A Comparative Analysis of Romania, Yugoslavia and Greece', *Annals of the Assoc. of Am. Geographers*, Vol. 57 (December, 1967), pp. 637–66 and 'Thessaloniki: The Impact of a Changing Hinterland', *East European Quarterly*, Vol. 2 (March, 1968), pp. 1–27.

HUEY LOUIS KOSTANICK

Author of the chapter 'Significant Demographic Trends in Yugoslavia, Greece and Bulgaria', is Professor of Geography, The University of California at Los Angeles. He has written 'The Resettlement of Bulgarian Turks in Turkey, 1950–1953', monograph, *University of California, Publications in Geography*, Vol. 8 (1957), pp. 65–164 and contributed 'Geopolitics of the Balkans', in *The Balkans in Transition*, edited by Charles and Barbara Jelavich (Berkeley and Los Angeles: University of California Press, 1963), pp. 1–51 and written 'Eastern Europe–Retrospect and Prospect', Presidential Address, *Yearbook of the Association of Pacific Coast Geographers*, Vol. 19 (1957), pp. 5–12.

THOMAS M. POULSEN

Author of the chapter 'Administration and Regional Structure in East-Central and Southeast Europe', is Associate Professor of Geography, Portland State University. He has published *Planning for Transportation in the Portland Metropolitan Area* (Portland: City Club, 1968).

NORMAN J. G. POUNDS
Author of the chapter 'The Urbanization of East-Central and Southeast Europe: An Historical Perspective', is University Professor of Geography and History, Indiana University. He has published *The Ruhr* (Bloomington: Indiana University Press, and London: Faber and Faber, 1952), *The Upper Silesian Industrial Region* (Bloomington: Indiana University Press, 1958), *Political Geography* (New York: McGraw Hill Book Co., 1963), *Eastern Europe* (London: Longmans Green and Co., and Chicago, The Aldine Press, 1969), *Poland Between East and West*, Searchlight book No. 22 (Princeton, N.J.: D. Van Nostrand Co., 1964) and has written 'The Spread of Mining in the Coal Basin of Upper Silesia and Northern Moravia', *Annals of the Assoc. of Am. Geographers*, Vol. 48 (1958), pp. 149–63.

JACEK I. ROMANOWSKI
Author of the chapter 'Geographic Research and Methodology on East-Central and Southeast Europe', is Assistant Professor of Geography, University of Washington. He has written 'The State Farm as a Vehicle of Agricultural Modernization in Poland'. Paper presented at the Far Western Slavic Conference, May 1–2, 1970. Mimeographed, 25p.

DEAN S. RUGG
Author of the chapter 'Aspects of Change in the Landscape of East-Central and Southeast Europe', is Professor of Geography, University of Nebraska, and served between 1961 and 1967 as Geographic Attaché assigned to the U.S. Embassy in Bonn, Germany. He has written 'Selected Effects of Planning on Urban Development in the Federal Republic of Germany', *Economic Geography*, Vol. 42 (October, 1966), pp. 326–35 and 'Die Nachkriegsentwicklung der Kartographie in der Bundesrepublik Deutschland', *Allgemeine Vermessungs-Nachrichten*, Vol. 74 (June, 1967), pp. 221–34, also published in English translation.

ORME WILSON, JR
Author of the chapter 'The Belgrade–Bar Railroad. An Essay

in Economic and Political Geography', was Desk Officer, Yugoslavia, Office of East European Affairs, U.S. Department of State and from the summer of 1970 is U.S. Consul General in Zagreb, Yugoslavia. He has been with the Department of State since 1950 with assignments in Frankfurt a/M, Southampton, Athens and the Air Force War College.

Biographical Sketches of Commentators

PAUL B. ALEXANDER. Associate Professor of Geography, University of Montana. Author of *Land Utilization in the Karst Region of Zgornja Pivka, Slovenia* (New York: Studia Slovenica, 1967); 'The Evolution of Karst Research in Slovenia, Yugoslavia', *The Rocky Mountain Social Science Journal*, Vol. V (1968), pp. 49–54; 'The Reka-Timavo River System of the Yugoslavian and Italian Karst', *Yearbook of the Association of Pacific Coast Geographers* (to be published during 1970).

S. EARL BROWN, Professor of Geography, The Ohio State University. Author of 'Grouping Tendencies in an Economic Regionalization of Poland', together with C. E. Trott, *Annals of the Assoc. of Am. Geographers*, Vol. 58 (1968), pp. 327–42.

IVAN CRKVENČIĆ, Professor of Geography, University of Zagreb, Yugoslavia. Author of 'Die Folgen der Urbanisierung in Jugoslawien am Beispiel der sozial-ökonomischen Struktur der Pendler und des Stadtrandes von Zagreb', *Münchner Studien zur Sozial- und Wirtschaftsgeographie*, Vol. 4 (1968), pp. 57–65; 'Conceptions modernes de l'études géographique des installations humaines', *Cvijićev zbornik* (Belgrade), 1968, pp. 345–55; 'Examples of Changing Peasant Agriculture in Croatia, Yugoslavia', *Economic Geography*, Vol. 33 (1957), pp. 50–71.

KAREL J. KANSKY, Associate Professor of Geography, University of Pittsburgh. Author of *Structure of Transportation Networks: Relationships between Network Geometry and Regional Characteristics*. Department of Geography Research Paper No. 84 (Chicago: The University of Chicago, 1963); 'Travel Patterns of Urban Residents', *Transportation Science*, Vol. 1 (1967), pp. 261–85.

JERZY F. KARCZ, Professor of Economics, University of California, Santa Barbara. Author of *Soviet Agriculture Marketing and Prices, 1928–1954* (Santa Monica, Cal.: Rand Corp., 1955); editor, *Soviet and East European Agriculture* (Berkeley, Cal.: University of California Press, 1965), 'Seven Years on the Farms: Retrospect and Prospect', *New Directions in the Soviet Economy*. U.S. 89th Congress, Second Session, Joint Commission on Foreign Economic Policy, Pt. II-B (Washington, D.C.: Government Printing Office, 1966), pp. 383–450.

ROGER E. KASPERSON, Associate Professor of Government and Geography, Clark University. Author of *The Dodecanese: Diversity and Unity in Island Politics*. Department of Geography Research Paper No. 108 (Chicago: The University of Chicago, 1968); with Julian V. Minghi, eds. *The Structure of Political Geography* (Chicago: Aldine Publishing Co., 1969).

LESZEK A. KOSIŃSKI, Associate Professor of Geography, University of Alberta, Canada. Author of 'Changes in the Ethnic Structure in East-Central Europen 1930–1960', *Geographical Review*, Vol. 59 (1969), pp. 388–402; 'Population Growth in East-Central Europe in the years 1961–65', *Geographia Polonica*, Vol. 14 (1968), pp. 297–304; *The Population of Europe. A Geographical Perspective* (London: Longmans, to be published in 1970).

IAN MURRAY MATLEY, Professor of Geography, Michigan State University. Author of 'The Marxist Approach to the Geographical Environment', *Annals of the Assoc. of Am. Geographers*, Vol. 56 (1966), pp. 39–54; 'Transhumance in Bosnia and Herzegovina', *Geographical Review*, Vol. 58 (1968), pp. 231–61; with Arthur E. Adams and William O. McCagg, *An Atlas of Russian and East European History* (New York: Frederick A. Praeger, 1967); *Romania*. European Profiles (New York: Frederick A. Praeger, to be published in 1970).

GUIDO G. WEIGEND, Professor of Geography, Rutgers University. Author of chapter 5: 'Western Europe', in *A Geography of Europe*, edited by George W. Hoffman (New York: The Ronald Press Co., 3rd edition, 1969), pp. 197–271; 'Some Elements in the Study of Port Geography', *Geographical Review*, Vol. 48 (1958), pp. 185–200; with James H. Street, *Urban Planning and Development Centers in Latin America* (New Brunswick, N.J.: Rutgers University, 1967), 97 p.

JOSEPH VELIKONJA, Associate Professor of Geography, University of Washington. Author of 'Territorial Identification and Functional Relations in Yugoslavia', Abstract, *Annals of the Assoc. of Am. Geographers*, Vol. 57 (1967), p. 193, 'The Socialist City in Yugoslavia', Paper presented at the Far Western Slavic Conference, May 1-2, 1970. Mimeographed, and a number of publications on the Italian migration to the United States and Canada.

BARBARA ZAKRZEWSKA, Professor of Geography, University of Wisconsin–Milwaukee. Author of 'The Changing Face of Small Towns in Poland', First Congress of Scholars and Scientists convened by the Polish Institute of Arts and Scientists in America, *Abstract of Papers*. Columbia University, 1966, pp. 68-9; 'Trends and Methods in Land Forms Geography', *Annals of the Assoc. of Am. Geographers*, Vol. 57 (1967), pp. 128-65.

List of Participants at the Conference on East-Central and Southeast European Geography
Austin, Texas April 18–20, 1969

Dr Paul B. Alexander, *Associate Professor of Geography, University of Montana*

Dr John L. Baxevanis, *Professor of Geography, East Stroudsburg State College*

Dr S. Earl Brown, *Professor of Geography, Ohio State University*

Mr Guy E. Coriden, *Director, European Office, Bureau of Educational and Cultural Affairs, U.S. Department of State*

Dr Ivan Crkvenčić, *Professor of Geography, University of Zagreb, Yugoslavia*

Dr Fred E. Dohrs, *Professor of Geography, Wayne State University*

Dr Edwin H. Draine, *Associate Professor of Geography, University of Illinois, Chicago Circle*

Dr Stephen Fisher-Galati, *Professor of History, Editor, East European Quarterly, University of Colorado*

Dr Jack C. Fisher, *Associate Director, Center for Urban Studies, Wayne State University*

Dr Muhamed Hadzie, *Member, Executive Council of the Socialist Republic Bosnia and Herzegovina, Professor of Political Economics, University of Sarajevo, Eisenhower Fellow. Since 1970 Federal Secretary of Foreign Trade, Yugoslavia*

Dr F. E. Ian Hamilton, *Lecturer in Geography, The London School of Economics and Political Science, Visiting Associate Professor, Portland State University and Northwestern University (1968–70)*

Dr Chauncy D. Harris, *Samuel N. Harper Professor of Geography, University of Chicago*

Dr Jordan A. Hodgkins, *Professor of Geography, Kent State University*

Dr George W. Hoffman, *Professor of Geography, The University of Texas at Austin*

Dr Koloman Ivanička, *Professor of Geography, University Komenského, Bratislava, Czechoslovakia*

Dr Karel J. Kansky, *Associate Professor of Geography, University of Pittsburgh*

Dr Jerzy F. Karcz, *Professor of Economics, University of California, Santa Barbara*

Dr Adolf Karger, *Associate Professor of Geography, University Tübingen, Germany, Visiting Associate Professor, Western Michigan University (1968–69)*

Dr Roger E. Kasperson, *Assistant Professor of Government and Geography, Clark University*

Dr Leszek A. Kosiński, *Associate Professor of Geography, The University of Alberta, Edmonton, Canada*

Dr Huey Louis Kostanick, *Professor of Geography, University of California, Los Angeles*

Dr Ladis Kristof, *Research Associate, Department of Political Science, Stanford University*

Dr David E. Kromm, *Assistant Professor of Geography, Kansas State University*

Dr Ian M. Matley, *Professor of Geography, Michigan State University*

Dr Theodore C. Myers, *Assistant Professor of Geography, University of Colorado*

Dr Ralph E. Olson, *Professor of Geography, University of Oklahoma*

Dr G. Etzel Pearcy, *The Geographer, U.S. Department of State; 1969–70 on California State College at Los Angeles*

Dr Thomas M. Poulsen, *Associate Professor of Geography, Portland State University*

Dr Norman J. G. Pounds, *University Professor of History–Geography, Indiana University*

Dr R. John Rath, *Professor of History, Rice University*

Dr Jacek I. Romanowski, *Assistant Professor of Geography, University of Washington*

Dr Dean S. Rugg, *Professor of Geography, University of Nebraska*

Dr Irwin T. Sanders, *Vice-President, Education and World Affairs, New York;* 1969–70 *on Professor of Sociology, Boston University*

Dr Kenneth Thompson, *Professor of Geography, University of California, Davis*

Dr Joseph Velikonja, *Associate Professor of Geography, University of Washington*

Dr Jack R. Villmow, *Professor of Geography, Northern Illinois University*

Dr Guido G. Weigend, *Professor of Geography, Rutgers University*

Mr Orme Wilson, Jr., *Country Director, Yugoslavia, Office of East European Affairs, U.S. Department of State; Summer* 1970 *on Consul General, Zagreb, Yugoslavia*

Dr Barbara Zakrzewska, *Associate Professor of Geography, The University of Wisconsin, Milwaukee*

Tables

3.1	Romania: The Variation of Selected Indicators of Regional Development	112
5.1	East-Central and Southeast Europe: Employment in Industry 1938–1939; 1950; 1968	176
5.2	East-Central and Southeast Europe: Indices of Increased Production and Employment, 1950–1967	178
6.1	East-Central and Southeast Europe: Summary of Changes in Administrative Regions, 1945–1951	230
6.2	East-Central and Southeast Europe: Characteristics of Provinces and Districts, 1948–1950	233
6.3	East-Central and Southeast Europe: Summary of Post-1951 Changes in Administrative Regions	244
6.4	East-Central and Southeast Europe: Characteristics of Provinces and Districts, 1959–1960	246
7.1	Role of the Hungarian Private Plots	284
8.1	Yugoslavia: Tourism: The Example of Slovenia	315
8.2	Yugoslavia: Regional Distribution of Bank Credits for Tourism	316
8.3	Yugoslavia: Matrix of Credit Flows of Investment in Tourism, 1967	317
8.4	Yugoslavia: Bank Credits Received (millions of new dinars)	318
8.5	Yugoslavia: Summary of Developments in Four Functional Areas	325
10.1	Europe: Annual Rate of Increase, Birth and Death Rates	397
10.2	Yugoslavia: Vital Statistics (per 1000 inhabitants)	401
10.3	Yugoslavia: Population by Republics (present-day territory)	403
10.4	Yugoslavia: Vital Statistics by Republics	405
10.5	Yugoslavia: Population in Cities subdivided into Communes	408

xxvi *List of Tables*

10.6	Greece: Vital Statistics (per 1,000 inhabitants)	411
10.7	Greece: Urban, Semi-Urban and Rural Population	413
10.8	Greece: Per Cent of Population Change, 1940–1951, 1951–1961	414
10.9	Bulgaria: Vital Statistics (per 1,000 inhabitants)	418
10.10	Bulgaria: Population, Male-Female Ratio and Urbanization	419
10.11	Bulgaria: Population by Administrative Divisions	420
10.12	Bulgaria: Population of Cities, by Censuses	421
11.1	Bulgaria and Greece: Comparative Economic Growth, Pre-war, 1950's, mid-1960	436
11.2	Bulgaria and Greece: Percentage Distribution of Gross Fixed Investments	445
11.3	Greece: Regional Differences	467

STATISTICAL SUMMARY TABLES

1	East-Central and Southeast Europe: Population Distribution, Pre-World War II and Post-war	492
2	East-Central and Southeast Europe: Cities of over 100,000	493
3	East-Central and Southeast Europe: Proportion of Active Population by Economic Sector (%), 1960's	495
4	East-Central and Southeast Europe: Land Use	496

Maps

1.1	East-Central and Southeast Europe, physiographic divisions	6–7
1.2	Agricultural Regions of East-Central and Southeast Europe	34–5
1.3	The German Democratic Republic	36
1.4	Poland	37
1.5	Czechoslovakia	38
1.6	Hungary	39
1.7	Romania	40
1.8	Bulgaria	41
1.9	Yugoslavia	42
1.10	Albania	43
1.11	Greece	44
2.1	Roman roads and urban-style settlements in the Balkan Peninsula	49
2.2	Hrady in the Czech lands	55
2.3	Cities in East-Central and Southeast Europe during the later Middle Ages	62
2.4	Urban development in East-Central and Southeast Europe, about 1965	67
3.1	Village area of Csepreg, Hungary	86
3.2	Collective farm in the German Democratic Republic	88
3.3	Theoretical structure of the socialist village	91
3.4	East Germany: Socialist Planning: The village of Marxwalde	92
3.5	Romania: Industry, 1938	103
3.6	Romania: Industry, 1966	106–7
5.1	Distribution of Major Industrial Plants, 1944–65	180–1
6.1	Changing boundaries of Stara Zagora Province, Bulgaria	236–37
7.1	Hungary: Average annual precipitation, 1900–50	275

7.2	Hungary: Agro-Climatic Moisture Regions	276
7.3	Hungary: Irrigation areas, 1966	277
7.4	Hungary: Depth of Water table	279
8.1	The Socialist Republic of Slovenia, Yugoslavia	302
9.1	Yugoslavia: Pattern of Freight Flow: Railways and Ports, 1967	368
11.1	Territorial Growth of Bulgaria	437
11.2	Territorial Growth of Greece	438
11.3	Distribution of urban and rural population in Bulgaria, 1959 and 1965	461
11.4	Distribution of urban and rural population in Greece, 1951 and 1961	462

I

Regional Synthesis: An Introduction*

GEORGE W. HOFFMAN

INTRODUCTION

East-Central and Southeast Europe comprise nine countries with an estimated mid-1969 population of 136 million and an area of 542,230 square miles. These countries are bounded by the Soviet Union and Turkey in the east and the Federal Republic of Germany (West Germany), Austria and Italy in the west, and in addition are bordered by the Baltic, Adriatic, Black and Aegean seas. The east–west distance varies from 650 miles between the Adriatic and Black seas to 240 miles between the Vienna Basin and the nearest border of the Soviet Union. The longest north–south distance, i.e., from the Baltic to the Sea of Crete, is over 1,300 miles, and it is 1,000 miles to the southern tip of the Albanian-Greece border. The region is centrally located on the European peninsula and its transitional character is exemplified by its varied physical and cultural characteristics. The individual political units vary greatly both in area and population. Poland with 120,000 square miles and 32 million people is the largest political unit, and Albania with 11,099 square miles and about 2 million people is the smallest. Like most of the customary European divisions this is an arbitrary one, without overall physical and historical unity, except perhaps the unity of location. In addition, the area has been given many different names, though the precise number of countries included with each regional name varies and is

* The author has drawn freely from his writings, 'Eastern Europe', in *A Geography of Europe*, edited by George W. Hoffman (New York: the Ronald Press, Co., 3rd edition, 1969), pp. 431–44; *The Balkans in Transition*, Searchlight book No. 20 (Princeton, N.J.; D. Van Nostrand Co., 1963), pp. 9–19; 'Eastern Europe: A Study in Political Geography', *Texas Quarterly*, Autumn (1959), pp. 57–61, and 'The Problem of the Underdeveloped Regions in Southeast Europe: A Comparative Analysis of Romania, Yugoslavia, and Greece', *Annals of the Assoc. of American Geographers*, Vol. 57 (December, 1967), pp. 637–66. The publishers' permission is appreciated.

related to the preference and emphasis of individual authors.[1] General location of the area is indicated by such terms as 'Eastern Europe', 'East Europe', 'Mid-Europe'. The term used in this book for the northern tier of countries, East-Central Europe, includes the German Democratic Republic (East Germany), Poland, Czechoslovakia and Hungary. The southern tier, Southeast Europe, comprises Romania, Bulgaria, Yugoslavia, Greece and Albania. Emphasizing its character as a transitional zone, the whole region has been called 'The Eastern Marchlands of Europe', an area where Western civilization gradually merges into a non-Western culture. Names such as 'Shatter Belt', the 'Devil's Belt', or for the southern part of the area, the 'Balkans' or 'Balkan Peninsula' emphasize its fragmentation into many political units resulting in instability and insecurity of its people. It should be mentioned right away that the word 'Balkans' is a corruption of the Turkish word *balak* meaning mountain. Names such as 'Iron Curtain', 'Cordon sanitaire', satellite or captive countries, express the political position or function. Eastern Europe as a general name is perhaps the most common and in this book is often used as an abbreviated form for the more precise description East-Central and Southeast Europe.

Few areas in the world show such great areal contrasts in terms of their physical and cultural features and their social change as do the countries of East-Central and Southeast Europe. The diverse and rugged relief often hinders interregional and local communications. Especially in the Southeast European highlands this condition encouraged particularism and isolationism, making it difficult to establish national cores, preventing unification, and generally contributing to the large inner fragmentation of the region. On the other hand, easy outer accessibility also left a profound impact on the cultural and political geography of the whole area. It enabled more powerful neighbours, such as the Romans and later the Venetians, Italians, Germans, Hungarians, Russians and the Ottoman Turks to occupy and control large sections of the region at one or another time. The neighbouring Empires of Russia, Prussia and, after 1871, a united Germany, Austria-Hungary, Turkey and others, struggled for mastery of important parts of East-Central and Southeast Europe and tried to impose their ideology, including their social and political systems. Certainly the strategic location of the area is paramount in explaining many of these developments; as Moodie wrote, this 'geographical character provides the key to understanding the strategic, economic, and social conditions which have developed there, and goes far to explain their politico-territorial structure.'[2]

While political and economic unity never has existed for the whole region, the present-day domination by the Soviet Union, or, as in Yugoslavia and Albania, by national Communist governments, brings unity of a certain type to the region. Pounds clearly defined the scope of any work on Eastern Europe (he excludes Greece) when he stated that 'none of these descriptions or definitions of Eastern Europe has any precision except the last (Soviet or Communist domination) and it is this which is used to define the scope of the book.'[3] The exception here is Greece, but its inclusion in the region is based on its intimate historical ties with the other countries of Southeast Europe which affect many of the present-day problems.

Cultural influences in Central Europe, from the Mediterranean region, from Russia and from the Orient, as well as the interaction of natural and social forces have made an impact on every one of the countries, complicating relationships with neighbouring countries as well as with the various regions within each national territory. Centuries of invasions and occupation by many peoples and cultures have resulted in diffusion, and have left a deep mark, on the present cultural landscape. As R. V. Burks clearly stated, 'Despite the continuing influx of competing foreign influences, or perhaps because of it, the peoples of Eastern Europe have developed distinct national personalities and highly articulated national cultures.'[4] The struggle against foreign domination was complicated by ethnic and religious conflict, which resulted in the ardent nationalism for which so many of the peoples of the region are so well known. There is no country in East-Central and Southeast Europe in which internal differences and contrasts brought about by past history have not been expressed in cultural and economic diversity. They have played a decisive role in developmental processes, and are shown especially in the spatial differences between and within the many political units. The geographical impact of spatial differences, especially of the problem of inter-regional contrasts within a country, often focused on economically underdeveloped or lagging regions, has received special attention in recent years. The economic and social progress of the countries of East-Central and Southeast Europe since their liberation from foreign occupation during the nineteenth and early twentieth centuries has been slow. Lack of progress, in part, resulted from the historical associations of various regional units of each state. Similar social and economic development characterized all the countries of the region before the outbreak of World War II. The standard of living was extremely low, the illiteracy rate, although showing considerable regional fluctuation, ran as high as 80% of those over ten years of age in some areas of the region, e.g. southern Yugoslavia, northwestern Greece, northeastern Romania, Albania, and most of Slovakia and eastern Poland. The population depending upon income derived from an agricultural surplus labour force was estimated at 35 to 45%, and this resulted in very low productivity and *per capita* yields.[5] The poverty of a large percentage of the peasants resulted, to a great extent, from the fragmented land holdings, industry's insignificant role in the overall economic activity, and the resultant agrarian surplus population. Industrialization was just beginning in most of these countries and it was largely confined to those parts formerly under Austrian and German control in southwestern and southern Poland (based on the coal fields of Upper Silesia and adjacent areas of non-ferrous mining), western Bohemia and northern Moravia in Czechoslovakia (iron ore and coal), western Romania, the Banat (iron ore) and northern Yugoslavia, Slovenia (coal, iron ore). For the most part it was government induced and controlled with much foreign capital invested.

World War II and changes in the political geography of the region left a deep impact on each of the countries of the region. With the exception of Greece all countries went through a political revolution which brought about basic social changes, old institutions were transformed and the static society with its traditional social structure was radically changed as new generations became indoctrinated in the advantages of socialism. Greece took the path of evolution (after a bloody civil war) and began to modernize some of its institutions along traditionally western lines, but the long-time impact of the 1967 military revolution is still difficult to assess. Regional contrasts expressed in social and economic developments have not been overcome in the relatively short period since independence was won by these countries, although in numerous instances they have been reduced to manageable proportions. Often these contrasts were made even more acute by the flow of investment capital to the more developed regions or to areas where development was easiest. On the other hand, an intensive application of modern technologies, the creation of a basic infrastructure, including essential modernization of transportation facilities, and most of all the development of a regional approach in planning the many spatial complexities, provide the ties which bind together countries and regions, whether they be multi-national or uni-national.

From the foregoing discussion it is clear that the countries included in our deliberation have many common geographic and cultural features, have had many common historical experiences and at the outbreak of World War II had many similar social and economic characteristics, several of which will be discussed later in this book. Many regions resisted the spread of innovation and improvements were, more often than not, slow to come about. Because of certain common historical experiences, many problems and characteristics have remained, in spite of diverse physical, social and political elements present during the last 25 years, and it is therefore possible to analyse most of the problems of the region under the same umbrella.

To provide the general reader with a better understanding of many of the problems mentioned, or questions discussed by individual authors, a brief regional synthesis based on a broad outline of the physical environment follows. Within each major division discussed there exists a variety of physical conditions, climatic influences, and minerals available which bring about regional differentiation of the environment which offers man a variety of economic opportunities. It is the author's hope that this summary will provide the conceptual framework for a better understanding of the following papers which discuss problems of the region's many contrasting social, demographic and economic processes.

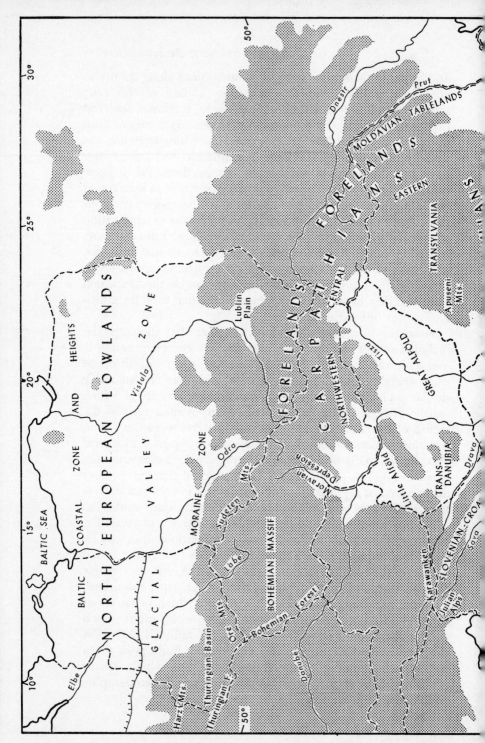

FIGURE I.I — EAST-CENTRAL AND SOUTHEAST EUROPE

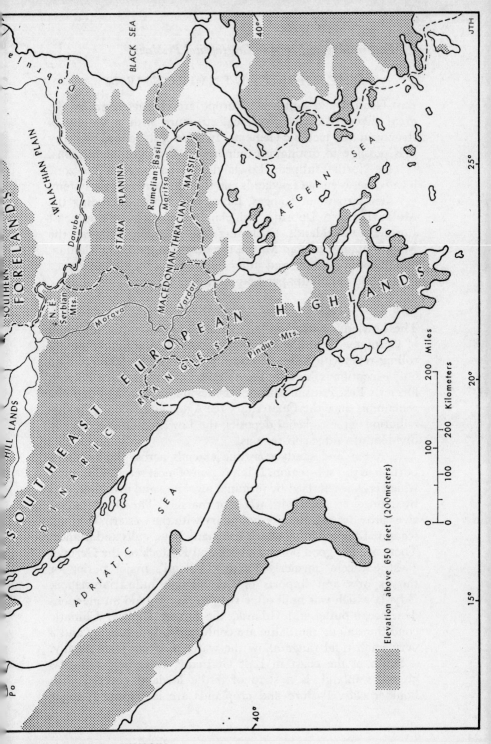

PHYSIOGRAPHIC DIVISIONS

THE PHYSICAL ENVIRONMENT

East-Central and Southeast Europe are characterized by the great diversity of its geological structure with a variety of structural and tectonic elements. In spite of structural diversity it is possible to distinguish four major physiographic regions: (1) the North European Lowlands, (2) an intermediate zone located between the Lowlands and the Alpine mountain system, (3) the Carpathian Ranges, Basins and Plains, including the Moldavian and Dobruja Tablelands, and (4) the Southeast European Highlands, including the Dinaric Ranges, the Macedonian–Thracian Massif, the ranges of the Stara Planina, the Adriatic Littoral and other Southeast European lowlands (for details see Endpaper map and Fig. 1.1).

THE NORTH EUROPEAN LOWLANDS

The Lowlands extend over most of the Democratic Republic of Germany (East Germany) and Poland, a region of gently rolling relief with hills seldom rising to more than 500 feet. The only exception is found in the moraines of northeastern Poland, formerly East Prussia. Much of the region is covered by deposits remaining since the Quaternary Ice Age, and based on the distribution of these glacial deposits, the Lowlands are commonly divided into three sub-regions:

1. A flat and sandy coastline extends across the northern portion of this sub-region. It is a region of most recent glaciation which is characterized by terminal moraines and drumlins, and by depressions filled with water or marshes. The Oder (Odra) river provides a distinctive boundary with the western, German section of the coastline which has many bays, gulfs and islands. There are few good ports: Lübeck and Rostock on the German coast, Szczecin (formerly Stettin), Poland's major outlet for the exports and imports of the Silesian industrial region, Gdynia which was built after 1919 to give Poland an independent ocean outlet and Gdańsk, the former Danzig. Climatic conditions along the Baltic are continental. Low salinity and a very small tidal range allow the water to freeze over readily.

South of the coast in East Germany, extending as far as 50 miles inland, is a strip of fertile lowland covered with boulder clay. Pasture and croplands are interspersed; sugar beet, rye, wheat and potatoes are grown here. Further south irregular hills mark a halt in the ice's movements. This zone is known as the Baltic Heights or Baltic Lake Plateau and is characterized by its many lakes, undrained hollows, sand, gravel, boulder clay and coniferous forests on the steeper slopes. South of the narrow Polish coastline are high sand dune ridges and south of these a broad undulating ground moraine which reaches an average height of 150 feet. The outwash deposits which were formed during a stable stage of the last glaciation reach over 1,000 feet southwest of Gdańsk. Some of the steeper slopes of the terminal moraine and sandy outwash plains are again covered by coniferous forests. Land is cultivated on the boulder clay of the more gentle slopes and numerous lakes are found in the valleys between the moraines, especially in the Masurian lake plateau of former East Prussia. The whole region is sparsely settled, with few and scattered urban concentrations. Agriculture is poorly developed and minerals are lacking.

2. Between the morainal heights and the uplands there is a series of parallel rather broad valleys alternating with somewhat higher ground consisting mainly of sandy soils. These northwest glacial valleys which were formed by the retreat of the melting ice are known as *Urstromtäler* in Germany and *pradoliny* in Poland. These glacial spillways extending in an east–west direction all across the North European Lowlands have a strong influence in the present drainage pattern and are inter-connected by diagonal valleys which offer an excellent opportunity for connecting the river systems of East Germany and Poland, and canals crisscross the area. The need for locks is almost obviated by the slight gradient from east to west. The most important of these canals is the Midland which links the Ems with the Elbe and is extended eastward by various canals to the Oder. It is continued by the Warta–Noteć–Vistula and Bug rivers and canals and the Warta–Bzura–Noteć–Vistula–Bug further south.

The river valleys have loamy soils, scattered worn-down morainic ridges, patches of wind-blown sand with intermittent small lake-filled depressions or marshes which are remnants of a more extensive system of glacial lakes. Most of these lowlands were originally covered by forests, but only an area east of the

Nisa (Neisse) river, stretching towards the middle Odra, is still heavily forested. The rivers of the Lowlands rise in the mountains of the Uplands and the Carpathian Ranges and benefit from the heavy rains, and the springtime melting of the snows which provides ample discharge. The Elbe, Odra and Vistula are the three most important rivers of the Lowlands in terms of their traffic. All three rivers are navigable for most of their course, but the Vistula plays the least important role.

3. A zone of transitional lands between the glaciated lowlands and the German-Bohemian and Polish Uplands consists of a belt of thinly layered morainic material and highly fertile soil. Further east it is characterized by a distinctive surface caused by post-glacial erosion. The areas included in these transitional lands are the Magdeburg borderlands and the Halle–Leipzig bay in East Germany and the Southern Moraine zone between the Nisa and the Silesian Plain which extends southward in the valley of the Odra. Much of the original rock is covered by alluvium or by the loess which was deposited during the inter- and post-glacial periods. Thanks to efficient agricultural practices this is a rich farming area with wheat, sugar beet, barley and vegetables predominating. In East Germany important mineral wealth such as the low-grade iron ore and petroleum in the Harz area between Hannover on the West and Magdeburg in East Germany, and important potash salts found in parts of the Saxony and Thuringian basins, in combination with valuable farm land supports a dense rural and urban population. In Poland scattered deposits of lignite extend from the Zielona Góra województwo towards Wrocław and eastwards towards Wielkopolska.

Climatic conditions in the North European Lowlands, especially the Polish part, are severe and increase from west to east with average January temperatures below 30°F even on the coast. Snow lies on the ground for 50 to 80 days every winter. Nevertheless with better drainage, even in the central parts of Poland, the land is well suited for agriculture with wheat, rye, sugar beet, and potatoes predominating. The country west of the Vistula is generally more densely settled than that in the east with a number of important cities – Poznan, Bydgoszcz, Toruń, Łódź and Warsaw – located at the points of contact with higher moraines and/or amidst the most productive agricultural soils.

THE TRANSITIONAL ZONE

A transitional zone of great diversity stretches from west to east between the Lowlands and the Alpine system. It consists of rolling dissected and forested hills with flat and rounded summits, granite massifs reduced by glacial action, old volcanoes, and basins and plateaus. The hills of this region are often steep-sided and rise to heights between 3,000 and 4,500 feet. This region can be divided into numerous physiographic units: the German group consisting of the Harz Mountains, the Thuringian Forest (Thüringer Wald) and the Thuringian Basin, and the Central Saxon Uplands; the Bohemian Massif (also called Basin) and its surrounding heights; and the Polish Uplands and Carpathian Forelands. The three distinct parts of Thuringia offer great contrasts. The forest is a densely populated, wooded mountain range, whose population is engaged mainly in manufacturing, demanding skilled workmen; agriculture is limited. In contrast the Basin contains rich agricultural lands. The Harz Mountains in higher elevations are covered with forests while elsewhere crops such as hay and potatoes are grown. Even today, the Harz Mountains remain a transportation bottleneck in an area of heavy traffic. South of the Halle–Leipzig Bay, the forested Ore Mountains reach an altitude of over 4,000 feet. These mountains drop off abruptly on the Bohemian side but slope gradually towards the bay (the Saxon Foothills); they support some agriculture, but also contain important mineral wealth, which includes uranium and lignite. The Ore Mountains, which run in a southwest–northeast direction, are separated from the Sudeten, which trend in a northwest–southeast direction, by the Elbe river which cuts through a sandstone range (*Elbesandsteingebirge*) in a deep canyon. The Transitional zone of East Germany is not a climatic barrier. Snow cover is abundant and snowfalls continue as late as March. Rainfall intensifies with increasing altitude.

The very ancient structures of the Bohemian Massif consist of a series of wooded mountains with a few peaks over 4,000 feet, several longitudinal depressions in the southwest and east, some basaltic uplands in the northwest, and a very fertile loess-covered intermontane basin (Bohemian Basin). The Bohemian Forest forms the boundary in the southwest, the Czecho-Moravian Uplands towards the east and heavily eroded horst-type mountains of the Sudetens with elevations to 5,200 feet in the northwestern part. The higher northwestern parts of the Sudetens contain the headwaters of the Labe (Elbe) river, and the lower eastern part those of the Odra and Morava rivers. Towards the northwest are the crystalline Ore Mountains where important uranium ore (pitchblend) is found near Jachymov. The Massif itself contains a number of depressions which are often rich in Tertiary lignite, mineral springs, and other important raw materials. Of note are the springs and kaolin of Marianske Lazne and Karlovy Varý, lignite of Repliče, and iron ore in the Berounka Basin. These resources, together with an old agricultural tradition, are the basis for an old and flourishing industry and present-day agriculture.

The Polish Uplands consist of several clear-cut sub-regions which were intermittently covered with fertile loess deposited during the Ice Age. (1) The Sudeten Mountains, largely composed of old rocks, an area covered with loess and an important cereal-growing region. (2) The Hercynian Uplands of the Silesian Plateau, east of the Odra river and overlooking the Silesian Plains, contain one of Europe's largest coal fields, as well as other minerals. The main centre for mining is the Upper Silesian coal basin, including its southern extension to Ostrava-Karvina in Czechoslovakia. (3) The incline eastward from the Silesian Plain to the Góry Świetokrzyskje (Holy Cross Mountains) is slight, with the high Plateau consisting of Cretaceous limestone and some of the area covered by loess. The old rocks of this area contain copper and iron ores and the main deposits are located in the Kielce-Radom district and near Częstochowa. (4) Between the upper valleys of the Vistula and San rivers lies an important triangular depression made up of alluvial deposits coming from the Carpathian slopes: Kraków-Sandomierz-Przemyśl. On its southern margin the depression is covered with fertile loess. (5) East of the Vistula and San rivers and south of Lublin is the Plateau of Lublin, consisting mainly of limestone covered with a heavy layer of loess. This is a treeless region, with wheat and sugar beet the main crops.

South of the Polish Uplands and east of the Bohemian Massif is another transitional zone, the Carpathian Forelands. This is a very important agricultural region and where loess is not present forests abound. The Moravian depression, between the Danube river (Vienna Basin) and the low gap of the Moravian Gate near Moravska Ostrava, forms part of the Carpathian Forelands. The Czecho-Moravian Uplands border the fertile longitudinal Moravian depression towards the west, and the Carpathian Ranges towards the east. The depression is one of Europe's important routes, connecting the North European Lowlands with the Danube Valley, and the Baltic with the Black Sea drainage. Agriculture, encouraged by mild winters, warm summers and sufficient rainfall, is dominant, with sugar beet, wheat and maize most widely distributed.

THE CARPATHIAN RANGES, BASINS AND PLAINS, INCLUDING THE MOLDAVIAN AND DOBRUJA TABLELANDS

The third major physiographic region of East-Central and Southeast Europe is dominated by the arc of the Carpathian Ranges and their forelands which extend from the Vienna Basin to the Iron Gate for a distance of over 1,000 miles. The Carpathian Ranges have a continuous outer sandstone belt of Tertiary rock and a discontinuous central zone of crystalline rocks and limestone. Towards the basin of the central zone there is a volcanic inner belt. Within the Carpathian Ranges various divisions are possible. Five subdivisions are customary: Northwestern, Central, Eastern, Southern (also called Transylvanian Alps) and Western.

The Carpathian Ranges

(1) The Northwestern Carpathians extend from the Danube and Morava rivers (known as Bielé and Malé Karpaty) to the Dukla Pass at an elevation of 1,640 feet in eastern Slovakia. The Ranges are widest here and are penetrated by several longitudinal rivers which facilitate transportation and settlement. The most important is the Váh river, which separates the High from the Low Tatry and offers a most valuable connection between western and eastern Slovakia. The High Tatry, with its highest peak Gerlachovskýštít (8,375 feet), is a zone composed of crystalline rocks which was glaciated during the

Ice Age and is the highest elevation in the Carpathian Ranges. South of the High Tatry are the Low Tatry, composed of gneiss and granite. South towards the Carpathian Basin (also called Pannonian Basin) are a series of young volcanic intrusions, known as the Slovakian Ore Mountains. These stretch towards the Hornád river and contain small deposits of iron ore, copper, gold and silver. Košice, on the Hornád river, is the most important town of eastern Slovakia and is located on important north–south and east–west routes. The area between the Tatry and Dukla Pass has many deep valleys and small intermontane basins and is heavily forested. The basins of the northwestern Carpathians, on the whole, are well cultivated with the greatest densities on the southern slopes of the Ranges.

(2) The Central Carpathians (also called Forest Carpathians) extend west from Dukla Pass, but their eastern border is poorly defined and is generally considered as located between the sources of the Tisza river near Tartars Pass. The width of the mountain ranges is about 60 miles and a number of transverse valleys offering low passes provide easy access between the Carpathian Basin and the upper Dniester. This area served historically as an invasion route for people from south Russia into the Carpathian Basin and west into the Danube river valley.

(3) The Eastern Carpathians, located entirely within Romania, extend southwards from the Tartars Pass to the Prahova Valley south of Braşov. This is the most rugged and inaccessible part of the Carpathian Ranges and has three clearly distinguishable zones; the parallel ridges of the eastern sandstone belt, a central limestone and crystalline zone, and a western zone of young volcanic material. Peaks vary in height from 5,000 to 7,000 feet; the slopes are heavily forested with deciduous trees on the lower, and coniferous forests on the higher, slopes.

(4) The Southern Carpathians or Transylvanian Alps swing abruptly to the west and extend from the Prahova Valley to two tectonic corridors, Timiş–Cerna and Bistra–Strei, to the west and northwest. Porta Orientalis offers easy access, including a railroad line between the Basin and the Walachian Plain. Numerous peaks of the Transylvanian Alps exceed 8,000 feet, and various glacial features gave the name 'Alps' to these

ranges. Moldoveanu Peak, 8,343 feet in the Făgăras Massif (located between the Prahova and Olt valleys), has the highest elevation. Only a central limestone crystalline zone and the foreland are represented. Typical features of this region are flat-topped plateaus and terraces, as well as numerous longitudinal depressions containing rivers. The present relief is largely the result of uplifting and warping and later stream erosion. Several important routes traverse the Transylvanian Alps at low altitudes, connecting the interior basin with the Walachian Plain. All of these routes are of long-standing historical importance: Predeal Pass (3,400 feet), Turnul Roşu (1,155 feet), and the Vulcan Pass (5,000 feet).

(5) The Western Carpathians (sometimes referred to as the Bihor Mountains after the highest elevation, 6,066 feet) is considered a separate branch of the Carpathians and stretches from the Danube river north to the Barcău Valley and is in general lower than the other Carpathian Ranges. Three main regions of different heights are distinguished: the Apuseni Mountains with the Bihor Massif, the metal rich Poiana Ruscă with a few peaks of 4,600 feet south of it, and the Banat Mountains are located between the Danube and the Timiş corridor. The Western Carpathians were not affected by the Quaternary glaciers and for this reason the typical glacial features are missing. They resemble the Hercynian massifs of the Harz Mountains in East Germany. On the other hand, extensive erosion surfaces with settlements reaching up to 3,900 feet are plentiful. Varied karst forms are distributed all through the region. Mention has already been made of the Poiana Ruscă, which is isolated from the other mountains of this region and is lower in elevation, but its metal and mineral wealth – the iron ore deposits on the eastern side – the marble of the southern slopes, and numerous minerals on the western side, make it one of the most important regions of the Carpathian Ranges.

Brief mention should also be made of the Carpathian Forelands (sub-Carpathians) which form a continuous belt and contain important petroleum deposits in the flysch formations between Bilteni-Ticleni, the Argeş, and Ploieşti regions to the area between Trotuş and Tazlău rivers in the Bacău region. This is a densely settled region, with rich agriculture and important subsoil resources.

Morariu and his colleagues discuss the specific character of the Carpathians and find 'extensive division into fragments, caused by numerous inner depressions and numerous passes, so wide, at places, as to become corridors of easy traffic', and secondly, 'the multitude of the high-altitude platforms, a fact which has made possible the use of the greatest part of the mountainous zones as permanent or seasonal settlement areas.'[6]

By their numerous depressions and passes, by the opportunities of settlement and economic exploitation afforded by their wide mountain platform (cattle-breeding and even high-altitude farming), the Romanian Carpathians have polarized an intensive human activity inside their area, where the national custom of the most authentic character, and the purest Romanian idiom have been preserved, and where the great treasures of Romanian folklore are to be found.[7]

The Carpathian Basin

The Carpathian Basin is sub-divided into the Transylvanian Basin (sometimes referred to as Tablelands) located in Romania and the Hungarian Basin or Plain, extending beyond the present boundaries of Hungary. The whole Basin is enclosed by the arc of the Eastern Alps, the Carpathian Ranges and the Dinaric Ranges, but it is not uniform in relief. It is divided into minor basins, hill lands and mountainous regions. Much of the Basin was covered by an inland sea until rivers from the surrounding mountains filled it with alluvial deposits; sections of it later were covered by loess. The Basin is drained exclusively by the Danube which enters by cutting across a spur of the Eastern Alps near Vienna and leaves by the gorges known as the Iron Gate, where a huge hydroelectric power plant and a dam across the Danube are being completed at the present time. The whole Carpathian Basin is usually divided into a number of broad regions, with the three western located largely in the Hungarian part of the Basin. (1) Little or *Kissalföld*, between the foothills of the Eastern Alps and the southeast slopes of the Bakony Forests and north across the Danube to the foothills of the Northwestern Carpathians.

The central portion, southeast of Bratislava, consists of a large alluvial fan of gravel and silt. Neusiedler Lake, at the Austro-Hungarian border, is surrounded by swamps and mud flats and is one of the remnants of the inland sea; (2) south and west of the Little Alföld are the rolling dissected and very productive hill lands of Trans-Danubia, the part in Yugoslavia being known as the Slovenian-Croatian Hill Lands. The hills are covered by loess and their greatest height is reached at about 3,300 feet. The Danube to the east and the Dinaric Ranges to the south border this sub-region. Lake Balaton, with a maximum depth of only 35 feet, is a shallow fresh-water lake remaining from the inland sea. The Slovenian-Croatian Hill Lands are very fertile, loess-covered lowlands interspersed with limestone and crystalline hills, and patches of forests alternate with densely cultivated lands; (3) the Great or *Nagyalföld* stretches irregularly east and north of the Danube to the foothills and volcanic belts of the Northwestern, Central and Western Carpathians. The Alföld is a perpetually settling block which has been covered by Tertiary and Quaternary strata and is dissected by the meandering Tisza river. The western part is an elongated, partly sand-covered region often referred to as 'Mesopotamia'. Southeast of Kecskemet is the *puszta* (waste), a region of soil saturated with salt. Near Debrecen is Hortobágy, a steppe region covered with sand which has a high soda content. Most of the Banat between the Tisza and the Transylvanian Alps is covered with fertile black soil; (4) the Transylvanian Basin is also a depression between the Western Carpathians and the Carpathian arc, which was uplifted and dissected and its Tertiary clay strata now reach elevations of up to 2,000 feet. The folding and uplifting of the Basin produced flattened vaults or domes where rich deposits of methane gas have been found.

The Carpathian Basin is transitional with regard to climatic conditions, with the greatest transition in the Hungarian part of the Basin. Rainfall declines from west to east, as, for example, between Bratislava (total precipitation 28 inches) and Szeged (22 inches) and also from north to south. Most of the rain falls in early summer, with smaller but still sufficient amounts in the latter part of the summer. Unfortunately, the summer rains occur during a period of high temperature and high evaporation which lessen the effectiveness of the precipitation. The results are often devastating droughts. Temperature differences between summer and winter are considerable, but by no means extreme. The Carpathian Ranges form a certain barrier between the humid and cool air masses of oceanic origin in the west and the continental air masses from the Eurasian continent with their icy winds causing huge snow drifts at times in Moldavia and the northern foothills of the Carpathians. The Basin is drained largely westward towards the Tisza and Danube. The latter crosses the Hungarian Basin from north to south and cuts through the Carpathian Ranges between Baziaş and Turnu Severin, flowing eastward towards the Black Sea where many braided meandering channels, formed by alluvial deposits, cut broad valleys below the general river level. The Tisza river with its sources in the central and eastern Carpathians has a more constant water regime largely due to the ample supply of snow in the source regions of its numerous tributaries. The Mureş river is most important from the point of view of volume of water carried, and also plays an important role as an economic and traffic axis.

The Hungarian Basin has numerous fuels sufficient in quantity to be mined. These include coal near Pécs and Komló on the southern slopes of the Mecsek hills, petroleum fields in southwestern Hungary and southeast of Budapest, a recently discovered oil field in the Szeged basin, and important oil fields in the south-central part of the Croatian Hill Lands and the Banat. Bauxite is the only metalliferous mineral produced in the Basin centred on the Bakony-Vertes hills. Lignite is found in the same general region and iron ore is available in small quantities in the Sajo Valley near the Czechoslovakian boundary. The mineral wealth of the Carpathian Ranges and the Transylvanian Basin in Romania has already briefly been indicated. The petroleum of the Carpathian flysch formation with the largest concentration between Ploieşti and south of Piteşti and natural gas with a very high methane content and only few other hydrocarbons are well distributed throughout the central part of the Transylvanian Basin. Bituminous coal reserves are mined largely in the Jiu Valley (north of Tirju-Jiu) and the Banat, together with iron ores. Lignite is plentiful, especially in the upper Jiu Valley. Iron ore is not plentiful but some mines are located in the Poiana Rusča and the Banat mountains.

Moldavian and Dobruja Tablelands, Walachian Plains

Between the Carpathian Forelands and the Prut river, from the Suceava Tableland in the northwest to the Danube in the southeast are a series of hills, slightly sloping towards the southeast, the Moldavian Tablelands. These hills and valleys descend from a maximum height of close to 1,800 feet in the northwest to less than 300 feet in the Jijia depression in the southeast. With the exception of a central plateau, the entire region is a hilly, loess-covered steppe; recurrent droughts and disastrous floods do much damage here. Further south, between the Danube and the Black Sea, are the Dobruja Tablelands which have two distinct parts, the northern heavily eroded old mountains and the southern, lower part with a more even relief. The steppe plateau of the Dobruja Tablelands is an important agricultural region with relief just under 600 feet. The plateau blocks the straight course of the Danube river and forces it into a northward direction. The river forms the eastern boundary of the Walachian Plain, sometimes referred to as the Danubian or Romanian Plain. Its northern boundary is the Carpathian Forelands and the lower basin of the Siret river and the western and southern boundary is formed by the Danube river. The Plain is a depression which in the Tertiary was a gulf of the Black Sea, but which by now has been entirely filled by alluvial deposits from the Transylvanian Alps and loess-type of material. The whole depression has been uplifted and is now about 150 feet above the sea level. The plain actually extends from the southern foothills of the Transylvanian Alps through the piedmont plateau of the Stara Planina. The valleys that cross the Plain 'run through intra-fluvial areas looking like wide, slightly grooved fields, sporadically filled, on the eastern part, by lakes, of which some are salt-water lakes'.[8] The Plain is one of Romania's most important agricultural regions, and when irrigation systems are more widespread they will control the rather frequent droughts and will contribute to a more intensive, perhaps industrially oriented agricultural production.

THE SOUTHEAST EUROPEAN HIGHLANDS

In structure and relief the Highlands are a complex mountain region with the lithic composition of the various ranges varying from the calcareous to the more resistant crystalline rocks. Volcanic intrusions indicate instability, but the various tectonic lines have become mineralized during the igneous activities, a contributing factor to widespread mining activities especially in Yugoslavia. On the other hand, they indicate a great structural instability which is clearly expressed by the annual large number of tremors, some of which reach a force of over VIII on the Richter Seismic scale.[9] The numerous steep mountains and small intermontane basins, heavily eroded, especially in the south, and the barrier which the mountain ranges cause, are formidable handicaps to communication over the whole region. A large part of the Highlands is composed of karst, containing all characteristic karst features.

Professor Josip Roglić in a number of studies analysed the 'appearance and reality' of the regional term 'Balkan', which has during this century so easily established itself, especially by American social scientists, and even by such an outstanding geographer as Jovan Cvijić.[10] Also the term 'Southeast European Peninsula', or 'Balkan Peninsula' has some popularity. The term 'Balkan' was derived from the Bulgarian Stara Planina (Old Mountains) which often is called by the corrupted generic Turkish name 'Balkan Mountains', and for a long time it was thought that the Stara Planina, the Hellenic 'Haemus', was a mountain range extending from west to east. When it was realized that the Balkan Mountains were not the mountain range after which the whole area could be named, some authors (especially the German, Theodor Fischer – 1893) used the term 'Southeastern European Peninsula' for the area south of the rivers Sava and Danube. It was, and still is, argued that this term had justified historical relations because these two rivers, in the nineteenth century, were border rivers of the Ottoman Empire. Besides this unrealistic division, rivers generally, as Professor Roglić rightly points out, 'cannot be borders of geographical units because by their natural properties and social significance they are units and do not separate'.[11] Further it is an important fact that nowadays geographers do sustain the notion that Europe is a continent and therefore Southeast Europe a peninsula of Europe. In the generally accepted notion of Eurasia, Europe is distinguished by its peninsular nature with Southeast Europe standing out with very specific characteristics, not the least of which is its generally mountainous character and its socio-economic isolation. Depending on the arbitrary decision of individual authors, Southeast Europe comprises Yugoslavia, Albania and Bulgaria, and at times also Romania, Hungary and Greece. There should be no place in modern terminology for such a misleading term as Southeast European Peninsula and especially the more familiar term 'Balkan Peninsula'.

In any physiographic division of Southeast Europe, three and possibly four mountain ranges stand out clearly: the Karawanken chain of the Eastern Alps forming the northern boundary; the limestone Julian Alps with their continuation in the Dinaric Ranges and their southward extension the Pindus Ranges all the way to the southern Aegean, to Crete; the ranges of the Stara Planina which form a continuation of the Carpathian Ranges with the link between the two, south of the Danube to the Timok river known as the Northeast Serbian Mountains, and the Macedonian–Thracian Massif, also called Rhodope Mountains, trending from southern Yugoslavia, crossing the Morava and Vardar depression and broadening southeastward into a series of uplands and dissected hills, broken by the Maritsa river northwest of Edirne (Turkey). In the turbulent history of this region, this mountainous core played a dominant role in the social, political and economic life of the people of the Highlands until very recently. Only after the agricultural revolution, with the introduction of suitable crops, especially maize, and improved farming techniques, and their inclusion in a more stable political framework, did the economically important and more easily accessible plains, depressions and basins, become more attractive to settlement. It is a relationship similar to that found between the Carpathian Ranges and the Walachian Plains and the Transylvanian Basin.

These lowlands, depressions and basins form the centre of present-day economic activities. Perhaps the only exception of a mountainous region participating in this process are the northern and north-central valleys of Bosnia, with their varied minerals which contribute to the economic activities of this region. The most important of the lowlands, depressions and basins are the lowlands of the Carpathian Basin (The Slovenian and Croatian Hill Lands and the Vojvodina) which have been mentioned earlier. Affording a short route between the Pannonian Basin and the Aegean Sea and its head port of Thessaloniki are the Transitional and Basin Lands of the Morava and Vardar depression. West of this depression are the foothills of the southeastern Dinaric Ranges, which consist of detached mountain blocks and basins connected by narrow passages. Other important lowlands are the agriculturally important Rumelian Basin (also called Maritsa Basin or the Plain of Thrace), located between the Stara Planina and the Rhodope Mountains, the Adriatic littoral between the Istrian Peninsula and the river Drin in Albania, small coastland regions in Albania and Greece, the Vardar (Axios) lowlands of northern Greece, the two tectonic basins of Larissa and Trikkala, the small coastal plain of Athens and the two plains on the Peleponnesus peninsula.

The Dinaric Ranges

Without doubt the most diversified highlands in Southeast Europe are the Dinaric Ranges. These parallel ranges of limestone of Cretaceous age, trend southeast from the Julian Alps in the Ljubljana Basin towards the Morava–Vardar depression. As the Prokletije Mountains they form the Albanian–Yugoslav border with an elevation of over 7,000 feet, continue in a series of parallel ranges southward to merge with the Epirus Mountains and finally continue as the Pindus Mountains into southern Greece. Smaller forks leave the main mountain chain and open to the north. The width of these ranges varies from 60 to 150 miles and the average elevation varies from 4,000 to 6,000 feet. The Julian Alps in Triglav reach 9,393 feet and the Dinaric Ranges in Durmitor 8,272 feet, their highest elevation. The various ranges are characterized by their diversity, ranging from the barren, dissected and waterless High Karst in the west to a series of parallel forested mountains and hill lands in the north and northeast. Soil erosion in the hills and sheet erosion in the basins is widespread in the southern region, especially in the Vardar depression and the Rhodope Mountains.

Access to the interior is blocked by the High Karst, a barren mainly Mesozoic limestone zone. The river valleys are very short and widely spaced because precipitation falling upon the limestone rocks sinks underground where it continues to flow. Rivers flow in deeply dissected valleys or through gorges which are difficult to traverse and they carry varying quantities of water and have considerable elevation differences between source and mouth. Their use for hydroelectric projects is now under way. The whole region consists of barren, rocky plateaus with a series of flat ridges, so-called *planina*; and longitudinal troughs, so-called *polja*, *dolinas* and *uvale*. The largest *polja* are those of Lika and Livno, each about 45 miles in length. The inner part of the Ranges, inner Bosnia and western Serbia, are less barren and rugged, sandstone and limestone predominant in a few places and crystalline rocks are visible. Narrow and open valleys are interspersed and extensive mining and logging activities are carried on throughout the countryside.

The Pindus Ranges

The limestone hills of the Prokletije and Epirus ranges are covered with *maquis* and offer few opportunities for agriculture; grazing and mountain agriculture are the main occupations of the people in this region. The Pindus mountain range is rugged and consists of blocks which are separated by narrow depressions which form transverse valleys, some of which are submerged. Regional differences in Greece are accentuated by the contrast provided by the narrow and widely scattered coastal plains, the relatively few upland basins, and the mountains. The latter consist of a series of folded limestone ridges of the central and western Pindus Ranges (6,800 to 8,000 feet). They form part of the Ionian Aegean drainage, which is characterized by karstic features, heavily eroded soils and a total lack of minerals. Transhumance is common. The Epirus, consisting of a series of limestone ridges, runs parallel to the western coast and ends at the Gulf of Arta (Amvrakokos), though some of its ridges are terminated in the mountains in the south bordering the Gulf of Corinth. The Epirus has ample rainfall, vegetation is dense and deciduous forests and the Mediterranean vegetation, if not cleared for cultivation and pasture, are substantial. The Ionian Islands, including Corfu, Levkas, Zante and others, are structurally similar to neighbouring Epirus. The islands have good soil, are densely populated, and due to their westerly location they have ample rainfall. The Pindus Mountains form a major divide and communication obstacle between the eastern and western parts of northern and north-central Greece.

The Northeast Serbian Mountains and the Stara Planina

The Carpathian mountains continue south and southeastward of the Danube as Northeast Serbian Mountains. East of the Timok river they are known as Stara Planina. The Northeast Serbian Mountains reach altitudes between 3,700 and 5,100 feet and are characterized by poor lines of communication, by karstic relief, and by valuable mineral resources. The Stara Planina is about 400 miles in length. It slopes gradually to the Danube and east towards the Black Sea. The Bulgarian Plateau, located between the Danube and the Stara Planina, is dissected by deep and broad valleys which are extremely fertile. The contrasts in relief and climate between the fertile valley plains and the plateau are great. The valleys have an abundant water supply, and are protected from the cold, dusty winter winds. The Stara Planina is easily crossed by two north–south railways, via the Isker Valley to Sofia and via the Shipka Pass, connecting the Danube port of Giurgiu (on the railroad to Bucharest) with the Rumelian Basin. South of the range is a transitional zone which stretches from western Bulgaria towards the Black Sea. It is a wide zone marked by a series of discontinuous and low ranges. Included are the Vitosa Mountains (7,510 feet) and the various parts of the Sredna Gora (Central Forest), sometimes referred to incorrectly as the 'Anti-Balkan range'. Between the Stara Planina and the Sredna Gora are a series of depressions which are characterized by numerous hot springs and also include the famous cultivation of roses in the Tundzha Valley.

The Macedonian–Thracian Massif

South of the Rumelian Basin is the Macedonian–Thracian Massif consisting of the Rhodope, Rila and Prim mountains. The Massif extends from the Transitional and Basin Lands of the Morava and Vardar depression in an easterly direction.

The highest part is Rila Planina (Mount Musala, 9,596 feet), an area of volcanic origin containing the headwaters of the Maritsa river. The Massif consists of a number of fertile river valleys and basins, e.g. Struma and Mesta valleys and the basins of Samokov, Pernik, Kjustendil. Passes give access to the valleys. This zone is rich in brown coal and lignite, and a variety of metals are found (especially in the central and eastern Rhodopes). The rivers of the Massif offer great hydroelectric potential and the forests of the region make an important contribution to the economy of the country. This region was opened up in the post-war period. The mild climate, especially in the valleys which open towards the south and southeast, permit the growing of Mediterranean plants, as well as tobacco. Part of the Massif is in eastern Macedonia and Thrace in Greece, a region of faulted basins separated by mountain zones, and river deltas.

Lowlands, depressions and basins

The importance of the lowlands, depressions and basins has been mentioned earlier. Of the depressions, that of the Morava and Vardar is without doubt the most important one due to the short route (about 300 miles) between the lowlands of the north and the head of the Aegean Sea. The depression consists of numerous tectonic basins, alternating with steep highland belts lying along the Morava, Ibar and Vardar rivers.

Recent faulting of rock strata and large-scale erosions at different ages, often associated with the formation of the Aegean Sea, profoundly modified or altered the flow of rivers in the region between the Danube and Aegean Sea. These structural developments reversed the drainage pattern of existing rivers and created new lakes, marshy basins, and new or modified watersheds. Lakes covered most parts of the region and were joined through the Grzelicka Gorge with the great Pannonian Sea. The Macedonian–Thracian Massif became an island. With the sinking of the Aegean when it became part of the Mediterranean Sea, the old lake within the peninsula was drained, and many separate basins with connecting rivers were established. Kosovo Polje and the plains of Skopje are examples of these former lakes. The Macedonian lakes of Ohrid and Prespa, draining towards the Adriatic, are other examples. New drainage basins were established and in connection with these anomalies, Marion Newbigin in her classic study[12] asked: To which geographic region does Macedonia really belong? This uncertainty also plays a considerable role in the political geography of this region.

Another important depression and basin is the earlier mentioned Rumelian Basin, which is drained by the Maritsa river and opens towards the Black Sea, though the plain is interrupted by a number of low ridges. The fertile alluvial soil and abundance of water permit the growing of a great variety of crops, including cereals, tobacco, rice, vegetables (tomatoes play an important role in Bulgarian export to western countries) and fruit. The Basin is Bulgaria's most important agricultural area. The Sofia depression, a plateau with an elevation of 1,800 feet, is surrounded by mountains, but due to the special characteristics of the river system in radiating from this structurally complex area, an extremely favourable and unique nodality of the Sofia area was brought about. Were it not for the small size of the depression, this area from a purely locational aspect easily could have exerted its influence over a much larger region of Southeast Europe.

The numerous narrow and widely scattered coastal lowlands offer great contrast and vary considerably in their size and economic importance. They include the narrow coastal zone of the Adriatic of Yugoslavia (Dalmatian coast), with its many islands and arms of the Adriatic Sea which present a most picturesque landscape and which are now very popular with tourists from all parts of the world. The coastal zone and the terraced hillsides of the littoral have only a limited amount of soil, but Mediterranean crops such as olives, vines, figs and limited pasture land and cereal acreage play an important economic role. The Albanian coastline, which varies greatly from the Adriatic littoral, alternates between swampy depressions, marshy deltas and sand bars which enclose shallow lagoons, though most of these areas have now been drained and a variety of crops have been introduced.

The east and south coasts of Greece have a very irregular configuration which contributes to the ease of communication between the mainland and the sea. The basin of Thessaly with its two tectonic basins, those of Larissa and Trikkala, are

Regional Synthesis: An Introduction 27

among the largest lowland areas of Greece. Oceanic influences are largely excluded from the Thessaly lowlands by the surrounding mountains and as a result rainfall is low. But with the introduction of modern agrotechnical measures the plains now produce a rich harvest of wheat and tobacco. Still, additional irrigation is desperately needed and would offer greater returns, including a greater diversification of crops. The coastal plains of northern Greece are often flooded, but have been partially filled by a change in the sea level and by silt which has been deposited in the delta by the Vardar, Struma, Nestos, Maritsa and other rivers. Rising from the alluvial plains are the rounded *maquis*-covered foothills of the Macedonian–Thracian massif. The more elevated parts of the lowlands have been agriculturally important for centuries. The Drama plains, for example, specialize in world-famous tobacco cultivation and are one of Greece's most densely populated regions. This is an important region in terms of convergence of routes, in terms of producing essential livelihood, and as a meeting place for people from many different countries. The contrast between northern and central Greece is quite marked. Besides feeling the full influence of the Mediterranean climate, central Greece is the typical Greek landscape of a small arable plain, close to the sea, from which arid foothills covered with scrubby *maquis* rise and give way to towering mountains with scant vegetation affording pasture only for sheep and goats. Athens is located in the southeastern part of central Greece, on a small coastal plain between barren heights and the sea, and for all practical purposes is one city with its seaport, Piraeus. Southern Greece, the peninsula of the Peloponnesus, can also be considered an island since the completion of the Corinth Canal in 1893. The plains of Messenia in the west have the greatest rainfall and are the most productive. That of Laconia is the longest, but only a few miles wide. The plains are surrounded by rugged mountains. Some rise above 6,000 feet and there are great contrasts between the barren, east–south exposed slopes and those of the wetter and wooded west and northwest. Typical Mediterranean products of the peninsula are grapes for wine, olives, and of special importance, the sundried raisins (currants). Included in southern Greece are the many islands in the southern Aegean Sea; for the most part these are mountain tops which are remnants of the Pindus Range which have been partially submerged.

The importance of the Danube as a drainage basin has already been mentioned in this discussion and its southeast European tributaries drain northwestern Yugoslavia, the Dinaric Ranges and the Stara Planina. The heavy winter rainfall of the region contributes to the Danube's winter maximum flow. Only the Neretva and Drin are important rivers which flow in the Adriatic Sea, though several short rivers flow within a short distance to the sea in Albania. Of the other rivers in Southeast Europe only the Vardar of Macedonia (called Axios in Greece) and the Maritsa (called Meriç in Turkey and Evros in Greece) both flowing into the Aegean Sea are of more than local importance. None of the rivers of Southeast Europe, with the exception of the Danube, are navigable. On the other hand, many of these rivers have more than local importance, largely because their valleys constitute important routeways. The rivers of the Dinaric Ranges flowing into the Adriatic are of potential importance for the development of hydroelectric plants. Several are already used for this purpose.

Climatic characteristics

The Southeast European Highlands are included in the broad patterns of the European climate and are generally classified as 'transitional'. Mediterranean climates, with mild winters and dry, hot summers, predominate along the littoral and most parts of Greece and penetrate into southern Macedonia and Bulgaria. Continental climates with dry and hot summers are found in the northern peripheral lands and in eastern Bulgaria, and a transitional climate, cool and wet, prevails in the northwestern and south-central Dinaric Ranges, the Northeastern Serbian Mountains and the Stara Planina. A number of strong local winds are created by pressure and temperature gradients due to the passing of warm depressions adjacent to cold areas. These are the *kosava* of eastern Vojvodina, and the *vardarac* in the Vardar Valley and the *bura* which blows heavily down the coastal slopes, isolating islands and local communities. Temperatures clearly prove the transitional character of the region with continental conditions predominant. Transitional conditions are prevalent in most of the Dinaric Ranges as maritime influence is blocked by the relief (Titograd has January temperatures of 32°F, while nearby Bar has 48°F). Vojvodina and eastern Bulgaria have the greatest extremes between January and July temperatures, and the highest July temperatures are recorded in southern Bulgaria and Macedonia. Rainfall tends to diminish from west to east, and annual seasonal precipitation varies regionally. Predominant winter rainfall is characteristic in much of the littoral; winter freeze is usually of short duration which permits cultivation up to 3,000 feet in all of Southeast Europe. The length of this cultivation varies regionally and depends greatly on the frost-free period together with factors of insolation and cloudiness. Southern Bulgaria and southeastern Macedonia as well as most of Greece, have the most frost-free days (275 to 320 days), or, like central and southern Greece, are not subject to winter freeze at all. The importance of late spring and summer rainfall is critical for the best agricultural regions of Southeast Europe, i.e. the southern margin of the Carpathian Basin, the Vojvodina to Macedonia, northern Greece and the Larissa Basin. Droughts are frequent and are responsible for great variations in crop output. Unfortunately, the same areas are often flooded, though drainage and irrigation works in some regions (Vojvodina, Axios delta) have meliorated the problem. In Yugoslavia alone floods have threatened roughly 20% of all farmland, and much re-cropping is at times needed.

Soil erosion

One problem of great concern in Southeast Europe is the gravity of soil erosion especially in southern Yugoslavia and Greece, affecting one-third of the agricultural area in the former, and one-half of Greece. The barren Karst, covering 10% of the area of Yugoslavia 'stands testimony to man's destruction of woodland to practice pastoralism and farming, to build ships and, during the Turkish occupation, to obtain clear views along the strategic southeastern frontier ridges overlooking the Adriatic'.[13] Modern tilling practices and the conservation of forests, both by limitation on felling of timber and the open grazing of sheep and goats, are just beginning in most Southeast European countries. The Alpine–Dinaric region with its wet climate and predominantly podzolic soils is the most important and extensive forest region of the Highlands.

Mineral and fuel resources

Mineral and fuel resources in Southeast Europe are widely scattered and are present in relatively small quantities; on the whole, Yugoslavia is best endowed as a result of its diversified geological history. Greek subsoil resources have been inadequately surveyed, but any sizeable development appears most unlikely. With the exception of some lignite (in northwestern Epirus) and bauxite on the north shore of the Gulf of Corinth, mining has been negligible. Northern Greece, especially Macedonia, is perhaps the area with the brightest prospects for successful mining, especially the exploitation of its bauxite reserves. The geological structure of Bulgaria shows small- to medium-sized and scattered deposits of ore reserves and a number of non-ferrous ores such as lead, zinc and copper and smaller reserves of chrome, manganese and molybdenum. Reserves of fuels are small with the exception of lignite. Metalliferous ores are available in the Macedonian–Thracian Massif, the western Stara Planina and the hinterland of Burgas.

Yugoslavia and Bulgaria have ample brown coal and lignite deposits. Albania has small deposits, but bituminous coal reserves are limited in all three countries. Bulgaria's largest fuel reserves are the lignite deposits northeast of Dimitrovgrad which are worked in several open-cast pits. Those of Yugoslavia are concentrated in Serbia (one-half) and north-central Bosnia (one-third). Yugoslavia, Albania and Bulgaria all have small oil and gas reserves, all insufficient for domestic needs. Yugoslavia's main natural gas fields are in the middle Sava valley and its main oil fields are located in the Vojvodina and the middle Sava valley. Petroleum occurs along the inner edge of the coastal plain of Albania between Vlorë and Elbasan with considerable reserves. Besides Romania it is the only country of East-Central and Southeast Europe exporting petroleum. Close to the southern margin of the petroleum fields are deposits of asphalt bitumen which have been mined since classical times. Petroleum was discovered after World War II in northwestern Bulgaria and the hinterland of Varna, but production has been small.

The main iron ore reserves are in central Yugoslavia (Ljubija and Vareš) where they support iron and steel works at Zenica and Sisak. Small deposits are widely scattered and those of northeast Serbia and western Macedonia are among the more important deposits. Bulgarian iron ores are in the Sofia Basin at Kremikovci, and at Krumovo. Smaller deposits are in the western Stara Planina and in the Rhodope Mountains. Copper is mined in Yugoslavia, Bulgaria and Albania. Yugoslavia is Europe's largest copper producer; the mines are located in the Northeast Serbian Mountains at Bor and especially Majdanpek. Bulgaria's copper ore is mined in the western Stara Planina and near Burgas. Albania has small mines of copper and chromite. Lead and zinc are mined in Yugoslavia with the most important mines in the Southern Dinaric Ranges at Trepča in the Kopaonik region of Kosovo, and in Bulgaria, in the southeastern part of the Macedonian–Thracian Massif at Madan and Rudozem.

Bauxite predominates in Yugoslavia with reserves throughout the western Dinaric zone and in central Bosnia. Antimony is mined in the mountains of northeast Serbia, and mercury (cinnabar) is found in the Carboniferous strata at Idrija in the Julian Alps. Other deposits obtained in small quantities are bismuth (as a by-product of the Trepča mines), pyrites at Majdanpek, chrome in southern Serbia, and manganese in northern Bosnia. Some of the mining activities, especially those in Serbia and Slovenia, date from the Middle Ages or even earlier.

* * *

These brief discussions of the major regional characteristics, especially the physiographic variations and sizeable, although mostly low-quality deposits, give an indication of the diversity of the landscape and resources available. The great diversity in both the cultural and the physical landscapes of the countries of East-Central and Southeast Europe leaves its impact on the economic activities to be discussed in the chapters which follow.

NOTES

[1] Several authors have used various names for this region: George W. Hoffman, 'Eastern Europe', in *A Geography of Europe*, edited by George W. Hoffman (New York: The Ronald Press, 3rd edition, 1969), pp. 431-2; Arthur E. Moodie, 'The Eastern Marchlands of Europe', in *The Changing World*, edited by W. Gordon East and Arthur E. Moodie (Yonkers-on-Hudson, N.Y.: World Book Co., 1956), pp. 111-13, and Harriet G. Wanklyn, *The Eastern Marchlands of Europe* (London: George Philip & Son, Ltd., 1941), pp. 1-3.

[2] Moodie, 'The Eastern Marchlands of Europe', *op. cit.*, p. 111.

[3] Norman J. G. Pounds, *Eastern Europe* (London: Longmans, Green & Co., Ltd and Chicago: Aldine Publishing Co., 1969), p. 1.

[4] Richard V. Burks, 'Social Forces and Cultural Change', in *The United States and Eastern Europe*, edited by Robert F. Byrnes, The American Assembly (Englewood Cliffs, N.J.: Prentice-Hall, Inc., 1967), p. 82.

[5] The Royal Institute of International Affairs, 'Agricultural Surplus Population in Eastern and Southeastern Europe', Committee on Reconstruction, Economics and Statistical Seminar (London: Chatham House, July 1943), Mimeographed, p. 15, and Paul N. Rosenstein-Rodan, 'Problem of Industrialization of Eastern and Southeastern Europe', *The Economic Journal*, Vol. LIII (1943), pp. 202-11.

[6] Tiberiu Morariu, Vasile Cucu, *et al.*, *The Geography of Romania* (Bucharest: Meridiane, 1966), pp. 26-8.

[7] Morariu, Cucu, *et al.*, *Geography of Romania*, *op. cit.*, p. 27.

[8] Morariu, Cucu, *et al.*, *Geography of Romania*, *op. cit.*, p. 37.

[9] The western Dinaric zone in the last 25 years has recorded eleven tremors of force VI to IX, the most recent tremors in Banja Luka (northwestern Bosnia) reached a force of close to VIII and the Skopje earthquake in 1963 reached over VIII. For an excellent account see F. E. Ian Hamilton, *Yugoslavia, Patterns of Economic Activity* (New York: Frederick A. Praeger, Co., 1968), pp. 60-1.

[10] Josip Roglić, 'Balkan Geography – Appearance and Reality', Paper read before the meeting of Balkans – Continuity and Change, University of California, Los Angeles, October, 1969, p. 31; also by the same author 'O Geografskom Položaju I Ekonomskon Razvoju Jugoslavije' (The Geographical Position and Economic Development of Yugoslavia), *Geografskog Glasnika* (Zagreb), Vol. XI–XII (1950), pp. 11–24. Jovan Cvijić, *La Péninsula Balkanique* (Paris: Armand Colin, 1918).

[11] Roglić, 'Balkan Geography', *op. cit.*

[12] Marion L. Newbigin, *Geographical Aspects of Balkan Problems* (New York: G. P. Putman's Son and London: Constable 1915).

[13] Hamilton, *Yugoslavia*, *op. cit.*, p. 71.

SELECTIVE BIBLIOGRAPHY ON EAST-CENTRAL AND SOUTHEAST EUROPE

ANDRÉ BLANC, *Géographie des Balkans* (Paris: Presses Universitaires de France, 1965).

JOVAN CVIJIĆ, *La Péninsula Balkanique* (Paris: Librarie Armand Colin, 1918).

PIERRE GEORGE et al., *Les Républiques Socialistes d'Europe Centrale*. Collection Magellan No. 15 (Paris: Presses Universitaires de France, 1967).

GEORGE W. HOFFMAN, *Balkans in Transition*. Searchlight book No. 20 (Princeton, N.J.: D. Van Nostrand Co., 1963).

—— 'Eastern Europe', in *A Geography of Europe*, edited by George W. Hoffman (New York: The Ronald Press Co., 1969, 3rd edition), pp. 431–524.

CHARLES JELAVICH, editor, *Language and Area Studies. East Central and Southeastern Europe. A Survey* (Chicago and London: The University of Chicago Press, 1969), specifically chapter 8, 'Geography', pp. 199–224.

BERNARD KAYSER and KENNETH THOMPSON (with collaboration of R. Vaternelle and Basil Coukis), *Economic and Social Atlas of Greece* (Athens: National Statistical Service, the Centre for Planning and Economic Research, and the Social Studies Centre, 1964).

WILLIAM H. McNEILL, *Europe's Steppe Frontier 1500–1800* (Chicago and London: The University of Chicago Press, 1964).

MARION L. NEWBIGIN, *Geographical Aspects of Balkan Problems* (New York: G. P. Putnam's Son and London: Constable, 1915).

R. H. OSBORNE, *East-Central Europe. An Introductory Geography* (London: Chatto and Windus and New York: Frederick A. Praeger, Inc., 1967).

NORMAN J. G. POUNDS, *Eastern Europe* (London: Longmans, Green & Co. Ltd, and Chicago: Aldine Publishing Co., 1969).

NICOLAS SPULBER, *The Economics of Communist Eastern Europe* (New York: The Technology Press of M.I.T. and John Wiley & Sons, Inc., 1957).

TRAIAN STOIANOVICH, *A Study in Balkan Civilization* (New York: Alfred A. Knopf, 1967).

H. G. WANKLYN, *The Eastern Marchlands of Europe* (London: G. Philip & Son Ltd., 1941).

Joint Economic Committee, Congress of the United States, *Economic Developments in Countries of Eastern Europe*, A Compendium of Papers submitted to the subcommittee on Foreign Economic Policy of the Joint Economic Committee, Washington D.C.: U.S. Government Printing Office, 1970, p. 634.

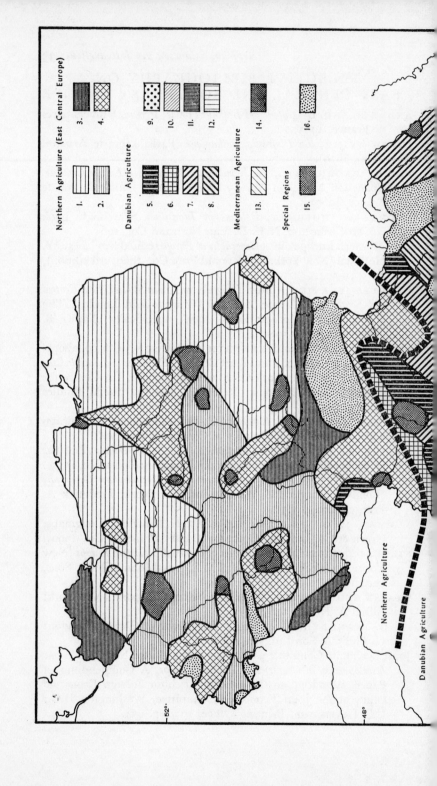

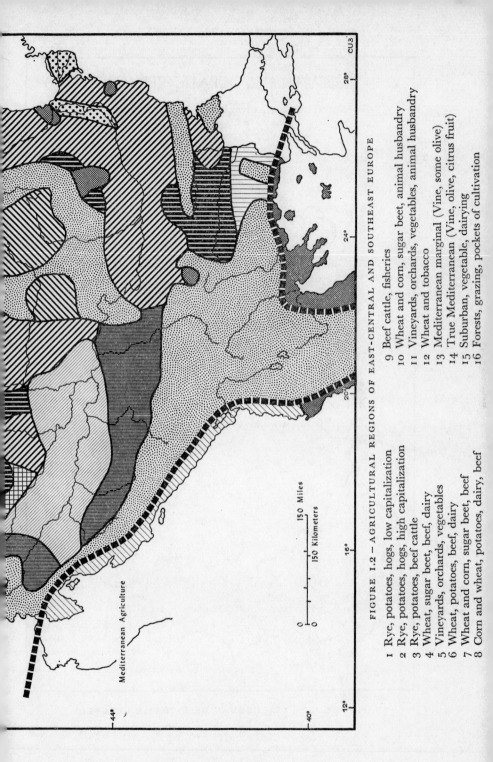

FIGURE I.2 – AGRICULTURAL REGIONS OF EAST-CENTRAL AND SOUTHEAST EUROPE

1 Rye, potatoes, hogs, low capitalization
2 Rye, potatoes, hogs, high capitalization
3 Rye, potatoes, beef cattle
4 Wheat, sugar beet, beef, dairy
5 Vineyards, orchards, vegetables
6 Wheat, potatoes, beef, dairy
7 Wheat and corn, sugar beet, beef
8 Corn and wheat, potatoes, dairy, beef
9 Beef cattle, fisheries
10 Wheat and corn, sugar beet, animal husbandry
11 Vineyards, orchards, vegetables, animal husbandry
12 Wheat and tobacco
13 Mediterranean marginal (Vine, some olive)
14 True Mediterranean (Vine, olive, citrus fruit)
15 Suburban, vegetable, dairying
16 Forests, grazing, pockets of cultivation

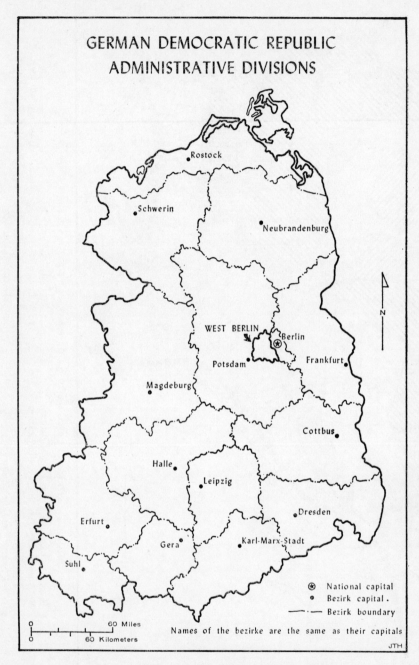

FIGURE I.3 — THE GERMAN DEMOCRATIC REPUBLIC

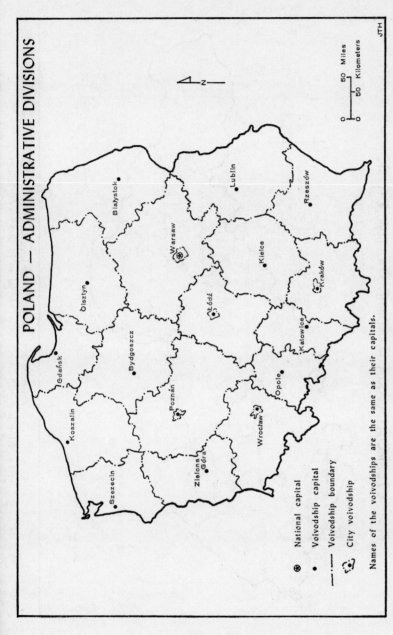

FIGURE I.4 – POLAND

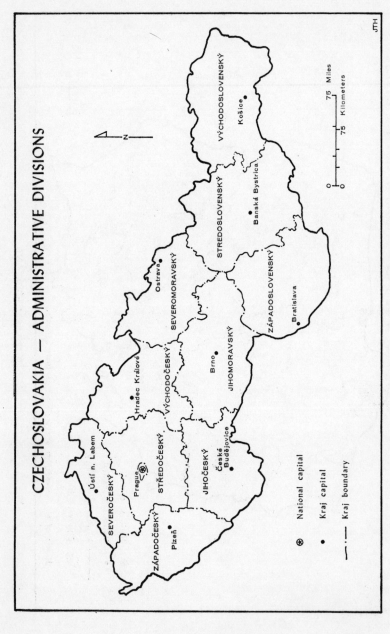

FIGURE I.5 — CZECHOSLOVAKIA
As of March 1970 administrative divisions in process of change

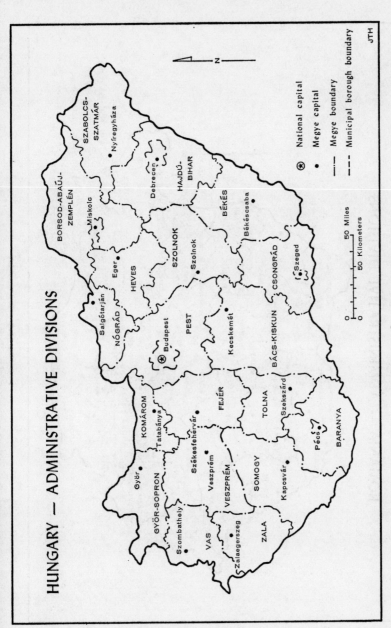

FIGURE 1.6 – HUNGARY

FIGURE I.7 – ROMANIA

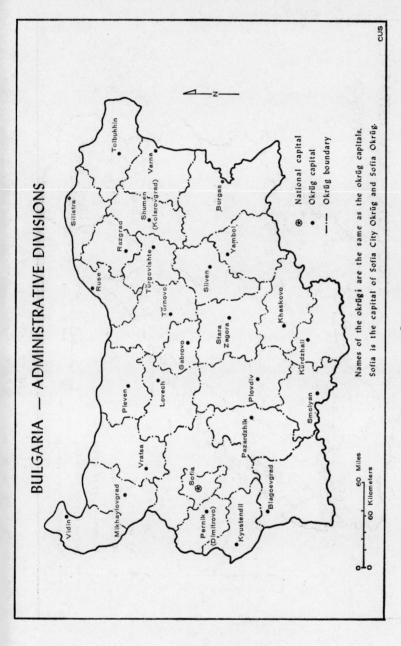

FIGURE 1.8 – BULGARIA

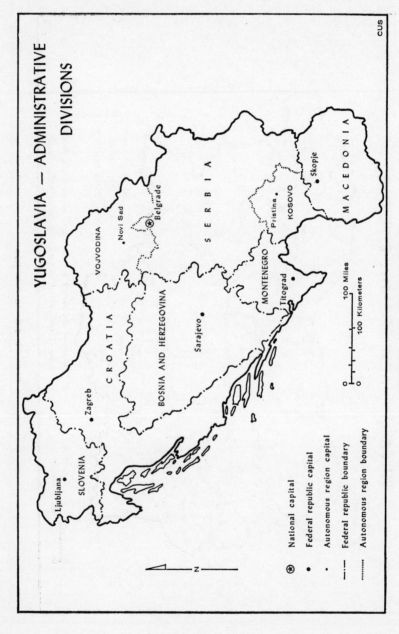

FIGURE I.9 — YUGOSLAVIA

FIGURE I.10 – ALBANIA

FIGURE I.II – GREECE

2

The Urbanization of East-Central and Southeast Europe: An Historical Perspective*

NORMAN J. G. POUNDS

In the period before statistics of national income began to be compiled and used, the foremost measure of the power and prestige of a state was the number, size and splendour of its cities. In them much of the surplus yielded by the land was invested in bricks and mortar. Their churches and civic buildings, the houses of their citizens and their walls and gates symbolized their wealth and prosperity. From the 'Order of Noble Cities' of Ausonius[1] and the fifteenth-century debate of the English and French heralds regarding the wealth of their respective countries[2] to the city atlases of the Renaissance and the travel narratives of the seventeenth and eighteenth centuries, cities were used as an index of affluence and power.[3]

In all such measures, Southeast Europe fared badly and East-Central only a little better. Cities were few, shabby and unpretentious. Fynes Moryson found even the 'better towns' of Poland built 'of timber and clay ... the houses being covered with straw, or tiles of wood, and the gentlemen's houses ... of no beauty'.[4] To a traveller at the very beginning of the nineteenth century, even Prague, probably the foremost at this time of all the cities of Eastern Europe, was 'forlorn and dreary: wide deserted streets, dirty Jews, and begging monks, ruinous palaces, and mouldering Gothic churches'.[5]

The purpose of this paper is to examine the ways in which the course of urban history, as it has been established for western and southern Europe, has impinged upon and influenced the historical development of East-Central and Southeast Europe, as it has been defined for the purposes of this volume.

* The author is deeply indebted for help, criticism and advice to Professors Ian Matley and Leszek Kosiński.

In the history of the Western World there have been three periods when the foundation and extension of cities was a major preoccupation of mankind. In each, urban living carried prestige and held economic advantages; the urban dweller thought of himself as a different kind of person from those who continued to live in the villages and the hamlets of the countryside.

The first of these periods was that of the classical civilizations of Greece and Rome. It began, perhaps in the seventh century before our era, in the Aegean region. It was diffused throughout the Greek world; was carried farther afield by Alexander, his successors and his imitators, and was adopted by the Etruscans and Romans. It was the Romans who gave to the classical city its greatest geographical extent. The foundation and growth of cities reached their climax in the second century of our era, after which very few were established and many underwent a long, slow decline and decay.

Though a few cities maintained urban institutions and some semblance of urban life through the early Middle Ages, they formed only a tenuous link between the classical period and the central Middle Ages, when the older cities were revived and a plethora of new towns was established. This period of urban revival was shorter than the classical period of urban ascendancy. According to how it is defined, it began between the eighth and eleventh centuries, culminated in the thirteenth, and waned in the middle years of the fourteenth.

The urban pattern established by this date underwent only minor changes before the development of the modern industrial and commercial city. In most instances the latter represented the continued growth of the larger cities of an earlier age, but some new towns were founded. The latter included the port cities of the lower Danube, the squalid alleys of Łódź and the tenement blocks of Katowice, as well as the 'socialist towns' of Nowy Tychy, Havířov, Dunaújváros and Victoria. Their purpose in every instance was clear; they were founded to house the workers in the docks and the 'dark, satanic mills', which the Industrial Revolution had called into being as its influence spread eastward through the region.

Between each of these periods of growth urban life and institutions stagnated or declined. Between the classical and medieval periods of urban growth, the decline was catastrophic.

Everywhere cities contracted, and many disappeared from the map. Rome was reduced to a shadow of its former self. The contraction of Constantinople was delayed, but at the time of its capture by the Turks, its population was probably no more than a tenth of that at the time of Justinian.[6] Between the fourteenth century and the nineteenth there was at best a long period of stagnation, when cities ceased to grow and none were founded. Locally, indeed, there was an absolute decline in their number and size.[7]

CLASSICAL URBANISM

The European city did not suddenly come into being, fully formed, like Athena emerging from the head of Zeus. There was, preceding each of the periods of urban growth, a proto-urban period when institutions existed capable of being fashioned into cities. The proto-urban nucleus of the classical city was, as a general rule, a fort of primitive design, situated on an elevated and naturally defensible site. The Cadmea, the Larissa, and the hill of Mycenae come to mind. Both archaeology and legend represent them as the seats of tribal leaders. If crafts and commerce were carried on within them, this was only because their princely rulers created a demand for military and exotic goods. The ἀκρόπολεις of the Achaean world of the Greeks differed from the Iron Age forts of southern Britain, the Celtic strongholds of Germany and France and the fortified enclosures of the early Slavs, more in the materials of which they were built and the climate and other physical conditions of their occurrence than in their function and purpose. For only limited areas of Europe is it possible either to map or to form realistic estimates of the numbers of these forts.[8] They were numerous, and in some areas very closely spaced. It is not unreasonable to assume that their density bore some relation to the size of the protohistoric population, and thus to the carrying capacity of the land. They were, from the point of view of the clan or community, places of refuge to which, in time of emergency, the local population could retreat. They were also – at least in East-Central and Southeast Europe – the seats of the clan leaders and the centres of religious cults. To only a very minor degree did they serve any economic purpose as centres of craft industries and foci of trade.

It was the achievement of the Greeks to take this proto-urban institution and to transform it into a place of beauty; not merely a place of refuge but a cultural centre. Thus, the city, which originated 'in the bare needs of life', continued, in the words of Aristotle, to exist 'for the good life'.[9] The good life consisted in conversation in the shady stoa of the αγόρα, in the colourful ceremony of the temple, and in visits to the theatre and the baths. Pausanias scornfully dismissed the city of Panopeus in Phocis, 'if city it may be called that has no government offices, no gymnasium, no theatre, no agora, no water conducted to a fountain . . .'[10] The true city was thus defined in cultural rather than economic terms.[11]

One has thus to ask, not whether the proto-urban city was spread through Eastern Europe, but how deeply the more refined, more sophisticated classical πόλις was diffused from the Aegean hearths of Greek culture into the Balkan backcountry. In this sense, the urbanization of the Balkan peninsula and Danube valley occupied most of the classical period. During the Greek period the πόλις made little progress beyond the coastal regions of Macedonia and Thrace, and even here, if the testimony of Libanius is to be believed,[12] many of the cities were somewhat marginal as judged by Greek standards. In Epirus, the Greek πόλις had as little influence as the Roman *civitas* was to have amid the hills of Lusitania.[13] Of the total of perhaps a hundred πόλεις which clustered along the northern shores of the Aegean from Kalchidike to the Hellespont and along the Black Sea coast as far as the *Ister*, a mere handful survived the classical period. The rest, however splendid their buildings, however cultivated the life that was lived in them,[14] succumbed to military or economic disaster, and for several not even their site is known with precision.

The Romans diffused the classical concept of the city very much more widely than the Greeks had been able to do, extending it not merely to the Rhine and Danube, but also to northern Britain and Transylvania (Fig. 2.1). But Roman urbanization was, at least in these peripheral regions of the Empire, imposed by the government and maintained for military and cultural rather than economic reasons. An open net

The Urbanization of East-Central and Southeast Europe 49

of roads was spread over Southeast Europe, with cities at strategic locations along them. One must conceive of each of these as the focus from which Roman ways of life and habits of speech were diffused into the surrounding areas. It is difficult to measure this diffusion. In Pannonia, where the terrain facilitated the acculturation process, it was probably very much

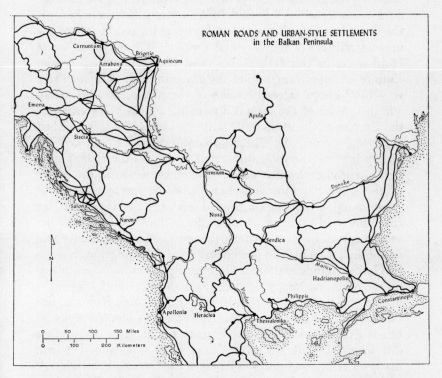

FIGURE 2.1 – ROMAN ROADS AND URBAN-STYLE SETTLEMENTS in the Balkan peninsula based on the Peutinger Table and Antoine Iter: Konrad Miller, *Itineria Romana* (Stuttgart: Strecker and Schröder, 1916)

more successful than amid the mountains of the Balkan peninsula. In any event it is almost impossible to measure it. The distribution of Latin inscriptions around the Danubian cities of Carnuntum (Petrolell), Aquincum (Obudá) and Poetovio (Ptuj) suggest the kind of cultural pattern which might have emerged.[15] The epigraphical evidence for the Balkan peninsula is, however, too tenuous for even this crude indicator of the role of cities to be used.

THE EARLY MIDDLE AGES

The city declined in importance – both economic and administrative – during the later years of the Western Empire and the dark centuries which followed. In Western and Southern Europe, the Christian Church proved to be the principal bearer of the urban traditions of the vanished Empire. 'The Papacy,' wrote Hobbes, 'is no other than the ghost of the deceased Roman Empire sitting crowned upon the grave thereof.'[16] In the same way the bishops were the heirs of the provincial and urban administrators of the later Empire. In almost every *civitas* of the Empire a Christian cell, with its bishop, had established itself, and during the Christian centuries which followed the bishop became the most influential as well as the richest local figure. His church assumed a prominent position in the centre of the city. Satellite churches gathered nearby,[17] and the whole religious complex attracted and sustained a small lay community of craftsmen and traders. There was no such development in most of Eastern Europe. Latin Christianity had established itself only in peripheral areas. The interior, overrun by Goths, Gepids and other Germanic peoples and then settled by the Slavs, no longer needed, nor indeed could it afford, the luxury of classical urbanism. In vain did cultured Romans, like Cassiodorus, plead for a restoration of urban life: 'let the wild beasts live in fields and woods: men ought to draw together into cities'.[18] Jordanes related how Pannonia, 'adorned with many cities', passed into the hands of the Germanic invaders,[19] and Constantine Porphyrogenitus described the devastation which resulted. 'The remaining cities, on the mainland and in the province (of Dalmatia), which were captured by the ... Slavs, now stand uninhabited and deserted, and nobody lives in them.'[20] The citizens of Salona found refuge within the walls of the palace which Diocletian had built beside the sea; those of Epidaurus (Cavtat) fled to the offshore island of Ragusa, and Narona and Dioclea were abandoned in favour of coastal settlements accessible to the sea-power of Byzantium.[21] In the interior the larger cities decayed, and the small towns and *vici* disappeared, leaving no trace on the landscape. Nowhere else in Europe was the process of de-urbanization as complete as in those parts of Southeast Europe which lay beyond the protection of the Byzantine emperors.

Within its boundary, however, the military danger served to perpetuate urban institutions. Procopius tells how the Emperor Justinian fortified the village of Lipljan, near Priština, where he had been born, 'with a wall of small compass ... in the form of a square, placing a tower at each corner'.[22] He also built 'an aqueduct and so caused the city to be abundantly supplied with ever-running water ... it is impossible to tell in words of the lodgings for magistrates, the great Stoas, the fine market-places, the fountains, the streets, the baths, the shops ...'[23] No doubt Procopius was flattering his emperor, but one cannot doubt that Justinian himself was expressing the classical admiration for urban living as well as taking steps to protect his subjects from invaders.

This system of fortified villages was extended over the frontier regions of the Empire, and in Greece itself, city walls were restored. Procopius credited the Emperor with fortifying 44 sites in Epirus, and with restoring the defences of 49 others; in Macedonia he repaired or established 46; in Dardania (northwestern Greece) no less than 69, and in Thessaly, 7.[24] Most of the names in Procopius' list were no longer identifiable. Doubtless they failed in their primary military objective, were overrun by the Slav or Bulgar invaders, and were destroyed. If they developed any distinctly urban functions – and there is evidence that some of them did – these were short-lived.

Stobi was one of the fortified sites listed by Procopius; indeed, it is the only one at which one can compare Procopius' description with reality. It lay in the Vardar Valley, 15 miles southeast of Titov Veles, where the modern road to Ohrid branches to the west. No settlement occupies the site today, but excavation has revealed a small town of the fifth and sixth centuries.[25] A complex defensive wall enclosed an area of about 20 hectares. Within were paved streets, lined in all probability by arcaded stoas. Crafts, particularly those associated with the manufacture of textiles, were carried on, and there were large, peristyle houses of Hellenistic design, inhabited probably by the wealthier citizens. Stobi was, in effect, a small, Hellenistic city, which flourished in the later years of the Roman Empire,[26] was sacked by Theodoric in 479, but was restored to enjoy a brief period of prosperity in the sixth century before its final extinction.

One must assume that many, if not most, of the sites which Justinian fortified had a similar history. A few probably were continuously inhabited during the period of invasion and Slav settlement, though probably no more populous or prosperous than comparable cities in France at this time. Among them was Thessaloníki, which had succeeded in holding off the attacks of the Slavs, though not, if one may credit the *Vita* of St Demetrius, without the help of a miracle or two.[27] The lists of bishops provide some evidence for the continued existence of cities, since an episcopal see without a city was almost a contradiction in terms.[28] At the councils held in the seventh century, representation was almost exclusively from the coastal regions of the Aegean.[29] The Balkan hinterland went unrepresented, for here no bishop served as ghostly successor to either Roman administrator or Thraco-Illyrian chief. By contrast, the Asian provinces of the Byzantine Empire were still 'covered with a network of cities',[30] each the *cathedra* of a bishop and the commercial centre of its local region.

The recovery of the Empire in subsequent centuries was reflected in the greater representation at the Councils of the Church. Bishops were present representing sees, and presumably cities, in the Rhodope and Maritsa regions, but most of the new foundations were in Thrace, Macedonia and Greece where the urban institutions of the classical period had suffered the smallest degree of interruption.

To a very small extent the invaders themselves participated in the process of creating cities. After the Bulgars had established themselves on the platform of northeastern Bulgaria they founded their first capital at Pliska. Excavation has revealed a very large enclosure which included not only churches and a small community of craftsmen but also a considerable body of farmers. Within this larger area was a smaller fortified enclosure, covering only about 700 square metres, doubtless the seat of the Bulgar prince.[31]

The seat of the Bulgarian kings was removed from this site during the tenth century to nearby Preslav.[32] Under the Second Bulgarian Empire the latter was also abandoned, and declined to a village, while the rulers built their formidable castle within the meanders of the Jantra at Trnovo. At the same time a small urban community developed to the north of the fortress. In all these instances, the proto-urban forms and institutions appear to have resembled both those which were at the same time being developed in the northern Slav lands and also those which evolved more than a thousand years earlier in the Aegean lands.

At the end of the eleventh century the crusading hordes came through the Balkans by a variety of routes on their way to Constantinople and the Middle East. The fullest account of the First Crusade is that of William of Tyre, based on the narratives of a number of informants, none of whose accuracy was above reproach. The separate groups of Crusaders, led by Walter the Penniless and Peter the Hermit both converged on Semlin (Zemun) and Belgrade. The chronicler relates that the Bulgars, 'who had little confidence in the defences' of the latter, retreated 'with their flocks and herds ... far into the dense and secret recesses of the forest'[33] – presumably the Šumadija.

A journey of eight days took Peter the Hermit from Belgrade to Niš, which the chronicler described as 'strongly fortified with a wall and towers'. The Crusaders, who had a marked aptitude for provoking violence, retaliated against the stubbornness of the local people by burning 'seven mills which were revolving near the bridge over the river (Nišava or Morava) ... These were quickly reduced to ashes. ... They set fire also to the houses of certain people which lay outside the walls.'[34]

At Belgrade, Walter the Penniless had demanded 'the privilege of a market'[35] from the local chief – presumably permission to purchase supplies. A similar request was made by Bohemond of Kastoria, where 'markets were not provided by the city for people who were passing through. Hence they were compelled to seize by force flocks and herds and other things necessary for food...'[36] On the other hand, a market was provided at Stralicia, 'a beautiful city of Dacia Mediterranea',[37] possibly to be identified with Sredec, or Sofia.

The evidence of the Crusading narratives, scanty as it is, nevertheless corroborates other sources, and suggests that there was very little evidence of urban life outside the frontiers of the Byzantine Empire itself; that in so far as the fabric of a few imperial cities had survived, they might have been inhabited by small colonies of agricultural and even pastoral peoples, and that specifically urban functions were rarely to be found.

THE PROTO-URBAN NUCLEUS IN SOUTHEAST EUROPE

The simple proto-urban nuclei of Southeast Europe had disappeared, either replaced by cities of classical concept and form or abandoned by the tribes which had formerly inhabited them. This was not so in those territories lying to the north of the Danube, which had never come within the boundaries of the Roman Empire. Both archaeological evidence and the writings of early chroniclers and travellers confirm that they were both numerous and important during the centuries which followed the collapse of the Roman Empire in the West.

These proto-urban nuclei belonged mostly to the late Iron Age. They were the seats of clan leaders; they probably contained the religious symbols of the clan, and they provided a place of refuge in time of war. Their function, at least in their earlier manifestations, was local, and such trade as may have been carried on was probably intermittent and restricted to a few luxury goods.[38] They tended, as in the Celtic-settled areas of Western Europe, to occupy easily defended sites, but in the Polish plain, where topography gave them little help, they relied more often on the protection of lake and marsh. On the basis of the archaeological evidence Kostrzewski enumerated no less than 355 such *grody* in Wielkopolska alone, of which 251 occupied low-lying or marshy sites.[39] They were no less numerous in Małopolska, and in the Prussian regions of the Baltic, where the 'Bavarian' geographer claimed that there were over 3,000 *civitates*.[40]

In the Czech lands the *hrady* usually occupied the summits of hills, and many of them were so well defended by nature as to be almost impregnable. Turek has published a map of Dark Age *hrady*. They form two clusters (Fig. 2.2) in northern Bohemia and southern Moravia respectively.[41] From the latter they extended eastward into Slovakia and, no doubt, at one time southwards across the Danube into the territory of Austria and Transdanubian Hungary. In both Poland and the Czech and Slovak lands the density of such forts can be roughly correlated with soil fertility, and, even if we admit that not all were in use at the same time, they none the less connote rather densely settled population.

Excavation has not made sufficient progress for more than a very small number of the *grody* and *hradiště* to be dated. The evidence suggests that most were built, in the main, between the seventh and the tenth centuries, but it is probable that some – especially the smaller of them – originated many centuries earlier.[42] Several of them were large, like Staré Město,

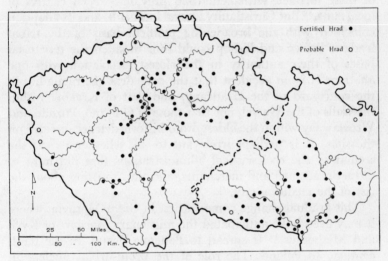

FIGURE 2.2 – HRADY IN THE CZECH LANDS, after Rudolf Turek, in *Slavia Antiqua*, Vol. 2 (1949–50)

the reputed capital of the 'Great Moravian' state, and Mikulcice, which covered from 10 to 20 acres. All were well defended with walls of earth, wood and masonry. Several of those examined were found to have rows of huts, which were permanently inhabited, as well as workshops and masonry-built churches. In several of them some form of trade was carried on, primarily, it would appear, in luxury goods. In all, these *hrady* suggest a cultural and economic level not very different from that represented in the ἀκροπόλεις of Achaean Greece.

By the ninth or tenth centuries, some of these fortified settlements had become more wealthy and politically more important than the rest. Tribal areas of increasing size came gradually to be controlled from them, until a rudimentary state took shape around them. It was at this time that the church set its seal upon these large and more important centres, by establishing a bishop or at least conferring some higher ecclesiastical status upon them. The church, in the sense of a physical structure of relative opulence and sophistication, became, along with the castle of the prince, the outward symbol of the settlements' pre-eminence. It is probable that many, if not most, of these fortified settlements had previously been centres of local cults.[43] But Christianity was no local cult, and its adoption coincided with the broadening political basis of the tribal leaders. Prince and bishop could thus broaden the territorial bases of their authority in the closest association with one another. It is no accident that the cathedral of Prague lies in the courtyard of the Hradčany, and that of Kraków within the walls of the Wawel, nor that those of Gniezno, Poznań and Wrocław lay within the oldest proto-historic defences erected on the sites. It is perhaps irrelevant to ask whether it was the secular or the ecclesiastical administrations that did most to attract craftsmen and merchants, and thus to bring about the rise of the medieval city.

This gradual change came earliest in Great Moravia, where it may even have anticipated the missionary journeys of Kyril and Methodius.[44] It spread to Bohemia,[45] and, in the tenth century, to Poland. The role of the proto-urban nucleus in attracting administrative and economic functions is more clearly demonstrated in Poland than elsewhere. It came a century or more later than in Moravia, was more adequately documented, and the *grody* sites have themselves been the object of a more thorough and prolonged excavation. The institutions and processes revealed closely resemble those demonstrated in Moravia. From a large number of fortified settlements, a few gradually separated themselves, attained a greater size and importance than the others, acquired churches, and became centres for ecclesiastical as well as secular administration.

The pre-urban nuclei may have originated in the eighth or ninth centuries, but their most rapid development was in the tenth and eleventh.[46] The smaller and less important may at this date have disappeared from history. Others began to acquire a two-fold structure, consistent with the growing feudalization of the period. The fortified enclosure, in origin a refuge for the local communities by whose labours it had been built, became gradually the seat of the prince, the *gród*, *hrad*, or castle, while an adjoining enclosure, or *podgrodzie*, developed to house and protect retainers, craftsmen, merchants and also the institutions of the church.[47]

The morphology of such settlements is now fairly well known. They resembled very closely those settlements which in northwestern Europe are known as 'mottes-and-baileys': a rounded enclosure, deriving from the original *gród*, with an appendant enclosure of rounded or crescent shape. The plan of pre-urban Poznań, as reconstructed from the excavations, illustrates the kind of complex that was emerging.[48] Similar structures have been demonstrated at Kalisz,[49] Gniezno,[50] Wrocław,[51] Sulecin,[52] Kraków.[53]

It is evident both from the literary sources and from the archaeological evidence that these pre-urban centres were beginning to assume the features and functions of true cities. Their local regions were yielding a growing surplus, which their rulers could employ in war or in conspicuous consumption. They attracted merchants and supported craftsmen.[54] Wolin was described as 'a most noble city . . . (and) a very widely known trading centre for the barbarians and Greeks (*sic.*) who live round about . . .' and, with some exaggeration, as 'the largest of all the cities in Europe'.[55] The nearby *Schinesghe* or *Schinesne*, undoubtedly the modern Szczecin (Stettin), also figured prominently in the early records, and was 'urbem . . . populosiorem et opulentiorem'. The anonymous *Gallus* also referred to the forts (*castra*) of Nakło and Miedzyrzecz,[56] and the Arab traveller of an earlier period, Ibrāhīm Ibn Ja'kūb, described Prague as 'built of stone and lime and . . . the richest in goods. Russians and Slavs come here with their merchandise from Kraków . . . and also Moslems, Jews and Turks . . .'[57]

Before leaving the question of the pre-urban nuclei of the northern Slav lands, some mention must be made of their military role. The life of Otto of Bamberg mentioned Uzda (Uście nad Noteć) as lying right on the frontier of Poland;[58] other early writers also mentioned *grody* or *castra* in such frontier locations, and some historians claim to detect a grand design on the part of Mieszko I (d. 992) or Bolesław Chrobry (992–1025) to erect a chain of forts or *grody* to protect Wielkopolska, the heartland of the first Polish state.[59] If this was indeed their policy, it must have borne a certain resemblance to that of Justinian in his building of fortified towns in the southern part of Southeast Europe to protect the Empire. It is a fact that many of the *grody* were well located to command the crossing of the marshy *pradoliny* of Wielkopolska. The chronicles record that some of them actually played a military role, but their continued existence in this proto-feudal state must have been predicated more on their economic than their military importance.[60]

The course of urban development, at least until the eleventh century, was markedly different in the lands lying north of the Danube from those to the south. In the latter, the primitive fortified enclosure of the tribe or clan had yielded to the classical concept of the city, and the latter had been largely destroyed during the long periods of almost continuous war.

Paradoxically, those areas which had never been subjected to Rome were less disturbed, and some of their earlier forts had slowly evolved by a kind of selective process into proto-urban centres and thus into true cities with economic as well as administrative functions. Thus, the more northerly and barbaric lands overhauled the more southerly and sophisticated, and have maintained their lead to the present day.

THE MEDIEVAL CITY

EAST-CENTRAL EUROPE

In the eleventh and twelfth centuries the city in the northern Slav lands was, in the main, an autonomous development. It may have owed something to both the example of the western states and the influence of the Christian church, and its growth was undoubtedly stimulated by traders from all directions and of most ethnic origins. But it was the native peoples of East-Central Europe, and principally the Slavs, who developed towns as the central places of distinct areas and as the foci of long-distance trade. It is, however, not easy to distinguish between this native movement and the wave of urbanism which spread eastwards from the Rhineland in the twelfth, thirteenth and fourteenth centuries. The latter is sometimes represented, especially by the German historians, as the imposition of a higher, more urbanized, more specialized culture and economy on the Slavs, who had hitherto lived a predominantly agricultural life. Nothing could be farther from the truth than this simplistic view.

It should be noted that in very many instances the Germans came in response to invitations from Slav princes, and that the non-agricultural component of this movement was probably quite small. The eastward migration of Germans arose in the first instance from overcrowded rural conditions in the Rhineland. In the rural areas the Germans settled, probably as small, distinct communities *among* the Slavs, as the fourteenth-century register of Sorau (Żary) in the *Województwo* of Zielona Góra clearly shows.[61] In the towns they formed a small élite of merchants and craftsmen among an urban population that remained predominantly Slav and mainly agricultural. The example of Reval in the sixteenth century, though outside both the area and the period at present under discussion, may nevertheless throw some light on the composition of the cities of East-Central Europe.[62] On the basis of personal names and the slight evidence of origin, no more than 30% of the city's population was found to be German. Almost all the others were native Esths from the nearby villages.

One might expect that the population of the larger Hanseatic cities would be overwhelmingly German. Their leadership unquestionably was, but at Danzig (Gdańsk) considerable numbers of the actual citizens came from the nearby and predominantly Slav regions of Pomorze Prussia and the Vistula Valley.[63] The proportion of immigrants who came from Germany never rose above 31% in the fourteenth century,[64] and later declined relative to that of the Slavs.

Such detail cannot be established for other cities of East-Central Europe. Prague, for a prolonged period the seat of the German Emperors, was strongly influenced, both culturally and economically, by Germans. There was probably a larger immigration of Germans to Prague than to the port cities of the Baltic, but everywhere else the impact of German settlers was small. A town of regular plan did not imply that its citizens were of German origin; only that the plan had been conceived by a local authority – presumably feudal, and executed by someone – perhaps German – who was aware of contemporary developments in this field in Central Europe.

The planned town, such as we find in late-medieval Poland, was not merely a Central or East-Central European phenomenon; it was common to all Europe. In lay-out and probably also in function it resembled the 'new towns' of England and the *bastides* of France.[65] It was a late medieval way of pioneering in empty or thinly populated land. Though urban in form, the new towns in both East-Central and Western Europe remained rural in function. They were a means of attracting rural population, not specifically from Germany, but from anywhere, by offering to the migrants the attraction of urban status and privileges.

The eastward diffusion of the town is well known. The spread of the 'German' town has been described by Rudolf Koetzschke[66] and others; it has been illustrated in numerous maps, most recently and elaborately in the *Atlas Östliches Mitteleuropa*.[67] It must be emphasized, however, that this map is not one of the diffusion of urbanism as such, but only of the spread of certain individual urban institutions. Nor did areas engulfed by this wave of 'German' urbanism change their economy. There were market centres and small urban agglomerations in the western Slav lands long before the so-called German towns made their appearance, and the latter remained more often than not merely very pretentious villages long after their establishment.[68]

Nor were the 'new towns' regularly established on virgin sites. In most instances the antecedents of the planned town are not known. In a few instances they were clearly preceded by villages, and in the most familiar instances they were grafted on to pre-existing Slav towns. In these latter cases it is usually easy to discern, on the ground, the several stages of urban growth. The topography of modern Wrocław, Poznań and Kraków shows in each case a fourfold division into the small area of the primitive *gród*, the early medieval *podgrodzie*, the later medieval planned city, and the modern development of industrial, commercial and residential areas peripheral to the medieval settlement. At Kraków these elements assumed an almost linear pattern; at Poznań, Opole, Wrocław, and above all at Prague, the later city grew up across the river from its earlier nucleus. At Łęczyca,[69] Gdańsk[70] and Kalisz none of the components of the later city were contiguous, and were in some instances quite widely separated.[71] In the course of time the Slav nucleus tended to be abandoned in favour of the more spacious settlement of the later Middle Ages. In some instances the latter grew and totally engulfed the older settlement; in a few, the towns have been so drastically rebuilt that their primitive zonation is no longer apparent in their architecture and street plan. Most often, however, the successive stages in the growth of the towns can be studied in the architecture of their churches: the fragments of romanesque in the oldest quarters, the gothic town church of the later medieval settlement, and the eclectic styles of the modern suburban growth.

The whole of Central Europe was over urbanized, in so far as there were too many settlements with the legal institutions and outward forms of towns. Within the limits of the German Reich there were as many as 3,000,[72] and Hektor Ammann counted no less than 88 in the basin of the Swiss Rhine.[73] The majority are classed as *Klein-* or *Zwergstädte*. They had urban pretensions, but their functions differed scarcely at all from those of large villages. They were market centres for surrounding villages, for which they supplied certain simple services. If this was the situation in the relatively well developed and commercially active areas of South Germany, it was even more the case in Eastern Europe. Much of Poland is in consequence plagued today with too many 'small towns' whose scanty functions have been eroded by modern transport facilities and replaced by those of a smaller number of more suitably located and larger towns.

The number of cities in East-Central Europe which served an area extending more than a market journey from their walls was very small, as indeed it was in all Central Europe. It is very difficult to assign size ranges to a given scale of functions for the cities of Central and Eastern Europe. Perhaps it might be suggested that towns with no more than 1,000 inhabitants (200–250 households) can be regarded as predominantly agricultural with exclusively local importance; that up to a size of 2,500 persons they were of regional but still restricted importance; that cities of 5,000 must have had an important

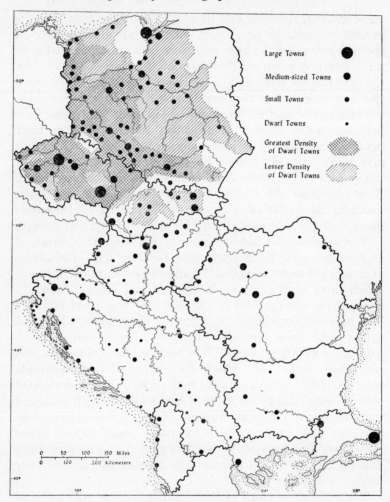

FIGURE 2.3 – CITIES IN EAST-CENTRAL AND SOUTHEAST
EUROPE DURING THE LATER MIDDLE AGES

basis in commerce and crafts, serving a very much larger area, and, finally, that the few cities of continental importance may have exceeded 10,000 people (Fig. 2.3).

This is quite clearly giving too great a precision to the size-function relationship of late medieval towns, but this classification, which is derived from the better documented urban history of Western Europe, may serve until fuller evidence becomes available.

Reliable estimates of population can be formulated only for the larger cities. Keyser has, on the basis of the city's *Bürgerbuch*, put the population of Danzig (Gdańsk) at about 10,000 in the mid-fourteenth century;[74] Rostock may have had about the same;[75] Kraków was probably similar, and Prague may have been somewhat larger. Brno is said to have had 12,000,[76] and Cheb (Eger), about 7,000 in 1390. It is doubtful whether Kaliningrad (Königsberg), Poznań, Toruń and Wrocław had very many more than 5,000 inhabitants.

In the sixteenth century very few towns of the present German Democratic Republic and western Poland had more than 500 hearths (2,000–2,500 persons), and some had less than a hundred. Frankfurt-on-Oder had only 757 hearths (about 3,000 persons) in the sixteenth century, despite its not inconsiderable importance as a river port at this time.[77] This small average size is amply demonstrated by the generally reliable hearth list of 1495 for Mecklenburg.[78] The towns would have been no larger in the fourteenth century and many may well have been smaller. It is to be presumed, furthermore, that their average size would have diminished towards the east, and that the towns of Poland and the Czech lands would, category for category, be smaller than those of Germany.[79]

There is for most of the smaller towns not a shred of statistical evidence, and the only measure of their comparative size is the area which they occupied within their walls. Fortunately, this can in most instances be derived from modern topographical maps, on which the line of the medieval walls, if not the walls themselves, can generally be traced with the help of early cadastral plans and the woodcuts of the early topographical writers, such as the Merians.[80] The map of late medieval towns in East-Central Europe is thus highly conjectural, and is based on literary and cartographical evidence which it is extremely difficult to quantify.

SOUTHEAST EUROPE

The extension of this map to the Pannonian Plain, Romania and the areas south of the Danube faces very great difficulties.

Literary evidence is slight, and topographical and archaeological little explored and in some instances virtually obliterated. The urge to establish cities and to endow them with rights and privileges was felt only in a few restricted areas, such as Transdanubia,[81] Transylvania, Slovenia and neighbouring parts of Croatia. More specialized mining towns were founded, notably in the Krušne Hory, where Schneeberg, Annaberg, and Altenberg were established in the later Middle Ages;[82] in Slovakia, and in Bosnia.[83] Mining settlements are apt to be short-lived, and relatively few of those established during the later Middle Ages actually enjoyed a long period of prosperity. Many declined into villages and some have disappeared as completely as ghost-towns of the American West.

The Tartar invasion of 1241 may have contributed to urban growth in so far as it emphasized the advantages of living within the shelter of a city wall. Buda took shape in part for this reason, but Pest was already a trading community of some importance.[84] In western Hungary, Sopron (Ödenburg), Pécs (Fünfkirchen), Győr and Bratislava were regional centres of 3,000 to 5,000 people. The German cities of Transylvania, particularly Kronstadt (Braşov) and Herrmannstadt (Sibiu), may have been somewhat larger.[85] It is doubtful whether the twin settlements which at this time made up Zagreb, could together have had much more than 2,500 inhabitants,[86] and Ljubljana (Laibach) was probably no larger.

Urbanization made very little progress in Southeast Europe; indeed, it further decayed during the latter Middle Ages over most of the area. A few of the urban sites of the late classical period continued to be occupied, but the most significant of them lay on the coast of the Adriatic and Black Seas, where they had developed under the protection of Italian naval and commercial power. The cities of the interior had never recovered from their devastation during the period of the invasions. Constantinople and Thessaloníki were diminishing in size and importance, and for others there is absolutely no statistical evidence.[87] Only the mining settlements of Bosnia, Serbia and Montenegro constituted an exception. The Romans had exploited the minerals of this area, but their operations did not give rise to urban settlements. In the later Middle Ages a number of mining towns grew up, of which the most important were probably Zvornik, Srebrnica, Rudnik and Novo Brdo. Some of these mining centres, particularly Kucevo and Rudnik, were in fact colonies established by the merchants of Dubrovnik. All, however, were small, and few survived the Turkish conquest.[88]

Thus by the fourteenth century a complete reversal of the earlier pattern of urban development had been achieved. The northern part of the East European region was now the most developed, while city life had disappeared from large areas of the Danube basin and Southeast Europe. Yet there were certain similarities between the nature of urban development in the classical and medieval periods. Both consisted in the growth of a small number of widely spaced towns of large size and diversified function, which in some way served the commercial needs of very extensive areas. Below these, hierarchically arranged, were greater numbers of cities of regional and local importance. The latter were smallest in size and most numerous. They were local centres where, in the words of Xenophon,[89] 'the same workman makes chairs and doors and ploughs and tables, and often this same artisan builds houses, and even so he is thankful he can only find employment enough to support him.' They remained more rural than urban in their function; they were essentially unspecialized and, in commercial terms, they provided little more than a medium of exchange between surrounding village communities.

THE MODERN CITY

The pattern of cities which had been established in East-Central and Southeast Europe by the mid-fourteenth century underwent little change until the nineteenth. More small towns were created in the fifteenth and sixteenth, especially in eastern Poland, which had not been greatly influenced by the medieval process of urbanization. War and resettlement changed radically the urban pattern in the Pannonian plain;[90] there was an urban revival in parts of Southeast Europe under Turkish rule;[91] centralized administration led to the growth of a number of capital cities, but, by and large, there was little growth or change before Eastern Europe felt the impact of the Industrial Revolution, and the nineteenth century inherited a basically medieval urban pattern.

Eastern Europe began to experience the revolution in industrial technology, which had taken place in northwestern Europe in the closing years of the eighteenth century. Its acceptance was gradual, largely because of the lack of an entrepreneurial class and the poverty of the local market. It was not until the middle years of the nineteenth century that it was reflected in any considerable urban growth. Much of East-Central and Southeast Europe did not experience the 'take-off into sustained growth', which is the feature of an industrial revolution, until after World War II.

Urban growth during the past century and a half has consisted partly in the foundation of new towns, but chiefly in the selective expansion of urban settlements already existing (Fig. 2.4).[92] New towns were very few during the nineteenth century, and most were associated with the exploitation of minerals, in particular coal. Most towns of the Upper Silesian industrial complex grew from villages at this time, including the largest of them, Katowice.[93] In Moravia, new towns also developed on the Czech extension of the Silesian coalfield. Away from the coalfield the most significant new town was Łódź, which was developed during the nineteenth century as a focus of the textile industries.[94] A number of small industrial centres, engaged mainly in the textile, footwear and ceramic industries, grew up in the hills which ring the Bohemian and Moravian plains, and others developed in Bulgaria, on the northern flanks of the Stara Planina, and in Slovenia and Croatia.[95] But urban population, which grew steadily during the half century preceding World War I, tended to move to the older urban centres. The cities which were already large during the later Middle Ages experienced the most vigorous growth, and their development was far more rapid than at any previous time. Fourteen of the major cities of East-Central Europe, for example, together increased almost exactly tenfold between 1819 and 1939.[96]

Urban development was less marked in Southeast Europe. The impact of industrialization was weaker and longer delayed than in East-Central. Political instability and the relative absence of clearly established urban centres and transport facilities deterred both the entrepreneur and the investor. Yet it would be a grave error to regard Southeast Europe as a

The Urbanization of East-Central and Southeast Europe 67

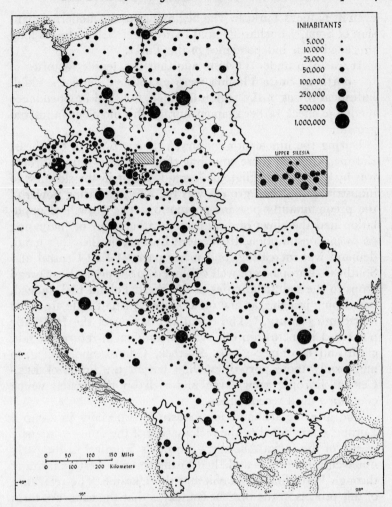

FIGURE 2.4 – URBAN DEVELOPMENT IN EAST-CENTRAL AND SOUTH-EAST EUROPE, ABOUT 1965. Based on maps in Norman J. G. Pounds, *Eastern Europe* (London and Chicago: Longmans Green and Co., and Aldine Publishing Co., 1969)

region in which urban and industrial growth was reduced to negligible proportions. The Turks themselves encouraged the growth of many cities for both administrative and economic purposes. Sarajevo was substantially an Ottoman creation, and even its name is Turkish. The beginnings of the modern expansion of Sofia,[97] Bucharest[98] and many other Balkan towns long preceded their independence of the Turks.

It was also under Turkish rule that the textile factories of the northern Stara Planina were established, and the small industrial towns of the region founded. Political independence tended to check rather than encourage this incipient industrial growth.

During the nineteenth century, urban growth in Southeast Europe, while nowhere equalling that in East-Central Europe, was by no means negligible. A major factor in the slowness of industrial and urban growth was the low purchasing power of the predominantly peasant population. The growth of population had the effect of diminishing rather than of increasing *per capita* consumption. In Western Europe, on the other hand, demand was increasing, and many parts of East-Central and Southeast Europe were well equipped to supply at least bread crops to the industrial cities of the West. Poland had, indeed, been doing just this, at first on a relatively small scale, from the fourteenth century.[99] The essential function of the Hanseatic ports had been to ship out rye and lumber in return for salt, cloth and consumer goods. Rostock, Gdańsk and Elbląg, to name only those port-cities lying within the limits of East-Central Europe, grew large and rich on this trade, which continued until recent times.

The growing demand for foodstuffs, especially in Central Europe, contributed to the reclamation of the sandy or marshy wastes of the Great Alföld[100] – Europe's 'frontier' in the early nineteenth century – and then to the spread of estate farming through Walachia, Moldavia and the Ukraine. The relevance of this process in the present context is that the export of farm products necessitated the construction of transport facilities, the building of docks, and the increase in both the number and the activity of markets. It contributed to the development of agriculture-oriented industries in the towns and to a general revival of urbanism. Access to the sea acquired a new importance, and the ports of the Black Sea, the lower Danube and the head of the Adriatic – points where the grain-growing lands of the Danubian and Southeast European countries had the easiest access to the avenues of world commerce – underwent a considerable expansion. Other products of East-Central and Southeast Europe shared in this process, notably the petroleum of Romania and the coal of Upper Silesia, with the consequent expansion of service centres on which they depended.

The urban growth of the nineteenth century was interrupted by two world wars and by a social and economic revolution which removed initiative from the entrepreneur and gave it to the state. This latter development was not altogether new. Both the Prussian and the Austrian governments had participated in economic developments which they regarded as militarily desirable. Between the two wars the Polish government embarked on a project for the industrial and consequently the urban development of its Central Industrial Region, with a more thorough-going organization of space and a more realistic approach to the problems of urbanism.

Other chapters in this volume examine the planning processes in the countries of the area. Among the objectives of the planners is the conscious and deliberate choice of areas for industrial and urban development. Their decisions are not necessarily more rational than those of the entrepreneurs of the nineteenth and early twentieth centuries, and many of the sites chosen may prove to have diseconomies as great as those of the stillborn towns of the Middle Ages. But the state has been able to bring to bear on the process of urban growth powers unknown to both the modern entrepreneur and the medieval *locator*. It can command the capital and other resources necessary for the rapid implementation of its plans; it can insure employment for the new citizens and a market for their products – at least, it can do so unless there is an error or miscalculation in its linear programming. In a certain sense, an accidental factor has been largely removed from the foundation and growth of cities, and one can at least expect that a new town or an expansion of an older town will be justified on social as well as economic grounds.

The urban growth of the past twenty years is here regarded as part of that more general process of urbanization which is associated with the adoption in East-Central and Southeast Europe of modern, large-scale, factory industry. The first century or more of this process was dominated by the western-style capitalist entrepreneur, and the last twenty or so years by socialist planning. This has given a twofold form to the modern town and to the modern extensions of older towns: the basically unplanned or only feebly controlled spread of a Łódź or a Katowice, and the closely controlled construction of a Nowy Tychy, a Dunaújváros, or a Novi Beograd. Towns of the latter type are few in number and large in size. They are to be met with from Poland to Bulgaria. They consist of tall apartment blocks, and they represent a highly concentrated, highly disciplined form of living, which is totally subordinated to the factories which these towns exist to serve. And as older towns come gradually to be rebuilt – like the devastated Muranów sector of Warsaw or the one-storied peasant huts of Bucharest – they are replaced by the tall blocks of apartments, rising, white and sanitary, over the relics of the past.

NOTES

[1] Ausonius, *Ordo Urbium Nobilium*.

[2] *Le Débat des Hérauts d'Armes de France et d'Angleterre suivi de, The Debate between the Heralds of England and France by John Coke*, ed. Léopold Pannier and Paul Meyer (Paris: Société des Anciens Textes Français, 1877); the French herald claims that 'pour une ville fermée que vous avez, nous en avons plus de douze bien peuplés, tant de gens mecaniques que autres . . .' *cap.* 116–19.

[3] See, for example, William Lithgow, *The Totall Discourse of The Rare Adventures and Painefull Peregrinations* (Glasgow: J. MacLehose and Sons, 1906), especially p. 362.

[4] Fynes Moryson, *An Itinerary* (Glasgow: J. MacLehose and Sons, 1907), Vol. 1, p. 133.

[5] Adam Neale, *Travels through some parts of Germany, Poland, Moldavia and Turkey* (London: 1818), p. 82.

[6] A. Andréades, 'La Population de l'Empire byzantin', *Actes du IVe Congres International des Études byzantines* (Sofia, 1935), pp. 117–26, suggests 800,000 to one million at its period of greatest prosperity. See also John L. Teall, 'The Grain Supply of the Byzantine Empire, 330–1025', *Dumbarton Oaks Papers*, No. 13 (1959), pp. 87–139. At the time of its capture the population was about 75,000; see H. G. Beck, 'Konstantinopel: zur Sozialgeschichte einer früh-mittelalterlichen Hauptstadt', *Byzantinische Zeitschrift*, Vol. 58 (1965), pp. 11–45, and Ömer-Lufti Barkan, 'Quelques observations sur l'organisation écono-

mique et sociale des villes ottomanes des XVIe et XVIIe siècles', *Recueil de la Société Jean Bodin*, Vol. 8 (La Ville), part 2 (Brussels, 1955), pp. 289–311.

[7] There is no work on the East European city even remotely resembling the *Deutsche Städtebucher* (Stuttgart, 1954 – in progress), edited by Erich Keyser, for West Germany. The best documented country is Poland, for which there exists *Slownik Geografii Turystycznej Polski*, 2 vols. (Warsaw: Komitet dla Spraw Turystyki, 1956).

[8] *Map of Southern Britain in the Iron Age* (Chessington, Ordnance Survey, 1962).

[9] Aristotle, *Politics*, I, 2.

[10] Pausanias, X; in *Pausanias's Description of Greece*, trans. and ed. J. G. Frazer, Vol. 1, p. 503.

[11] Norman J. G. Pounds, 'The Urbanization of the Classical World', *Annals of the Association of American Geographers*, Vol. 59 (1969), pp. 135–57.

[12] Glanville Downey, 'Libanius' Orarion in Praise of Antioch (Oration XI)', *Proceedings of the American Philosophical Society*, Vol. 103, No. 5 (1959), pp. 652–86, *cap.* 207.

[13] Nigel G. L. Hammond, *Epirus* (London: Oxford University Press, 1967), pp. 670 ff.

[14] Aristotle himself came from the small Kalchidic πόλις of Στάγερος. The capital of the Macedonian Kings, Pella, is now represented only by excavation.

[15] The map of Latin inscriptions in Pannonia is based on Andreas Mócsy, *Die Bevölkerung von Pannonien bis zu den Markomannen Kriegen* (Budapest: Verlag der ungarischen Akademie der Wissenschaften, 1959). See also Edith B. Thomas, 'Uber die römische Provinz Pannonien', *Das Altertum*, Vol. 2 (1956), pp. 107–19, and Pavel Oliva, *Pannonia and the Onset of Crisis in the Roman Empire*, (Prague: 1962). The standard work on Roman inscriptions is *Corpus Inscriptionum Latinarum*, ed. Theodor Mommsen. The Balkans are covered by Vol. 3, parts 1 and 2 (Berlin, 1873), and in the supplementary volumes. For the lower Danube valley see also Veselin Beševliev, 'Spatgriechische und spatlateinische Inschriften aus Bulgarien', *Deutsche Akademie der Wissenschaften zu Berlin, Institut für Griechische-Römische Altertumskunde, Berliner Byzantinische Arbeiten*, Band 30, 1964.

[16] Thomas Hobbes, *Leviathan*, Book 4, *cap.* 47.

[17] Jean Lestocquoy, 'Le paysage urbain en Gaule du Ve au IXe siècle', *Annales: Economies-Sociétés-Civilisations*, Vol. 8 (1953), pp. 159–172; Eugen Ewig, 'Kirche und Civitas in der Merowingerzeit', in *Le Chiesa nei Regni dell' Europa Occidentale e i loro rapporti con Roma sino all' soo* (Spoleto, 1960), Vol. 1, pp. 35–41.

[18] *The Letters of Cassiodorus*, ed. Thomas Hodgkin (London: Henry Frowde, 1886), Book 8, no. 31, p. 378.

[19] *The Gothic History of Jordanes*, ed. C. C. Mierow (Princeton: Princeton University Press, 1915), p. 127.

[20] *Constantine Porphyrogenitus de Administrando Imperio*, ed. Gy. Moravcsik, (Budapest: 1949), 139. On Roman urbanization of Illyria see Henri Cons, *La Province Romaine de Dalmatie*, (Paris: Ernest Thorin, éditeur, 1882).

[21] J. A. R. Munro, W. C. F. Anderson, J. G. Milne, F. Haverfield, 'On the Roman Town of Doclea, in Montenegro', *Archaeologia*, Vol. 55 (1896), pp. 33-92.

[22] Anna Comnena described Lipenium as 'a small fort lying at the foot of the Zygum': *The Alexiad of the Princess Anna Comnena*, IX. 4, ed. Elizabeth A. S. Dawes (London: Routledge and Kegan Paul, 1967). A number of small classical cities were able to survive on the Istrian coast; see Ernst Klebel, 'Über die Städte Istriens', *Studien zu den Anfängen des Europäischen Stadtwesens, Reichenau Vorträge*, 1955-6, Vol. 4 (1958), pp. 41-62.

[23] Procopius, IV, i, 17-18.

[24] Procopius, IV, iii, 27-9.

[25] Ernst Kitzinger, 'A Survey of the Early Christian Town of Stobi', *Dumbarton Oaks Papers*, no. 3 (1946), pp. 81-162; see also Charles Picard, 'Observations archéologiques en Yougoslavie', *Académie des Inscriptions et Belles-Lettres, Comptes Rendus* (1954), pp. 70-95.

[26] It appears as *Stopis* in the *Peutinger Table*; see Konrad Miller, *Itineraria Romana* (Stuttgart: Strecker und Schröder, 1916), columns 555-8.

[27] *Miracula Sancti Demetrii*, *Migne's Patrologia Graeca*, Vol. 116, columns 1284-93. See also Dmitri Obolensky, 'The Empire and its Northern Neighbours, 565-1018', *Cambridge Medieval History*, Vol. 4, part 1 (Cambridge, 1966), pp. 473-518.

[28] See also Charles Delvoye, 'Salonique, seconde capitale de l'empire byzantin, et ses monuments', *Revue de l'université de Bruxelles*, 9e série, Vol. 2 (1949-50), pp. 391-444, and Paul Lemerle, *Philippes et la Macédoine orientale à l'époque Chrétienne et Byzantine* (Bibliothèque des Ecoles Francaises d'Athenes et de Rome, Fasc. 158. Paris, 1945).

[29] George Ostrogorsky, 'Byzantine Cities in the Early Middle Ages', *Dumbarton Oaks Papers*, No. 13 (1959), pp. 45-66.

[30] *ibid.*, p. 58.

[31] Stamen Michajlow, 'Plisca - première capitale slave à la lumière des dernières fouilles' (in Russian), *Slavia Antiqua*, Vol. 6 (1957-8), pp. 323-68.

[32] Christo Gandev, 'Naissance et développement des villes et de leur quartiers artisanaux en Bulgarie depuis le VIe siècle avant notre ère jusqu'au XIXe siècle de notre ère', *Kwartalnik Historii Kultury Materialnej*, Vol. 10 (1962), pp. 559–63.

[33] William of Tyre, *A History of Deeds Done Beyond the Sea*, ed. E. A. Babcock and A. C. Krey (New York: Columbia University Press, 1943), Vol. 1, p. 100.

[34] *ibid.*, I, p. 20. It is interesting to note that Edrisi noted the flour mills at Niš, together with its gardens and vineyards: *Géographie d'Edrisi*, trans. P. Amédée Jaubert (Paris: Imprimerie Royale, 1836–40), Vol. 2, pp. 382–3.

[35] *ibid.*, I, p. 18.

[36] *ibid.*, II, p. 13.

[37] *ibid.*, I, p. 18.

[38] Lubor Niederle, *Manuel de l'Antiquité Slave* (Paris: E. Champion, 1926), Vol. 2, pp. 299–307.

[39] Josef Kostrzewski, *Les Origines de la Civilisation Polonaise* (Paris: Presses Universitaires, 1949), p. 105.

[40] Geograf Bavarski, *Monumenta Poloniae Historica*, Vol. 1 (Lwów, 1864), pp. 10–11.

[41] Rudolf Turek, 'Stav výzkumu staročeskych hradišť', *Slavia Antiqua*, Vol. 2 (1949–50), pp. 389–421.

[42] Josef Poulík, 'The Latest Archaeological Discoveries from the Period of the Great Moravian Empire', *Historica* (Prague), Vol. 1 (1959), pp. 7–70; *ibid.*, 'Some Early Christian Remains in Southern Moravia', *Antiquity*, Vol. 32 (1958), pp. 163–6; Libuše Jansová, 'Oppidum celtique de Hrazany sur la Vltava moyenne', *Historica*, Vol. 4 (1962), pp. 5–21; František Kavka, 'Der Stand der Forschungen über den Anfang der Städte in der Tschechoslowakei', *Kwartalnik Historii Kultury Materialnej*, Vol. 10 (1962), pp. 546–9.

[43] Václav Vaněček, 'Staré Čechy 8.–9. stol', *Slavia Antiqua*, Vol. 2 (1949–50), pp. 300–20; J. Poulík, in *Historica*, Vol. I (1959), pp. 7–70.

[44] Zdenek R. Dittrich, 'Christianity in Great Moravia', *Bijdragen van het Instituut voor Middeleeuwse Geschiedenis der Rijksuniversiteit te Utrecht*, Vol. 33 (1962).

[45] George Vernadsky, 'The Beginnings of the Czech State', *Byzantion*, Vol. 17 (1944–5), pp. 315–28; Joseph M. Kirschbaum, 'La Grande-Moravie et son héritage culturel', *Etudes Slaves et Est-Européennes*, Vol. 2 (1957–8), pp. 69–80; also F. Graus, 'Origines de l'état et de la noblesse en Moravie et en Bohême', *Revue des Etudes Slaves*, Vol. 39 (1961), pp. 41–58.

[46] Witold Hensel and Lech Leciejewicz, 'Villes et Campagnes',

Annales: E-S-C., Vol. 17 (1962), 209–22; Henryk Münch, 'Über den frühmittelalterlichen und mittelalterlichen Stadtgrundriss in Polen und seiner Erforschung', *Kwartalnik Historii Kultury Materialnej*, Vol. 10 (1962), pp. 346–59.

[47] On the general evolution of the Polish city see the numerous and repetitious papers of Aleksander Gieysztor, of which 'Les origines de la ville slave', *La Città nell' Alto Medioevo, Settimane di Studio del Centro Italiano di Studi sull' alto Medioevo*, Vol. 6 (1959), 279–303, is typical; Witold Hensel, 'Le développement des recherches archéologiques sur les origines de l'etat polonais', *Archaeologia Polona*, Vol. 1 (1958), pp. 7–56.

[48] Zdzisław Kępiński and Krystyna Józefowiczówna, 'Grobowiec Mieszka pierwszego i najstarsze budowle Poznańskiego grodu', *Przegląd Zachodni*, Vol. 8 (1952), pp. 370–97.

[49] Zdzisław Kaczmarczyk, 'Rola dziejowa Kalisza w wiekach średnich', *Przegląd Zachodni*, Vol. 7 (1951), pp. 32–46; Pierre Francastel (ed.), *Les Origines des Villes Polonaises* (Paris: Mouton and Co., 1960), pp. 179–89.

[50] Pierre Francastel, *op. cit.*, 127–36; Witold Hensel, *Najdawniejsze Stolice Polski* (Warsaw: Państwowe Wydawnictwo Naukowe, 1960).

[51] *ibid.*, pp. 191–215.

[52] Walenty Wojtkowiak, 'Geneza miasta Sulęcina', *Przegląd Zachodni*, Vol. 4 (1951), pp. 524–41.

[53] Roman Jakimowicz, 'Kilka uwag nad relacja o słowianach Ibrahima ibn Jakuba', *Slavia Antiqua*, Vol. I (1948), pp. 439–59.

[54] Adam of Bremen, *History of the Archbishops of Hamburg–Bremen* (New York: Columbia Univ. Press, 1959, II. 22), p. 66. See also sources cited in Tadeusz Lewicki, 'L'apport des sources arabes médiévales (IXe–Xe siècles) à la connaissance de l'Europe centrale et orientale', *L'Occidente e l'Islam nell' Alto Medioevo*, Spoleto, 1965, Vol. I, pp. 461–85; Herbert Ludat, 'Frühformen des Städtewesens in Osteuropa', *Studien zu den Anfängen des europäischen Städtewesens: Vorträge und Forschungen*, Vol. 4 (Lindau, 1958), pp. 527–53.

[55] Władysław Kowalenko, 'Starosłowiańskie grody portowe na Bałtyku', *Przegląd Zachodni*, Vol. 6 (1950), pp. 378–419; Ryszard Kiersnowski, 'Kamień i Wolin', *ibid.*, Vol. 7 (1951), pp. 178–225; Karol Górski, 'Upadek słowianskiego Wolina', *Slavia Antiqua*, Vol. 5 (1954–5), pp. 292–331.

[56] 'Galli Chronicon', *Monumenta Poloniae Historica*, Vol. I (Lwów, 1864), pp. 379–484.

[57] *Relatio Ibrāhīm Ibn Jàkub de Itinere Slavico*, ed. Tadeusz Kowalski, Monumenta Poloniae Historica, New Series, Vol. 1

(Kraków, 1946), pp. 48–54 (Polish text); 145–51 (Latin text). See also Ivan Borkovsky, 'Der altoböhmische Přemysliden-Fürstensitz Praha', *Historica* (Prague), Vol. 3 (1961), pp. 57–72.

[58] *Vita Ottonis Episcopi Babenbergensis*, Monumenta Poloniae Historica, Vol. 2 (Lwów, 1872), pp. 32–70.

[59] Notably Zygmunt Wojciechowski, 'Znaczenie Giecza w Polsce Chrobrego', *Przegląd Zachodni*, Vol. 8 (1952), pp. 410–16.

[60] Zygmunt Wojciechowski, 'Znaczenie Giecza w Polsce Chrobrego', *Przegląd Zachodni*, Vol. 8 (1952), pp. 410–16. Herbert Ludat, 'Frühformen des Städtewesens in Osteuropa', *Studien zu den Anfängen des europäischen Städtewesens, Vorträge und Forschungen*, Vol. 4 (Lindau, 1958), pp. 527–53.

[61] *Das Landregister der Herrschaft Sorau von 1381*, ed. Johannes Schultze, Veröftentlichung der Historischen Kommission für die Provinz Brandenburg und die Hauptstadt Berlin, Vol. 8, part 1, 1936.

[62] Heinz von zur Mühlen, 'Versuch einer soziologischen Erfassung der Bevölkerung Revals im Spatmittelalter', *Hansische Geschichtsblätter*, Vol. 75 (1957), pp. 48–69.

[63] Erich Keyser, 'Die Bevölkerung Danzigs und ihre Herkunft im 13. und 14. Jahrhundert', *Pfingstblätter des Hansischen Geschichtsvereins*, Vol. 14 (1924).

[64] *ibid.*, pp. 21–4; the tables are derived from the *Bürgerbuch* of the city of Danzig.

[65] Maurice Beresford, *New Towns of the Middle Ages* (London: Lutterworth Press, 1967); see also Henryk Münch, 'Geneza rozplanowania miast wielkopolskich XIII i XIV Wieku', *Prace Komisji atlasu historycznego Polski* (Kraków, 1946).

[66] Alfred Püschel, 'Das Anwachsen der deutschen Städte in der Zeit der mittelalterlichen Kolonialbewegung', *Abhandlungen zur Verkehrs- und Seegeschichte*, Vol. 4 (Berlin, 1910); Fedrerick G. Heymann, 'The Role of the Towns in Bohemia of the later Middle Ages', *Cahiers d'Histoire Mondiale*, Vol. 2 (1954), pp. 326–46; Rudolf Koetzschke, *Geschichte der ostdeutschen Kolonisation* (Leipzig, 1937).

[67] *Atlas Östliches Mitteleuropa* (Bielefeld: Velhagen und Klasing, 1959), pp. 59–61.

[68] See Henryk Münch, *Geneza rozplanowania miast wielkopolskich XIII i XIV Wieku*, Prace komisji atlasu historycznego Polski (Kraków, 1946).

[69] Andrzej Nadolski, 'Die Entstehung der Stadt Łęczyca', *Kwartalnik Historii Kultury Materialnej*, Vol. 10 (1962), pp. 502–9.

[70] Andrzej Zbierski, 'The Early-Mediaeval Gdańsk in the light of Recent Researches', *Kwartalnik Historii Kultury Materialnej*, Vol. 10 (1962), pp. 418–34.

[71] An analogous situation existed at Brandenburg, and, though for somewhat different reasons, at Braunschweig and Hildesheim, Hans Planitz, *Die Deutsche Stadt im Mittelalter* (Graz-Köln, 1965), pp. 69–70, 145–6.

[72] Heinrich Bechtel, *Wirtschaftsgeschichte Deutschlands*, Vol. 1 (Munich: G. D. W. Callwey, 1951), pp. 255–7.

[73] Hektor Ammann, 'Die schweizerische Kleinstadt in der mittelalterlichen Wirtschaft', *Festschrift Walther Merz* (Aarau: Verlag H. R. Sauerländer & Cie., 1928), pp. 158–215.

[74] Erich Keyser, 'Die Bevölkerung Danzigs und ihre Herkunft im 13. und 14. Jahrhundert', *Pfingstblätter des Hansischen Geschichtsvereins*, Vol. 14 (1924); the population doubled within a century; see Erich Keyser, *Bevölkerungsgeschichte Deutschlands* (Leipzig: Verlag von S. Hirzel, 1938), pp. 202–3.

[75] Karl Koppmann, 'Uber die Pest des Jahres 1565 und zur Bevölkerungsstatistik Rostocks im 14., 15. und 16. Jahrhundert', *Hansische Geschichtsblätter*, Vol. 29 (1901), pp. 43–63. See also Heinrich Reincke, 'Bevölkerungsprobleme der Hansestädte', *Hansische Geschichtsblätter*, Vol. 70 (1951), pp. 1–33.

[76] Berthold Bretholz, 'Eine Bevölkerungsziffer der Stadt Brünn aus dem Jahre 1466', *Zeitschrift für Sozial und Wirthschaftsgeschichte*, Vol. 5 (1897), pp. 176–84. The author finds this estimate rather high.

[77] Hearth-lists printed in J. Jastrow, *Die Volkszahl deutscher Städte zu Ende des Mittelalters und zu Beginn der Neuzeit* (Historische Untersuchungen, Berlin, Heft 1, 1886), pp. 189–202.

[78] Friedrich Stuhr, 'Die Bevölkerung Mecklenburgs am Ausgang des Mittelalters', *Jahrbuch des Vereins für mecklenburgische Geschichte*, Vol. 58 (1893), pp. 232–78; also Karl Hoffmann, 'Die Stadtgründungen Mecklenburg-Schwerins in der Kolonisationszeit vom 12. bis zum 14. Jahrhundert', *Jahrbücher des Vereins für mecklenburgische Geschichte und Altertumskunde*, Vol. 94 (1930), pp. 1–200.

[79] Henryk Samsonowicz, 'Das polnische Bürgertum in der Renassancezeit', *Studia Historica Academiae Scientiarum Hungaricae*, Vol. 53 (1963), pp. 91–6.

[80] Matthaeus and Caspar Merian, *Topographia Saxoniae*, etc. a series of volumes on the topography of Central Europe, published from 1643 onwards, and reprinted beginning in 1960.

[81] J. Szücs, 'Das Städewesen in Ungarn im 15.–17. Jahrhundert', *Studia Historica Academiae Scientiarum Hungaricae*, Vol. 53 (1963), pp. 97–164.

[82] Karlheinz Blaschke, 'Bevölkerungsgang und Wüstungen in Sachsen während des Späten Mittelalters', *Jahrbucher für Nationalökonomie und Statistik*, Vol. 174 (1962), pp. 414–29.

[83] Constantin Jireček, *La Civilisation Serbe au Moyen Age* (Paris: Collection Historique de l'Institut d'Etudes Slaves, No. 1, 1920), pp. 24–31.

[84] A. Kubinyi, 'Topographic Growth of Buda up to 1541', *Nouvelles Études Historiques* (Budapest), Vol. 1 (1965), pp. 133–57.

[85] Stefan Pascu, 'Les sources et les recherches démographiques en Roumania', *Actes du Colloque International de Démographie Historique*, ed. P. Harsin and E. Hélin, Liège, no date, probably 1964, pp. 283–303.

[86] Jack C. Fisher, 'Urban Analysis: A Case Study of Zagreb, Yugoslavia', *Annals of the Association of American Geographers*, Vol. 53 (1963), pp. 266–84, quoting evidence in Mijo Mirkovic, *Ekonomska Historija Jugoslavije* (Zagreb, 1958).

[87] Peter Charanis, 'A Note on the Population and Cities of the Byzantine Empire in the Thirteenth Century', *The Joshua Starr Memorial Volume, Jewish Social Studies*, Publications No. 5, pp. 135–148.

[88] Constantin S. Jireček, *Die Handelsstrassen und Bergwerke von Serbien und Bosnien während des Mittelalters* (Abhandlungen der königl. böhmischen Gesellschaft der Wissenschaften, Prague, 1879); George W. Hoffmann, 'Thessaloníki: The Impact of a Changing Hinterland', *East European Quarterly*, Vol. 2 (1968), pp. 40, 1–27.

[89] Xenophon, *Cyropaedia*, VIII, ii, 5.

[90] A. N. J. Den Hollander, *Nederzettingsvormen en problemen in de Groote Hongaarsche Laagvlakte* (Amsterdam, 1947), pp. 82–133.

[91] Ömer-Lufti Barkan, 'Quelques observations sur l'organisation économique et sociale des villes ottomanes des XVIe et XVIIe siècles', *Recueil de la Société Jean Bodin*, Vol. 7, *La Ville*, part 2, Brussels, 1955, pp. 289–311.

[92] Kazimierz Dziewoński, 'Procesy urbanizacyjne we wspólczesnej Polsce', *Przegląd Geograficzny*, Vol. 34 (1962), pp. 459–508, presents an excellent survey of modern urban development in Poland.

[93] Norman J. G. Pounds, *The Upper Silesian Industrial Region*, Indiana University, Slavic and East European series, Vol. 11 (1958), pp. 148–51.

[94] Adam Ginsbert, *Łódź*, Łódź, 1962; Ludwik Straszewicz, 'The Łódź Industrial District', *Przegląd Geograficzny*, Vol. 31 (1959), Supplement, pp. 69–91.

[95] Peter Sugar, *Industrialization of Bosnia–Hercegovina 1878–1918* (Seattle: University of Washington Press, 1963).

[96] *Das östliche Deutschland: ein Handbuch* (Würzburg: Göttinger Arbeits Kreis, 1959), p. 689.

[97] Herbert Wilhelmy, *Hochbulgarien* (Kiel: Buchdruckerei Schmidt u Klaunig, 1935).
[98] Florian Georgescu, ed., *Istoria Oraşului Bucureşti* (Bucharest, 1965).
[99] Wilhelm Koppe, 'Revals Schiffsverkehr und Seehandel in den Jahren 1378/84', *Hansische Geschichtsblätter*, Vol. 64 (1939), pp. 111–52; Georg Lechner, 'Die hansischen Pfundzollisten des Jahres 1368', *Quellen und Darstellung zur hansischen Geschichte*, neue Folge, Vol. 10 (1935). See also R. W. K. Hinton, *The Eastland Trade and the Commonweal in the Seventeenth Century* (Cambridge University Press, 1959).
[100] A. N. J. Den Hollander, 'The Great Hungarian Plain: A European Frontier Area', *Comparative Studies in Society and History*, Vol. 3 (1960–1), pp. 74–88; 155–69, especially pp. 159–61.

Comments

IAN M. MATLEY

Professor Pounds has given us an interesting and well-documented account of the three major periods of urban development in East-Central and Southeast Europe. In particular he has demonstrated that the periods of classical and medieval urban development were separated by a period of stagnation or decline in many areas.

Professor Pounds points to the de-urbanization of the parts of the Southeast Europe lying outside the borders of the Byzantine Empire compared with the maintenance of many of the old classical urban centres within the Empire. Although de-urbanization certainly took place in Pannonia and some areas of Dalmatia, as well as in Dacia, continuous occupation of some urban sites took place. For example, although the Roman city of Apulum (Alba-Iulia) in Transylvania was destroyed in the eighth century by migratory peoples, there is archaeological evidence of continued occupance of the site, including the construction of a fortified centre in the ninth and tenth centuries. During the thirteenth century, in spite of the destruction of the city in 1241 by the Tatars, the development of the city continued after its designation as the seat of the Roman Catholic episcopate of Transylvania. In the seventeenth century the development of the city was halted by further destruction by the Turks and Tatars, resulting in the rebuilding of the city on a neighbouring site.

Continuous occupance from Dacian times can be demonstrated for the sites of other Transylvanian cities such as Cluj and Turda. I am not arguing against the theory of periods of stagnation or decline in urban development in Eastern Europe but simply emphasizing that not all urban sites were completely abandoned during the period of the migration of peoples after the disintegration of the Roman Empire.

Professor Pounds' comments on the development of the East European medieval city by the Germans are interesting, suggesting as they do the relatively small number of Germans involved and the predominance of native population in the

German-founded cities. These comments apply mainly to cities in Poland but I am sure that the same situation applied to the so-called Transylvanian 'Saxon' cities of Kronstadt (Braşov), Hermannstadt (Sibiu), Schässburg (Sighişoara) and others, where the German population was probably outnumbered by Romanians in the Middle Ages. I would heartily agree that the extent and effects of German urban settlement in Eastern Europe have often been exaggerated.

I should like to mention the role of the Turks in the urban development of Southeast Europe. In spite of the destruction caused by the Turks to many urban centres they were also instrumental in the development of others, although in few cases can it be claimed that they were founders of cities. In particular, the development of Sarajevo, starting in 1462 with the establishment of the 'saray' or governor's residence, was due almost entirely to the Turks. By 1489, it was given the status of 'Seher' or large town and became one of the most important Turkish cities in Europe with a population of some 50,000 in the mid-sixteenth century. The decline of Sarajevo was linked with the decline in Turkish power and was hastened by the sacking of the city by Prince Eugene of Savoy in 1697 and by a series of floods and fires in the nineteenth century.

Other Southeast European cities such as Novi Pazar and Skopje, the ancient capital of Serbia, also saw their greatest development during the period of Turkish control.

As in the case of cities founded by the Germans, the population of these Turkish cities consisted predominantly of native peoples.

Professor Pounds states accurately that the industrial revolution of the nineteenth century resulted in few new towns being established in East-Central and Southeast Europe, but points out that many of the older cities received an impetus for continued growth. I should place Bucharest in this category rather than regard it as an urban centre which developed only after the mid-nineteenth century. When Bucharest became the capital of Walachia in 1659 a period of rapid growth began. By 1750 the city was described as a notable city with 10,000 houses. By 1831 there were 70,000 inhabitants and by 1860, 123,000. It is true that the first major industrial work-shops and the railroad station were not built until the 1860's, when

Bucharest became the capital of Romania, and that the population rose more rapidly, doubling between 1860 and 1900, but it is nevertheless not accurate to think of Bucharest as entirely a product of the late nineteenth century.

The growth of the city of Ploieşti in the late nineteenth century can be attributed in great degree to its development as a petroleum refining centre after 1857. However, it should also be remembered that Ploieşti was an important settlement on the routes from Bucharest to Transylvania and Moldavia and around 1700 was the third largest populated centre of the country.

Galaţi and Brăila, the most important Romanian ports on the Danube saw considerable development before the major influx of Western capital into the country. Galaţi was already a busy port for trade with Turkey and the Mediterranean at the beginning of the eighteenth century. Although the town suffered damage at times during the wars between 1769 and 1829, the trade of the port was increased considerably after the Treaty of Kuchuk-Kainardji in 1774 when the Turkish monopoly on the foreign trade of the Romanian principalities was limited and more goods started to arrive along the Danube from Austro-Hungary, Russia and the Western countries. After the Treaty of Adrianople in 1829, when the Turkish trade monopoly was abolished, trade increased further. The city was rebuilt on the lines of Odessa, shipyards were established, and foreign merchants settled in the city. Brăila saw similar developments in the same period. By the 1840's the grain trade of Walachia through Galaţi and Moldavia through Brăila was growing rapidly. At that period about 200,000 tons of shipping a year called at both ports. Galaţi and Brăila were thus not creations of the nineteenth century although they saw considerable development during that period.

Constanţa on the Black Sea on the other hand saw little growth before 1878. It was a small port before that time, used by Genoese seamen in the twelfth to fourteenth centuries and later by the Turks. In 1865 it had only 2,000 inhabitants.

My argument is that although many of the Romanian industrial and trade centres saw a period of accelerated growth after the influx of Western capital in the late nineteenth century, many of these urban centres had seen considerable growth at an earlier period.

3

Aspects of Change in the Landscape of East-Central and Southeast Europe*

DEAN S. RUGG

A visitor to the Soviet Union and Eastern Europe is impressed by certain characteristics which make the landscape rather distinct. The most notable forms are those which are socialist and reflect the impact of a particular political system upon the landscape. Associated with these forms are others which represent normal technological developments and those which are national in nature. The appearance, association and location of these landscape elements all intrigue the geographer who is tempted to ask if they can be analysed as a typology and utilized as keys to the understanding of the forces that are shaping the geography of Eastern Europe. Such a question is related to the overall problem of whether or not a landscape can reflect national policy and whether or not such a phenomenon as the 'socialist landscape' exists.

The primary problem in such an analysis is the difficulty in dealing with the term 'landscape'. Any landscape is the product of varying cultures operating through time, and the separation of these cultural elements is an imposing task. Sauer's early attempt to systematize the study of the landscape, while appealing, did not generate research work on the overall associations of forms in an area; instead, investigation has centred on the distribution and spread of certain forms.[1] As Hartshorne has pointed out, the problem in using the landscape as a frame of reference for geographical studies is that the term is used in so many different ways: some would restrict the investigation

* The author expresses his appreciation for comments received on earlier versions of this paper to the following persons: Brian Blouet, Chauncy Harris, Daniel Hafrey, David Hooson, W. A. Douglas Jackson, Ian Matley, Joseph Spencer and Norman Stewart. Ladis Kristof presented a very thoughtful critique of the paper presented at the Austin meetings. All these comments contributed to and assisted the author in the final preparation of his study.

to forms perceptible to the senses while others insist that invisible forces must be included in order to understand the forms.[2] German geographers, however, utilize the term *Landschaft* as a conceptual tool to analyse a restricted area in terms of the visible elements of the landscape and the processes which lie behind them. Karl Schröder, in his study of the changes in the agrarian landscape east of the Elbe River, utilizes this concept as he also tries to analyse the impact of Marxist–Leninist ideology.[3]

Hartshorne has stated that the function of the landscape in geographical study is to serve as a key to regional understanding.[4] The forms serve as the outward manifestation of the ideas and attitudes of a culture. Therefore, landscape analysis requires an investigation of the factors behind the appearance and location of the forms. In the Marxist countries, certain landscape forms appear to reflect the ideas of a particular political ideology. It is possible that the landscape can serve here as a key to understanding the Communist culture area. Certainly one notices that the post-1945 changes in the Eastern European landscape have created a degree of uniformity and distinctiveness that contrasts with the older peasant villages and the isolated pre-war centres of economic activity. This hypothesis supplements Whittlesey's statement that central authority, acting for the whole of its territory, 'tends to produce uniformity in cultural impress even where the natural landscape is diverse'.[5]

In Eastern Europe the landscape is derived from three main sources: (1) communist forces of industrialization, mainly socialistic; (2) Western forces of modernization, also socialistic but with capitalistic overtones; and (3) national and/or ethnic forces of the individual countries. In a state like Romania, these three forces are represented, respectively, by the new industrial town and complex of Gheorghe Gheorghiu-Dej, the Ploieşti oil fields and the peasant village. The purpose of this paper is to develop a tentative typology of the landscape elements that are found in Eastern Europe today and to utilize these forms as keys to the understanding of the ideas that are shaping the area. Such a study reflects Whittlesey's call in 1934 for a more detailed analysis of the impress of central authority upon the landscape.[6]

SOCIALIST FORMS IN THE LANDSCAPE OF EAST-CENTRAL AND SOUTHEAST EUROPE

RURAL LANDSCAPE

Collectivized agriculture is central to the communist plan for creating a 'socialist' society. Although a large-scale state industry can be built supported by a system of private agriculture, the regimes would then have to reward this group and be dependent upon it. Such an agricultural system might be more efficient but the possibility of systematically utilizing peasants for extracting capital and for future industrial labour would be made difficult. Even more important would be the continued threat to the Communist Parties by the sizeable peasant groups (30 to 80% of the population) with separate national, local and religious traditions. Therefore, the political factor became crucial as the communist authorities imposed a revolution from above in the social structure of the countryside. As Philip Mosely states:

> It was absolutely necessary, according to the communist program, to transform the village and the agricultural sector in the image of the urban and proletarian sector.[7]

In the Soviet Union and in all communist countries of Europe except for Poland and Yugoslavia, a large-scale, state-controlled agricultural system now exists with the collective and state farms as its major components.

The collectivization process is essentially a planning tool for the mechanization of agriculture. The pre-communist tenure systems of Eastern Europe were altered as the state took over control of all land, subdividing it into large units. The goal was to develop these units on a labour-intensive basis, thereby releasing peasants to the cities while maintaining and, if possible, increasing food output. Unfortunately, few geographical details exist on the typical structure of collective and state farms other than those of average size.

Perhaps the most distinct aspect of the collective and state farm is its relationship to the settlement pattern. Miskin states that 'A collective is not simply a single farm worked on cooperative lines but is often a large area comprising several villages.'[8] David Mitrany adds that the process of collectivization is part of an economy planned and controlled from urban centres.[9] The relationship between settlement and farm is also true of the state farm because the centre serves as residence and cultural focus for the salaried workers. The collective-state farm system represents an attempt to superimpose a rational pattern of large fields suitable for mechanization over an existing settlement pattern. Details are not available as to how the boundaries of

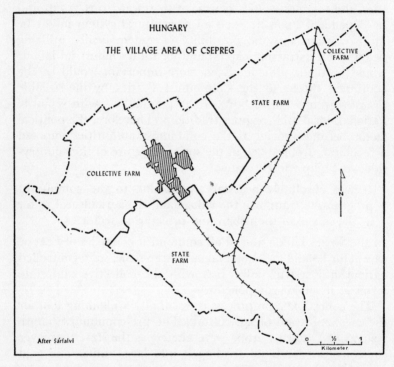

FIGURE 3.1. – VILLAGE AREA OF CSEPREG, HUNGARY

farms and fields are established. Apparently it differs by country and local community, depending on the rate of collectivization and the density of population. In portions of East Germany, Poland and Hungary pre-war large estates were simply converted into state farms. On the other hand, where small holdings were important and population density was higher, the collective farm became dominant.

A few studies of the collectivization process are given in

Geographia Polonica for 1965 (No. 5). In Eastern Europe the commune centred on a village is the common form of local political organization. As 'socialist' farms were organized, the area of the commune may have included parts of several collectives and state farms. This is true of the village area of Csepreg in western Hungary which has portions of two collectives, one state farm, and a state forest (Fig. 3.1). In other cases, a single collective or state farm incorporates several villages. Whatever the case, the total effect of collectivization on the rural landscape is impressive. Schröder has mentioned a few of these changes: land use in terms of cultivated plants and livestock rearing; marketing facilities; use of machinery; road network; and the forms and spatial relationships of settlements.[10] In this paper, three aspects of change in the rural landscape are discussed as part of an initial and sketchy attempt to establish a typology: fields, farm buildings and villages.

Large fields

As one travels by air from Czechoslovakia or Hungary to Austria, the contrast in field size on respective sides of the border is startling. In Austria the small strip pattern, derived from the medieval three-field system and subsequent laws of inheritance, predominates, as it does in much of Western Europe. In the communist countries large farms and fields are the rule. In the Soviet Union and in all of Eastern Europe except Poland and Yugoslavia, about 90% or more of the agricultural land is in the 'socialized' sector. The size of state and collective farms is large in all of these areas: in 1967 Romanian state farms averaged 15,000 acres and collectives about 4,750 acres, a contrast to the vast Soviet Union where in 1965 the state farms of the Virgin Lands, largely specializing in wheat, averaged 150,000 acres each.[11] Figures on sizes of fields are lacking but these would vary depending on the area. A typical layout of fields for a collective farm in East Germany is shown together with the pattern of fields before the changes were made (Fig. 3.2).

Changes in field size necessitated changes in fences. Again, little data are available on this subject but Schröder states, for example, that the traditional hedgerows of west Mecklenburg have been reduced.[12]

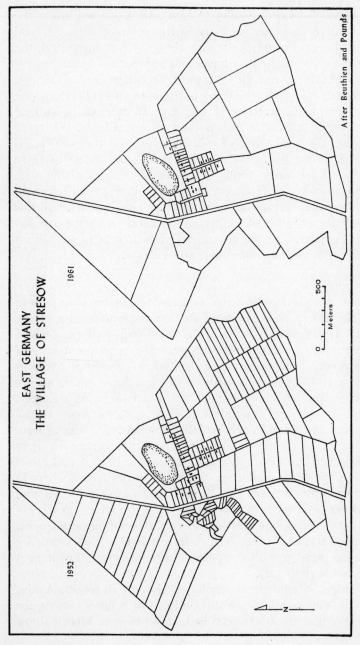

FIGURE 3·2 — COLLECTIVE FARM IN THE GERMAN DEMOCRATIC REPUBLIC. The Village of Stresow

Farm buildings

The visitor to Eastern Europe is impressed also by the number of post-1945 farm buildings that have been constructed. Although the farmer himself generally lives in the nearest village, the reorganization of farm operations on a large-scale basis requires certain new structures in both village and collective. Unfortunately, data regarding the type and distribution of such buildings are not available. General agreement among writers seems to exist, however, that a majority of the new farm buildings are related to livestock. In the peasant agriculture of Eastern Europe between the wars livestock rearing was of a subsistence and extensive nature with meat and dairy products uncommon in the diet.[13] World War II did not help the situation. When collectivization programmes began, the members were allowed to keep a few animals but most livestock was grouped together in an attempt to increase the quantity and improve the quality of production through large-scale methods. Since structures for large herds or flocks were lacking, new ones were constructed. For example, a United Nations report on Romanian agriculture states that in 1960, 40% of the capital stock of the collectives was in buildings.[14]

One of the few geographic studies including case studies of collective and state farms is Horbaly's work on Czechoslovakia.[15] He mentions the early emphasis in 1949–50 on the construction of feeding barns for pigs since propagation of this quick-maturing animal would help the meat shortage. At the 'Gigant Smiřice' state farm in eastern Bohemia, ten large pig-fattening structures were constructed, each 500 feet long 44 feet wide, 5 feet high at the walls and 14 feet high in the centre; each of these will accommodate 1,000 swine. Other 'socialist' building types mentioned by Horbaly include cow and poultry barns, calving sheds, storehouses, slaughter house and meat-processing plants, and dehydrating plants for potatoes and hay. René Dumont states that in Czechoslovakia.

> Cooperative stock-rearing had led to the construction of enormous collective cow houses, which were subsidized on condition that local materials were employed and used economically, and that the members themselves should supply practically all the labor.[16]

Structural relationships of farm and village

Under collectivization programmes, fields and even certain new farm buildings may be developed rather rapidly. However, villages take longer to change although Marxist–Leninist ideology includes plans for the transformation of the village. One important result of this transformation is a series of relationships between farm and village that result in new patterns of settlement. These relationships may be summarized as follows: the collective or state farm is incorporated out of village-commune areas; the village and farm structures, being collective in nature, are developed together, and the village serves as an administrative, shopping, industrial, cultural and residential centre for the state-owned farms. The first of these was mentioned above. In this section the second characteristic is discussed, and in the next section the service function of the village is described using the House of Culture as an example.

One of the principles of communist village planning in East Germany is the development of separate but adjacent areas of residence (*Wohnbereich*) and economic activity (*Wirtschaftsbereich*).[17] The theoretical pattern of the socialist village is shown on the map (Fig. 3.3). Although it is expressed as theoretical, Schröder states the process is apparent throughout the northern plain.[18] Placing residential and economic areas together follows the ideological principle of reducing differences between town and countryside as the *Wirtschaftsbereich* becomes a 'factory in the village', a place where all economic processes related to the collective or state farm are concentrated. Here are found all buildings for the handling of crops and animals plus the processing of products. The adjacent *Wohnbereich* is a residential area that varies in character from an old peasant village to a settlement that served a Prussian estate; within these types new single-family houses have appeared (even some of two or three stories) that represent attempts at modernization.[19] The village also possesses certain functions – political, economic, and social – that together determine its position in the socialist hierarchy of settlement. A 1961 source for East Germany stated that this hierarchy is based on centrality with the highest order of village including a Machine Tractor Station, collective and (possibly) a state farm, creamery,

Change in the Landscape of East-Central and Southeast Europe

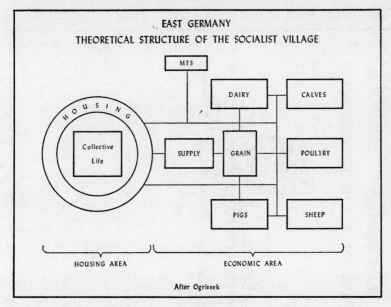

FIGURE 3.3 – THEORETICAL STRUCTURE OF THE SOCIALIST VILLAGE

Gemeinde administration building, House of Culture, school, department store, inn and post office.[20] In the lower orders of the hierarchy, the village functions that relate to administration and to the agricultural system decrease. In theory, the socialist hierarchy of central place functions in villages is greatly dependent upon planning decisions, especially those connected with a state-directed agriculture. This is reflected in the realignment of *Gemeinden* to fit economic reality in a socialist state.[21]

The hierarchy outlined for East Germany, however, is ideal and each village may differ according to local circumstances. A concrete example is provided by Marxwalde, formerly called Quilitz, a village about 20 miles north of Frankfurt/Oder (Fig. 3.4).[22] The map illustrates the degree to which this village corresponds to the ideal 'socialist' type. Designated as a 'Main Village', this old *Angerdorf* has been adapted to conform remarkably well to the theoretical plan for development of village with 'residential' and 'economic' areas. The former noble's house and park has been turned into a House of Culture

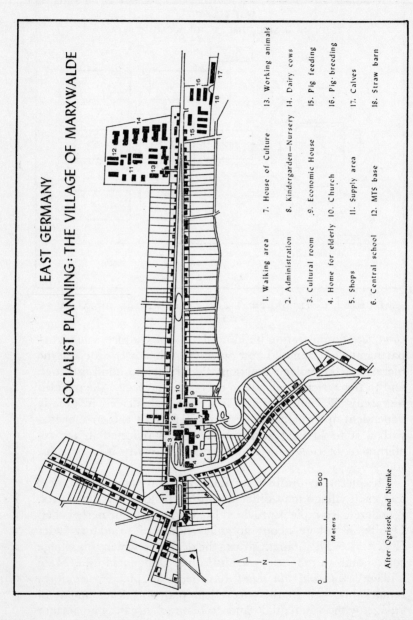

FIGURE 3.4—THE GERMAN DEMOCRATIC REPUBLIC: SOCIALIST PLANNING. The Village of Marxwalde

with adjacent athletic field and public park. Grouped around the wide portion of the 'village green' are administration buildings, school, shops, church and House of Culture. At one end of the village begins the 'economic area' of the collective farm. Dominant buildings are the elaborate stables for cows and pigs but others exist for oxen, breeding of pigs and storage of machinery for the farm. The stables for calves are located near the meadow land, and sheep and poultry installations lie farther in the country because of the disease problem.

The systematic relationship between village and farm that is developing in East Germany is not as apparent in the other countries of East Europe although a lack of data makes conclusions impossible. This problem seems to offer a fertile field for geographical research.

House of Culture

Within the village of socialist East-Central and Southeastern Europe many changes have taken place as indicated in the last section. Perhaps the most distinct 'socialist' structure in the village, however, is the House of Culture or, in smaller settlements, the club or reading room. These centres comprise a hierarchy of outlets for the communist cultural programmes and, therefore, act as foci of Party influence in the rural community.[23] The emphasis has been on the rural areas because of isolation and traditional resistance of the peasant to regime propaganda. The official goal is for the culture centre to replace the local church and tavern as a focus of rural life. In varying degrees, the institution is a part of the landscape of all countries of Eastern Europe, having been copied from the Soviet Union. In 1968 the writer viewed many of these in Romania, including new ones in construction, and was told that this structure, along with the school and police station, is typical of the 'villagescape'. A recent Romanian atlas indicates the wide regional distribution of this centre and its increase since the years after World War II.[24] A case study of Petrarch, Bulgaria indicated that a House of Culture with library, lecture hall and theatre was under construction in the late 1950's. George Hoffman, in a later study on Bulgaria, specifically mentions that 'nearly every village now has an impressive, centrally located culture center'.[25]

The organization and activities of these cultural centres illustrate their role in centralized Party policy. Unlike the prewar reading rooms and farmers' circles, which were truly local groups, the House of Culture and its related forms are run by the district or local councils. To the communist they are the visible, material symbols of the regime's care for the masses. To others they are the visible, material symbols of central authority, often being regarded as a kind of state office. Varying in size from sparsely equipped centres having only a library or a game room to lavishly equipped buildings with laboratories, gymnasiums, auditoriums, and even athletic fields, the activities serve several purposes, but in all of these the social and political goals are closely related. Although the purpose of the institution is to raise the cultural–educational level of the rural areas through presentations in the arts and technology, the political and economic campaigns of the regime are promoted. For example, the theme of increasing fruit production may be present in a play, folk singing or a lecture. Conflicts between local desires and regime planning may be resolved through discussions held in the House of Culture. In the past, Soviet 'culture' was disseminated through these community centres, but in recent years the revival of 'national' traditions has become more important, although with communist overtones. Finally, the purely entertainment purpose remains significant to the centre which sponsors cultural brigades that compete in acting, singing and dancing contests at the national level. However, even here the programmes have often been related to the 'building of socialism'. When such 'bourgeois' or 'capitalist' innovations as rock-and-roll are seen, strict warnings may come from the Party regarding the degeneration of Houses of Culture into centres of frivolity. Such a crude attempt to link leisure time with official political purpose has caused many to resist the influence of this institution. Some rural residents, especially the intelligentsia, feel alienated from their own local institutions and the village as a social unit suffers.

URBAN LANDSCAPE

Urbanization remains a key process in socialism since the centres of administrative and planning regions are cities that serve as foci for industrialization and ideological indoctrination programmes developed by the Party. Food, labour and industrial raw materials flow into the cities, and the products are distributed back through the region. Therefore, regional and city planning, in theory, are interrelated in communist countries and this integrated planning throughout a national area is a fundamental characteristic of the organization of 'socialist' space; however, differences in the degree of this integration may exist depending upon the degree of regional autonomy, e.g. Yugoslavia. This interrelationship of region and city is reflected in the often quoted goal of reducing the disparity between town and countryside. In the days of Khrushchev, there was much talk of developing agro-towns where the workers on state farms (the 'agricultural factory') would live in apartment houses like an urban worker.[26] At the present time, little visible evidence is available of progress towards this goal.

The separation of distinct elements of socialism in the urban landscape is difficult because so many of these forms are characteristic of Western cities. However, the basis for establishing a typology of urban forms in the countries of Eastern Europe rests largely on the strong centralized planning that is altering the structure of communist cities, supposedly in accordance with Marxist–Leninist principles. These states have the administrative power to plan cities in a way that few Western states and cities have. Changes in the townscape of Eastern Europe are most evident in the new 'socialist' cities or in new portions of older cities where the regime has attempted to have the urban pattern reflect the ideology of communism. In theory, the new socialist city should correct the defects of the capitalistic era, including the evils of unrestricted land speculation, and create a new pattern which illustrates the classlessness of society. All parts of the city should contribute to the spatial unit as a whole, that is, the emphasis should be on the integration of functional land uses rather than their separation into distinct areas. In the new urban construction of Eastern Europe the effects of such planning is evident in three aspects of the landscape: city centre, neighbourhood unit, and new 'socialist' city.

City centre
Theoretically, the role of the city centre in 'socialist' cities is completely different from that in 'Western' cities and a visitor

to Eastern Europe feels this is apparent in practice. In the 'Western' city, the Central Business District is the hub around which the other functions arrange themselves in a market economy where rent and transport costs are predominant locational factors. In socialist countries, the centre is developed as a focus for political and cultural activities.[27] Commercial areas are limited in size in comparison to Western countries because of restricted emphasis on the production of consumer goods. It is not surprising then that in large cities of Eastern Europe, where few automobiles and single-family houses are found, high urban densities place a tremendous burden on the stores and on public transport. An attempt is made, therefore, to decentralize commercial activities and thereby reduce the congestion of the centre. Shopping trips to the centre of the city are discouraged by state establishment of small service areas near housing areas. If factories are located near the housing areas, intra-urban traffic is further reduced. The visitor to Eastern Europe notices that large department stores, office buildings and land uses related to the automobile are missing from the central portions of cities. In fact, the relative absence of cars, service stations and land uses connected with or dependent upon the automobile is in direct contrast with American cities. However, this pattern is rapidly changing.

National policy under a planned economy requires a focus of activity which takes the place of the Central Business District and which serves to co-ordinate the entire urban complex, especially the commercial, industrial and residential functions that have been systematically decentralized. Such a focus is political and cultural. Included in or near the district are various form elements that are distinctly 'socialist' such as the following:

- East Berlin–Stalin Allee (now Frankfurter Allee) and Marx–Engels Platz
- Warsaw – Palace of Culture
- Prague – Klement Gottwald Mausoleum
- Budapest – Liberation Memorial on Gellert Hill (though this can easily be questioned)
- Bucharest – Scinteia publishing house and in the centre, the new concert and meeting hall

Sofia – Georgi Dimitrov Mausoleum and the Square of September 9

Belgrade – Marx–Engels Square with the Trade Union House, Central Committee Building and the office building of 'Borba'.

Tiranë – Skanderbeg Square with statue of Stalin.

Other common elements include buildings of the Communist Party or government, frequently adorned by a Red Star, the ever-present statue of the Red Army soldier and monuments relating to the accomplishments of socialism. Less permanent elements include the posted official newspaper, the loudspeaker for Party propaganda and posters or Party flags. On certain days such as May 1 and November 7, these temporary activities are intensified with flags and banners supplemented by pictures of leaders. In short, many form elements exist in the socialist urban landscape to remind one that a uniform ideology prevails in all the countries, although the degree of intensity may vary.

Certain other elements of the city centre in Eastern Europe seem to be characteristic although they are not distinctly 'socialist'. The uniformity of store fronts and the lack of neon signs contributes to a drabness that is noticeable to the visitor. Hotels of quality are constructed to provide the accommodations primarily for Western tourists and businessmen who will spend hard currency. Finally, the lack of consumer goods is partly compensated for by numerous cultural institutions that provide services – opera, concerts, theatre, circus, films, sport and books – at reasonable prices.

Neighbourhood unit concept

In the development of self-contained residential areas of apartment houses, which possess shops and services for the people living there, the communist countries are following a concept that is significant in 'Western' countries as well. However, in communist states the 'neighbourhood unit' is a social tool for planned urban development as the state establishes residential areas utilizing official norms for size and services without class distinction – at least in theory. At the same time, definite economies are possible in mass construction, and intra-urban transport is relieved as units become more self-contained. Neighbourhood units are made up of a series of apartment houses with dwelling units of different size, oriented around a service centre including shops and personal and cultural services. A series of such neighbourhood units, located around a political–cultural focus and accessible to industrial plants, represents the socialist model of spatial urban organization that retains the planning principles of uniformity and predetermined size.

Michael Frolic has mentioned two types of neighbourhood units in Soviet cities which represent different stages in the application of socialist principles to urban development.[28] The *mikrorayon* represents a transitional stage of collective living in that the neighbourhood possesses an organic unity of residential, cultural and service functions that is not always characteristic of separately planned subdivisions and shopping centres in Western areas. However, the *mikrorayon* lacks employment centres, and its relationships to other parts of the city are not systematic. A more theoretical city of the socialist future, envisaged by S. G. Strumilin, the Soviet economist, is based on the *commune*. This residential living unit of 2,000–2,500 people possesses all the services plus a factory offering employment. Most significantly, its social structure reflects the emphasis on collective life: all individualism is to be submerged in the social fabric of the community and group participation is to be extolled, whether working in a factory or eating in a large dining hall. In Eastern Europe, the *mikrorayon* stage is predominant although high degrees of urban self-containment are visible in the new socialist cities.

'Socialist' cities

Perhaps the most visible changes in the urban landscape of communist countries in Eastern Europe are the new 'socialist' cities constructed along Marxist–Leninist principles. Such a pattern is characteristic of the Soviet Union. According to V. Shkvarikov, between 1926 and 1963, over 800 new towns were built, over one-third of them being constructed on vacant land, e.g. Zaporozhje, Magnitogorsk and Komsomolsk.[29] After World War II, the Communist governments of Eastern Europe also established new cities, generally industrial centres, which were designated as show pieces for 'socialism' and often included a large factory as a working nucleus for the self-contained project. The main examples are:

Eisenhüttenstadt, East Germany
Nowa Huta, Poland
Nowe Tychy, Poland
Havířov, Czechoslovakia
Dunaújváros, Hungary
Kazincbarcika, Hungary
Tiszapalkonja, Hungary
Gheorghe Gheorghiu-Dej, Romania
Victoria, Romania
Dimitrovgrad, Bulgaria
Qytet Stalin, Albania
Velenje, Yugoslavia
Titograd, Yugoslavia

These new cities have risen 'out of the fields' to house the growing worker class, and departures from the traditional capitalistic pattern should be complete as no real 'urban' complex existed on the site. Jack Fisher states that the cities are usually built to accomplish a single specific purpose: to house the workers of a large plant (Nowa Huta), to relieve congestion in an adjacent industrial area (Nowy Tychy), or to serve as a regional administrative centre (Titograd).[30]

Nowa Huta is an industrial satellite town for Kraków, Poland. A great integrated iron and steel plant was constructed here after World War II to symbolize the new industrial era of socialism. Unlike the steel plants of Upper Silesia, which were built largely by German technology and capital, this factory was completely Polish and 'socialist'. The capacity of the plant is over two million tons of steel, a big help to Polish industrialization. Coking coal is supplied from Polish and Czechoslovak sources, but the iron ore come from the Soviet Union. Many of the worker-residents were of rural origin, for one of the locational factors in the decision to undertake the project here was apparently the rural overpopulation that has been characteristic of the area since it was part of the Austrian Empire as the province of Galicia. Nowa Huta is a self-contained new town of 125,000 and is organized along socialist principles around a central square with shopping and collective institutions located to serve the worker-resident of the six-storey apartment houses. The steel plant is adjacent to the housing area and the city is linked to Kraków by frequent train service.

What is 'socialist' about the city centre, neighbourhood unit or new town? As pointed out above, certain buildings, squares or monuments are distinctly socialist because they reflect the ideology through style, ornaments or personification of specific individuals. The city centre and neighbourhood units possess a certain association of forms and spatial organizations that appear different although admittedly such a statement may be hard to prove. Finally, the 'socialist' cities represent the attempt to apply socialist principles of city planning. However, can we go further and make remarks about socialist architecture? The author is inclined to agree with Yurick Blumenfeld who states that the skyscrapers of Moscow, Warsaw and Bucharest – wedding cakes frosted with pediments and pseudo-rococo ornamentation – have a distinctive period flavour.[31] Although it may be resented as a Russian import, the Palace of Culture in Warsaw may have more character than the more modern structures around it. To some, the newer architecture of Eastern Europe is not art but the reflection of a Party-directed mass culture devoid of originality and based on the plan above all: 'Dreary, grimy, monotonous blocks of brick or concrete are bereft of spiritual feeling or human content.'[32] Perhaps it is the lack of character that does symbolize the socialist city. Supplementing this idea are negative qualities of poor landscaping, overcrowded transportation, inadequate services and advanced air pollution; the latter is derived from low-grade fuel used in heating, industry and vehicles. On the other hand, certain positive qualities are present with the frequent presence of 'collective' institutions such as parks (of rest and culture), schools, nurseries or kindergartens, hospitals, youth clubs – all identified in some way with the regime. One concludes that the 'socialist' character of these cities is in part impressionistic but nevertheless worthy of trying to capture.

NON-SOCIALIST FORMS IN THE LANDSCAPE OF EAST-CENTRAL AND SOUTHEAST EUROPE

Earlier in this paper, the hypothesis was offered that the landscape of Eastern Europe is composed of three types of form elements: socialist, Western and national. In the preceding section an attempt was made to identify the more distinctive socialist forms. In this part of the paper, the Western and national forms are briefly mentioned in order to complete the landscape analysis.

WESTERN FORMS

The goal of the communist regimes is to create an industrialized society. The reorganization of the agricultural and village pattern of the East European countries is aimed at providing the food, labour and popular support for the vast programmes of industry that are established. Urban areas also must be reorganized or created from scratch to hold the factories that are constructed. New patterns of resource use are established to provide raw materials for the industry and the energy for the population centres. The landscape forms that result from these programmes of industrialization – mines and forests, power plants and electrification lines, transport lines and factories – lack distinct 'socialist' character and resemble similar elements of modernization found around the world. The development of such forms is illustrated by the case of Romania.[33]

Resources

The increasing exploitation of resources is evident in the landscape. Oil production is now more widely distributed between four areas in contrast to 99% from Ploieşti in 1938, thereby accounting for a doubling of production in spite of declines in the older area. Natural gas and oil now divide the the energy burden whereas before the war oil was over three times as important as gas; Transylvania exhibits an industrial base as natural gas supplements the non-ferrous minerals and timber of this area. Coal production is four times that of 1938, and use is being made of extensive and widely distributed brown coal and lignite sources, e.g. Comăneşti. Coking plants have been added to the industrial landscapes of Reşiţa and Hunedoara. Finally, the government has instituted a national forestry programme aimed at reforestation, balanced use of different species (including beech), watershed control, conservation and development of timber-based industries.

Electrification

Certain effects of a post-war electrification programme are apparent in the Romanian landscape. Over 20 new power plants have been constructed in the post-war period and a completely new transmission network links the various pre-war and post-war stations. The big hydroelectric installation at Bicaz on the Bistriţa is a major showpiece for socialism in Romania. However, no less significant is the gleam of lights in the countryside at night, as 5,046 villages that lacked electricity in 1945 now possesses it.[34]

Transportation and communications

The Romanian landscape also reflects certain post-war changes in transportation and communications. Only a few railroads have been built but these have improved access to Bucharest, the coal mines of the Jiu valley and the backward areas of Maramureş. Double-tracking is 75% greater than in 1938 which helps to explain the fact that tonnage of rail freight increased over four times between 1938 and 1965. The road network has been modernized. In many cases, new gas pipelines are visible since they are often constructed above ground; this network supplied 347 localities in 1963 in comparison to only 86 in 1950. Conveyance of people and goods along the Danube river has increased while new airports illustrate the fact that air passenger traffic is 30 times that of 1938. The net effect of all these changes is a more balanced means of access throughout the country permitting new patterns of areal organization.

Industrialization

The most significant aspect of the modernization trend in the Romanian landscape is the industrialization programme. Reflecting its variety of unexploited resources, this communist country had a gross industrial output in 1965 that was over nine times that of 1938 and in recent years its rate of production

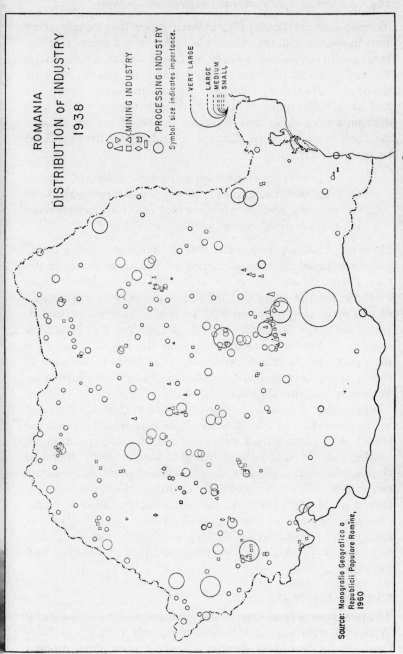

FIGURE 3.5 — DISTRIBUTION OF INDUSTRY IN ROMANIA, 1938

increase has averaged 13%, at least twice that of any other East European country. Although Romania in terms of employment is still only about 40% as industrial as Czechoslovakia, a diversification of the structure has begun to appear, both by sector and by area. By 1965 over 700 new industrial enterprises had been built, many of them by Western firms.[35] The wide distribution of these new factories is one of the most characteristic elements of the Romanian landscape. Christian Kind reports that the

> Planners are trying to distribute the new industrial centers equally all over the country, so as to make rational use of labor reserves and to create a net of strong points from which further development can be carried on.[36]

However, Hoffman points out that the relative position of the underdeveloped regions, in terms of an equal share in the total industrial production, has advanced only slightly.[37] Some evidence for the wider distribution is given by a comparison of two maps (Figs. 3.5 and 3.6). In 1938 industry was concentrated along the Bucharest–Ploieşti–Braşov line and in smaller clusters in Western Transylvania and the Banat. The 1966 map portrays the expansion of the older areas in addition to rather astonishing development of backward areas in Moldavia, Walachia and the Dobruja.

The policy of industrializing backward regions is illustrated in the landscape of Moldavia. In 1968 the writer viewed the Galaţi steel plant which will eventually double the national production. At Gheorghe Gheorghiu-Dej, a great chemical complex has been built consisting of a power plant and separate factories for oil, rubber and basic chemicals. Other new installations in Moldavia include the hydroelectric plant at Bicaz, chemical factories at Săvineşti and Roznov, a rolling mill at Roman, the ball-bearing factory at Bîrlad, the antibiotics factory at Iaşi, and wood working combines at Suceava and Comăneşti.

NATIONAL FORMS

The two types of landscape forms discussed above – socialist and Western – evolved, for the most part, only recently in East-Central and Southeast Europe and have been superimposed over 'national' forms that are derived from a much longer period of history. The term 'national' is difficult to use, primarily because so many culture groups settled in the area known as the 'Shatter Belt'. Independent states have not existed for much of recent history in Eastern Europe as external powers controlled much of the area. The forms derived from Rome and Byzantium are still visible. In the nineteenth century four Empires – Austria, Germany, Russia and Turkey – controlled major portions of the area and left their respective impact on the landscape. Dominated by these powers, the separate cultural groups of Eastern Europe had difficulty in expressing their individuality, especially in the landscape. Perhaps it is safe to say that the 'national' landscape forms of East Europe represent a mixture of external and internal cultures.

Rural landscape

Although altered by the impact of socialism, the rural landscape of Eastern Europe still reflects the peasant tradition which dominated the area until 1945. Now the large fields of collective and state farms have replaced the small strips of the past, except in Poland, Yugoslavia and isolated portions of the other countries. However, the village areas still reflect national culture or ethnic groups although the Communist Party has succeeded in making certain transformations. In a Romanian village it is still not unusual to observe a painted cottage with carved wooden gateway to the yard, a plot of maize and a flock of sheep nearby and a cumbersome slow-moving oxcart. Perhaps most characteristic of the rural areas of all East European countries is the sweep-well, a cross-cultural rather than a national form, yet one which conveys to this observer the pre-1940 essence of the area.

Urban areas

External influence in Eastern Europe has always been most evident in the cities, and the national forms, both now and from the past, reflect a mixture of foreign and local cultures. In Poland, German and Russian influences vie with the national taste. Poznan has a German character while Orthodox churches are common in the settlements along the Soviet border. Yet the

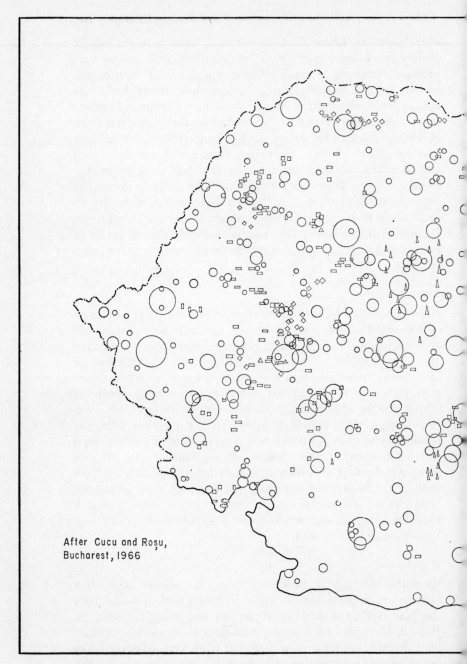

After Cucu and Roșu, Bucharest, 1966

FIGURE 3.6 — DISTRIBUTION

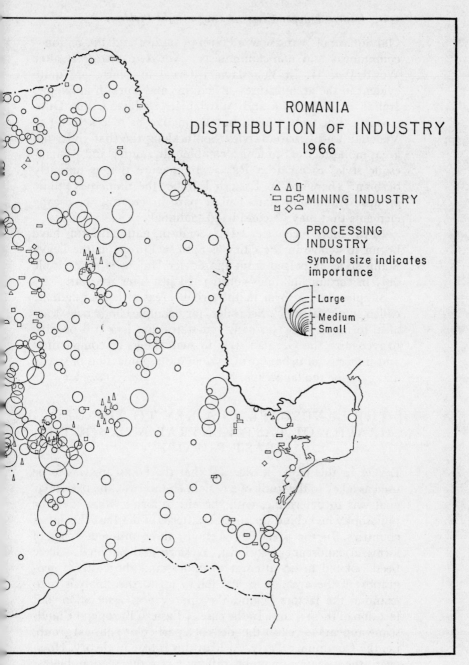

Old Square of Warsaw was Polish in nature and the nation – communists and non-communists – voted to restore it after World War II. In Yugoslavia external influences are quite evident in the architecture – Byzantine and Turkish in Skopje, Italian in Dubrovnik and Austrian in Ljubljana. But Dubrovnik is not truly Italian but Dalmatian as it is a blend of Venetian and Croatian styles. South along the Ibar river, the great monastery of Studenica combines Byzantine and Romanesque styles to create a Raška architecture that is uniquely Serbian. Therefore, in Eastern Europe, the numerous ethnic groups within one state had a part in creating landscape elements that may be considered 'national'.

In the more recent period of socialism, national forms have begun to evolve in the cities of Eastern Europe as the heavy Stalin style, imported from the Soviet Union, is now visible only in certain buildings constructed just after the war.

Supplementing form in the urban areas is what might be called 'national style'. Schröder, for example, finds new 'socialist' forms in East Germany but states that it is still possible to recognize the German style in such things as compactness and intensity of urban form, systematic organization of space, and care for the landscape.[38]

THE LANDSCAPE AS A KEY TO THE GEOGRAPHY OF EAST-CENTRAL AND SOUTH-EAST EUROPE

Earlier in this paper it was said that the landscape was to be used as a key to the study of a region. In this case, the particular goal was to determine, with the aid of forms, what ideas or philosophy lie behind the spatial patterns of the East European countries. In the preceding sections, three different types of forms in landscape – socialist, Western and national – have been isolated in an attempt to understand the cultural geography of the area. A logical follow-up to this analysis is to examine the factors behind the appearance, association and location of these forms. In the case of Eastern Europe, the landscape appears to reflect the ideas of a particular political group, i.e. the Communist Party. At least this is true in theory. However, the practice may be different. In the section below, several ideas are examined that appear to illustrate the impact of the Party on the landscape.

THE EVEN DISTRIBUTION OF ECONOMIC ACTIVITY

According to Marxist ideology, the locational decisions on economic activity, especially industry, are aimed at a more even distribution than has existed in the past. In contrast to capitalism, which Lenin stated was characterized by structural and spatial imbalances in the economy, socialism aims at planned proportional development of the national economy, which Soviet theorists state is the basic economic law of the ideology.[39] Through national plans, the state exercises deliberate control over the distribution and regional character of production at all levels. This planning not only stands for national efficiency but also for social justice in the sense of more equitable distribution of wealth.[40] Under such a social philosophy of planning, all parts of a communist country, *in theory*, would be affected and gradually the landscape would reflect the impress of central authority. Locational decisions would not reflect just profit or efficiency but the desire to lift backward areas, even if the programme was costly.

The theme of equality between regions in a communist state needs elaboration. In a sense, the goal of planning is not strict equality but a more rational distribution of the elements of production with each region contributing according to its resources. In the Soviet Union, attempts to even out regional development were based on certain principles of industrial location that seemingly contradict each other unless understood in terms of specific applications: locating certain industries closer to sources of raw materials; making regions self-sufficient; promoting specialized production within certain regions; and raising the level of backward or minority areas. Therefore, projects that seem unrealistic to capitalistic states – for example, the Urals–Kuznetsk exchange of resources and an integrated iron and steel plant in the Soviet Republic of Georgia – were carried through at much expense. The result has been a redistribution of the centres of economic activity in the Soviet Union with greater contributions from the eastern areas; however, Soviet planners recognize that the harsh environment will prevent Siberia and Central Asia from ever being 'equal' in terms of settlement. Soviet statements regarding the 'even' distribution of industry must be taken in this context.

> The more unconfined geographical distribution of [socialist] industry leads to its more *even* distribution over the territory of the country. However, what do we understand under the *even* distribution of industry? It does not mean, of course, that the same industrial enterprises of the same size and capacity must be established all over the country. Scientifically speaking, the even distribution of industry is such that the natural, labour, and productive resources all over the country be utilized along some very definite lines of planning so that the development of productive forces in an area (district, town, etc.) would not hinder the development of productive forces in other areas. The even distribution of industry implies the development of smaller towns, construction of new ones, the *scattering* of industry over the territory of the country leading to the *general increase in the production of public labour*.[41]

In spite of the emphasis on the even or rational development of resources and industry in the Soviet Union, there is much evidence to suggest that the fulfilment of a short-run goal often takes precedence over the socialist theoretical goal although 'social cost' continues to be a locational factor.

Similar statements regarding planning and the redistribution of economic activity are found in the communist states of Eastern Europe. In contrast to the Soviet Union, no large areas are unsuitable for settlement (like Siberia) and the problem is to utilize the resources of all regions in a rational manner. In the previous sections Romania was used as one example to illustrate how all sections of the country now appear to be contributing to the overall economy. A variability test on the relative contributions of the various Romanian administrative regions in certain selected products shows that in 1965 more balance existed between the areas than was found in previous years of communism, although they are still by no means equal (Table 3.1 on p. 112). It would appear that practice is *to some extent* following the ideology of even or rational distribution of economic activity as expressed above and in the following statements on East Europe and Romania:

The operation of the socialist laws of industrial location results in more even geographic distribution of industry within individual countries as industrial production is extended to new territories and leads to an industrial upsurge in areas that were backward under capitalism; this process tends to level the economic development of parts of each country.[42]

Only on the basis of doing away with the lag of some regions, of harmoniously developing the technical and material basis of socialism throughout the homeland's territory, can conditions be created for the forward march of socialist society, for the gradual transition to communism.[43]

All three types of landscape elements – socialist, Western and national – have been affected by the theory of planned proportional development. Certainly the wide distribution of collective and state farms is apparent and the same, to some extent, is true of the village through the House of Culture. The new socialist cities mentioned above often serve as growth centres in areas of underdevelopment. For example, the chief engineer in the Borzeşti chemical factory at Gheorghe Gheorghiu-Dej informed the writer that the decision to locate a new town of 40,000 and its accompanying industrial complex in the Trotus valley was stimulated by the backwardness of the area and the accessible resources of brown coal (Comăneşti), oil (Moineşti) and salt (Tîrgu Ocna). The city and factories at Gheorghe Gheorghiu-Dej, in many ways, resemble their Western counterparts but certain differences remain which are socialist – a point discussed above.

CREATION OF AN URBAN SOCIETY

Mosely states that

> the communist ideology propagates the idea of the superiority of an urban type of civilization and the necessity of reshaping rural life along the patterns of an industrial society.[44]

Mitrany supplements this when he says that

> the process of collectivization, as part of an economy planned and controlled from urban centres, is at the same time

TABLE 3.1 – *Romania: The Variation of Selected Indicators of Regional Development (Relative Percentages of the Totals for the Country)*

	Urban Population		Retail Sales		Movement of Goods		Coal Prod.		Brick Prod.		Gross Indust Prod.		Employees in Indust.	
	1959	1965	1959	1965	1955	1965	1950	1965	1950	1965	1950	1965	1959	1965
Argeș	2.9	3.0	4.1	4.9	6.59	8.12	5.76	6.26	.81	3.05	2.7	3.4	3.6	4.0
Bacău	4.6	4.8	5.1	4.9	6.64	6.69	4.55	2.63	2.15	11.71	5.4	6.5	5.6	5.7
Banat	8.8	8.5	9.0	8.4	7.06	7.39	6.65	2.85	18.33	13.81	11.8	9.2	11.1	10.3
Brașov	8.5	8.5	8.0	7.4	4.47	5.66	2.99	4.92	12.94	8.19	—	12.5	11.5	11.6
București	3.4	3.4	4.2	5.3	13.84	10.52	—	—	3.23	4.00	1.8	2.2	1.7	2.1
Cluj	5.9	5.9	5.9	5.7	4.64	6.29	2.22	1.19	6.47	5.33	5.0	4.2	5.9	5.7
Crișana	4.3	4.1	4.2	3.8	2.90	3.97	6.30	11.65	4.04	4.38	3.0	2.4	3.4	3.3
Dobrogea	1.5	1.6	2.2	2.4	7.03	4.96	—	—	.81	.86	0.8	1.1	1.4	1.5
Galați	4.9	4.9	5.0	5.3	7.80	6.28	—	—	1.08	1.62	4.4	4.3	3.8	4.0
Hunedoara	5.4	6.1	4.3	4.0	11.48	6.81	65.62	48.25	1.62	2.00	6.4	5.8	6.9	6.4
Iași	3.5	3.4	3.8	3.9	4.14	3.64	0.04	—	1.89	2.38	1.9	3.3	2.0	2.4
Maramureș	3.7	4.1	3.7	3.6	2.49	4.12	0.09	—	2.16	3.81	2.7	2.7	3.9	4.1
Mureș-Aut. Maghiară	3.3	3.5	4.3	4.2	3.98	4.27	—	—	7.01	9.24	2.9	4.1	3.9	4.2
Oltenia	4.3	4.5	4.7	5.6	7.31	9.10	0.04	11.41	1.89	3.71	3.0	4.2	3.2	4.0
Ploiești	8.3	8.1	7.1	7.2	4.72	7.70	5.74	10.84	9.43	14.86	14.2	10.4	10.1	8.9
Suceava	2.7	2.7	4.0	3.8	4.91	4.48	—	—	0.54	1.43	3.0	2.1	3.3	3.0
Orașul București	21.8	20.4	18.3	17.2	—	—	—	—	23.18	9.62	20.1	20.4	17.8	17.8
Orașul Constanța	2.2	2.5	2.1	2.4	—	—	—	—	2.42	—	0.7	1.2	0.9	1.0
Total	100	100	100	100	100	100	100	100	100	100	100	100	100	100
Variation Early	911.12		780.66		759.88		4490.60		1275.11		1021.38		888.26	
Late	859.68		743.26		684.38		2791.33		909.91		959.08		858.40	

Source: *Anuarul Statistic al Republicii Socialiste România* (Bucharest: Direcția Centrală de Statistică, 1966), Tables 46, 91, and 194.

binding the villagers together as they have never been before, on a professional and class basis.[45]

Such statements are consistent with the emphasis placed on reducing the differences between town and countryside. The apparent goal is the establishment of state farms as 'factories' in the country with the workers living in agro-towns and benefiting from the same cultural aspects of socialism as the city workers. In theory, this idea will reduce the traditional enmity between urban and rural areas as the formerly alienated peasants contribute to the building of socialism. The criterion of a rational distribution of industry – cited above – is related to this idea of an urban society since the network of growth centres in backward and other areas will facilitate the development of regional and local industries utilizing labour released by collectivization. The wide distribution of such centres prevents the overconcentration of economic activity and leads to a more equal distribution of wealth.

Do the forms in the landscape support this idea of urban control of society? Certainly the Party's central control of agriculture is a strong feature of the landscape through the collective-state farms and the village structure. Agro-towns have not yet made their appearance, however, and there is less mention of them in recent years. One might argue that the urbanization of the countryside in communist countries is part of a normal process found in Western countries although in the United States it certainly is not part of a uniform, state-run system of agriculture. Certainly the peasant villages of Eastern Europe still, for the most part, are national in their appearance and it is only in the appearance of new structures affected through central control – *Wirtschaftsbereich*, administrative and cultural buildings, and single-family houses – that the urban aspect is apparent.

EMPHASIS ON MINORITY GROUPS

One of the factors or objectives mentioned as affecting the distribution of economic activity in the communist countries is the emphasis on gaining the support of minority groups, often ethnically and culturally different from the primary national group. In the Soviet Union, these groups, for the most part, are distributed on the periphery of the country and it became an important object of State policy after 1918 to gain their voluntary support. The policy was first stated in 1929 in *Pravda*:

> In order to consolidate the alliance between the border regions and the center, it is first of all necessary to put an end to the estrangement and isolation of the border regions, to their patriarchal manner of life and lack of culture . . .[46]

In 1939 Molotov mentioned several projects which illustrated the application of this principle, among them the Mingechaur hydroelectric station in Azerbaidzhan, the drainage of the Colchis lowland in Georgia, the construction of the Zeravshan reservoir in the Uzbek SSR, and the completion of the Balkhash copper works in the Kazakh SSR.[47]

Evidence of the impact of ethnic groups on the location of economic activity is also found in Eastern Europe. Between World Wars I and II, East Europe's ethnic diversity was expressed through use of the term 'Shatter Belt'. Although Leszek Kosiński has shown that the diversity was reduced after 1945, several of the states have sizeable minority groups and their national policies reflect a desire to accommodate these groups.[48] The most significant examples of this ethnic diversity is provided by two federal states – Czechoslovakia and Yugoslavia – whose political structure reflects the presence of sizeable minorities. In Czechoslovakia, the relative weakness of the Slovak region *vis-à-vis* the Bohemian core area resulted in serious attempts to provide growth centres in the former area such as the East Slovakian Iron Works near Košice.[49]

The greatest political-ethnic fragmentation, however, is found in Yugoslavia where six states – Serbia, Croatia, Slovenia, Bosnia–Herzegovina, Macedonia and Montenegro – have certain rights which are protected within the legislative system by a Council of Nationalities. They also have certain possibilities for allocation of money within the federal budget. In addition, two special areas of Serbia – Vojvodina and Kosovo – have certain autonomous status because of ethnic differences; the former area is one of the most ethnically fragmented areas of Europe while the latter includes over one million Albanians. Fisher suggests that this multinational structure has complicated problems of economic development and that

Economic criteria have been less of a rationale for investment than the constitutional provision which suggests that equality must be shown to each of the six republics.[50]

Problems, therefore, arise as to the location of various developmental projects. For example, refrigeration plants located in the backward areas of Montenegro and Macedonia proved to be unsuccessful in terms of quality of product and distance from northern markets. The Party asks if it would be more rational to locate such plants in the more advanced areas of Slovenia and Croatia even though regional pride in the southern areas suffers.

In the other states the problem of sizeable minority groups is less evident. Romania has about 1·5 million Hungarians and the size and cultural influence of this group led to early formation of a Hungarian Autonomous Area; until recent years this area benefited from a Romanian policy that allowed for separate cultural institutions and for the establishment of economic projects. However, trouble in the region has been intermittent since 1956 (the Hungarian Revolution) and in 1968 the special province was abolished. It goes without saying that the Romanian government cannot ignore the presence of such a large cultural group.

One may conclude by stating that to a limited extent the appearance and location of forms in the East European landscape, whether Western, national or socialist, are based on the presence of minority groups.

VIEW OF RESOURCES

The socialist governments of Eastern Europe view resources within the context of planning policy. Therefore, the rational development of resources is closely tied to regional planning and even to the support of international projects of socialism, e.g. Romanian natural gas supplied to Hungary or the Iron Gate's hydroelectric power project. National planning in Eastern Europe provides the opportunity for handling resources as a Western country would in a war-time situation. A good example is provided by Romania where the government views resources in three categories: (1) those in large quantity including oil, natural gas, salt and timber provide a basis for trade and large-scale industry; (2) those in lesser amounts like lead,

116   *Eastern Europe: Essays in Geographical Problems*

zinc, copper, bauxite, gold and coal are exploited with the goal of achieving self-sufficiency; and finally, (3) those lacking, such as iron and coke, are supplied by imports; for example, the Soviet Union in 1965 supplied close to 50% of Romania's needs in iron ore, a degree of dependence that is also true of the large iron and steel plants of East Germany, Poland, Czechoslovakia, Hungary and Bulgaria. The forms in the Romanian landscape reflect this attitude towards resources: examples include the export of crude oil or oil-derived products from new producing areas like Argeș or Bacău; the expansion of coal mines in the Jiu river valley, and the import of coke and iron, via the Black Sea and Danube river, for the new iron and steel plant at Galați.

EGALITARIAN MODEL OF SOCIAL RELATIONS

Communist ideology emphasizes the goal of a classless society. The question, therefore, arises as to whether this type of idea is reflected in the landscape and geography of the East European countries. Perhaps the best test is in housing. The observer in communist countries feels that a certain degree of uniformity exists which may arise, however, as much out of a desire to maximize the effectiveness of limited investments as out of an ideological goal. Residences of the 'new' class – Party members, cultural leaders and technocrats – are certainly not very ostentatious. Fisher makes the following statement on socialist theory regarding cities:

> There is to be no sharp distinction or structural division among the various parts or components; the equality of the residential areas and their social composition is to be everywhere the same. No part of the city should ideally attract or repel certain classes. . . . All parts of the city of socialist man will be composed, in theory, of all people – a truly classless potpourri.[51]

This is supported by a Polish planner:

> The other characteristic [besides an orientation toward history] of the Polish planner's ideology may be found in his defense of social democratization and an egalitarian model of social relations. . . . At the core of these attitudes [of in-

> dustrialization, urbanization, mobility, and social services] is the simple truth that people are equal as in regard to their basic needs, desires, and aims and in their right to satisfy them. . . . The egalitarian outlook is reflected in the planner's conception of the neighborhood unit [where] social segregation should be avoided. . . . The egalitarian principle is also visible in housing standards. The difference between particular housing estates is practically negligible. Within a given neighborhood unit there is no difference in standards at all.[52]

The same planner admits that differences in housing do exist but the differences reflect competition among institutions, the main housing investors, and not between individuals. Differences also exist between sections of a country that possess varying historical backgrounds, e.g. Serbia and Slovenia. However, superimposed over these contrasts is a significant egalitarian idea:

> The belief that the state owes each family an adequate dwelling unit with minimum standards appears to be one of the pillars of the socialist creed that was universally accepted.[53]

Although the existing urban reality in Eastern Europe appears to be as yet only partly influenced by Marxist philosophy, the egalitarian idea has had some impact on landscape form and, therefore, on the geography of the area.

RENTABILITY

Ian Hamilton has pointed out that the most important recent criterion for investment in Yugoslavia is what he calls 'rentability' of the project.[54] 'Rentability' can be summarized as the relation of total income of a project (after the deduction of total costs) to the fixed and variable assets of the project. The system tends to act against locations of new economic activity in backward areas because these projects generally incur higher investment costs in infrastructure while those areas have little capital to contribute towards financing new projects. Although this principle was slow to gain headway, reforms made since 1965 in the direction of Yugoslav decentralization of economic development now make 'rentability' dominant over the more vague but once important criteria of 'social cost'. Hamilton states that these reforms may jeopardize the policy of developing the backward areas and encourage greater regional imbalances, for 'the greater the dependence for economic growth upon autonomous funds, the greater is the advantage to the developed areas'.[55]

The importance of the rentability principle in other East European countries is perhaps less because decentralization is not generally a feature of economic development. However, market and profit factors are more significant now in most countries, and it may be that the location of economic activity in the future will be based more on efficiency of operation than has occurred in the past. If this becomes true, then the backward areas – Białystok, Slovakia, Alföld, Dobruja and Macedonia to cite examples – may not receive as much attention as before.

THE VALIDITY OF THE CONCEPT OF A SOCIALIST LANDSCAPE

In this paper an attempt was made to utilize the landscape of Eastern Europe as a key to the understanding of the area. Forms in the landscape from socialist, Western and national origin were described in a tentative typology followed by an analysis of certain ideas that help to explain the appearance and location of these forms. However, these forms, regardless of their perceptible relationships to socialist, Western or national origins, are all controlled today by one Party which, through a national policy of planning, is able to exert a strong impress upon the landscape. Can one call a landscape that manifests such overall control a socialist landscape? There is little question that some forms are more uniquely socialist than others. Schröder seems to imply a socialist landscape when he contrasts the

> siedlungs-geographische Gegensatz zwischen gewachsenen Formen im Western [West Germany] und den Planformen im Osten [East Germany].[56]

In any case, he finds it difficult to discern a German character to the landscape east of the Elbe river. It may be that the socialist landscape is temporary and eventually the Western forms will become predominant as they are in most countries in the process of economic development. Certainly the idea of a socialist landscape is intriguing but its establishment will rest on continued research on specific forms, their origin and change through time.

NOTES

[1] Carl O. Sauer, 'The Morphology of Landscape', *University of California Publications in Geography*, Vol. II (1925), pp. 19–54.

[2] Richard Hartshorne, *The Nature of Geography* (Association of American Geographers, 1949), pp. 189–236.

[3] Karl H. Schröder, 'Der Wandel der Agrarlandschaft im ostelbischen Tiefland seit 1945', *Geographische Zeitschrift*, Vol. 52 (1964), pp. 289–316.

[4] Hartshorne, *op. cit.*, p. 167.

[5] Derwent Whittlesey, 'The Impress of Effective Central Authority upon the Landscape', *Annals, Association of American Geographers*, Vol. 25 (1935), p. 90.

[6] Whittlesey, *op. cit.*, pp. 85, 97.

[7] Philip E. Mosely, 'Collectivization of Agriculture in Soviet Strategy', in I. T. Sanders, *Collectivization of Agriculture in Eastern Europe* (Lexington: University of Kentucky Press, 1958), p. 58.

[8] Vladimir Soloukhin, *A Walk in Rural Russia* (New York: E. P. Dutton and Co., 1967), p. 7.

[9] David Mitrany, *Marx Against the Peasant* (London: George Weidenfeld and Nicolson Ltd., 1951), pp. 229–30.

[10] Schröder, *op. cit.*, p. 296.

[11] Central Statistical Board, *Statistical Pocket Book of the Socialist Republic of Romania* (Bucharest, 1968), pp. 138–9; D. Hooson, *The Soviet Union* (Belmont, California: Wadsworth Publishing Co., Inc., 1966), p. 68.

[12] Schröder, *op. cit.*, p. 302.

[13] Doreen Warriner, *Economics of Peasant Farming* (London: Oxford University Press, 1939), pp. 98–101.

[14] United Nations, 'Economic Development in Romania', *Economic Bulletin for Europe*, Vol. 13 (1961), p. 89.

[15] William Horbaly, *Agricultural Conditions in Czechoslovakia*, Research Paper No. 18 (Chicago: Department of Geography, 1951), pp. 55–94.

[16] René Dumont, *Types of Rural Economy* (London: Methuen & Co. Ltd., 1957), p. 506.

[17] Rudi Ogrissek, *Dorf und Flur in der Deutschen Demokratischen Republik* (Leipzig: VEB Verlag Enzyklopädie, 1961), p. 101.
[18] Schröder, *op. cit.*, p. 303.
[19] Schröder, *op. cit.*, pp. 303, 306.
[20] Ogrissek, *op. cit.*, pp. 101–6.
[21] Schröder, *op. cit.*, p. 308. In the *Kreis* of Grevesmühlen (Mecklenburg), 145 *Gemeinden* were consolidated into 38 large *Grossgemeinden*.
[22] Ogrissek, *op. cit.*, pp. 107–10.
[23] 'The House of Culture', *News from Behind the Iron Curtain*, Vol. 4 (1955), pp. 42–9. Much of the information in this section came from this source.
[24] Victor Tufescu, *Atlas Geografic – Republica Socialistă România* (Bucharest: Editura Didactică şi Pedagogică, 1965), p. 101.
[25] Micheline Billaut, 'La Collectivisation agraire en Bulgarie: l'exemple du village de Petrarch', *Annales de Géographie*, Vol. 69 (1959), p. 491; see also George W. Hoffman, 'Transformation of Rural Settlement in Bulgaria', *Geographical Review*, Vol. LIV (1964), pp. 45–63, specifically p. 63.
[26] 'The City of Socialist Man', *East Europe*, Vol. 10 (1961), p. 4.
[27] Jack C. Fisher, 'Planning the City of Socialist Man', *Journal, American Institute of Planners*, Vol. 28 (1963), p. 253.
[28] B. Michael Frolic, 'The Soviet City', *Town Planning Review*, Vol. 34 (1964), pp. 286–8.
[29] Vjacheslov Shkvarikov *et al.*, 'The Building of New Towns in the USSR', *Ekistics*, Vol. 18 (1964), p. 307.
[30] Fisher, 'Planning the . . .', *op. cit.*, pp. 255–9.
[31] Yurick Blumenfeld, *Seesaw – Cultural Life in Eastern Europe* (New York: Harcourt, Brace and World, Inc., 1968), pp. 60–1.
[32] Blumenfeld, *op. cit.*, p. 60.
[33] Mihail Haşeganu (ed.), *Wirtschaftsgeographie der Rumänischen Volksrepublik* (Berlin: Verlag die Wirtschaft, 1962); T. Morariu, V. Accu, *et al.*, *The Geography of Romania* (Bucharest: Meridiane, 1966); *Monografia a Republicii Populare Romîne*, Volumes I and II (Bucharest: Academia de RPR, 1960); Central Statistical Board, *Statistical Yearbook of the Socialist Republic of Romania*, various years.
[34] *A Report on Electric Power Developments in Rumania* (New York: Edison Electric Institute, 1964), p. 18.
[35] With the Soviet decision to use Western help, viz., Italy, in establishing an automobile industry, the smaller East European countries need feel no ideological restraints on turning to capitalistic countries in the interests of efficiency.

[36] Christian Kind, 'Rumania Today', *East Europe*, Vol. 17 (1962), p. 5.

[37] George W. Hoffman, 'The Problem of the Underdeveloped Regions in Southeast Europe. A Comparative Analysis of Romania, Yugoslavia and Greece', *Annals of the Association of American Geographers*, Vol. 57 (December, 1967), pp. 637–66, specifically p. 648.

[38] Schröder, *op. cit.*, pp. 310–11.

[39] W. A. Douglas Jackson, 'Comment' to article in J. Karcz, *Soviet and East European Agriculture* (Berkeley: University of California Press, 1967), p. 100.

[40] E. H. Carr, *The Soviet Impact on the Western World* (New York: The Macmillan Co., 1947), p. 27; a similar comment was made by F. E. I. Hamilton at the Conference on East European Geography, Austin, Texas, April, 1969.

[41] Y. G. Saushkin, *Economic Geography of the Soviet Union* (Oslo University Press, 1956), p. 49. Italics are his.

[42] N. D. Stolpov and N. F. Yanitskiy, 'Structural and Geographic Shifts in the Economy of the European People's Democracies, Members of CEMA, in the Period of Completion of the Construction of Socialism', *Soviet Geography: Review and Translation*, Vol. 6 (January, 1965), p. 18.

[43] Nicolae Ceauşescu, *The Five Year Plan* (Bucharest: Romanian News Agency, 1966), pp. 16–17.

[44] Mosely, *op. cit.*, p. 50.

[45] Mitrany, *op. cit.*, pp. 229–30.

[46] Iosef V. Stalin, 'The Policy of the Soviet Government on the National Question in Russia', *Pravda*, October 10, 1929, as cited in *Marxism and the National and Colonial Question* (New York: International Publishers, no date), p. 82.

[47] Vlacheslav Molotov, *Tretii piatiletnii plan razvitiia narodnogo khoziastva SSSR, 1938–1942 g. g.* (Leningrad, 1939), pp. 23–5 as cited in J. G. Rice, 'Ideological Theory Underlying the Distribution of Industry in the USSR', Discussion Paper No. 19, Department of Geography, University of Washington, May 8, 1959, p. 30.

[48] Leszek A. Kosiński, 'Changes in the Ethnic Structure in East-Central Europe', *Geographical Review*, Vol. 59 (1969), pp. 388–402.

[49] Koloman Ivanička, 'Problems Connected with the Research of Regions in Czechoslovakia', *Function and Forming of Regions*, Acta Geografica, Universitatis Comenianae No. 8 (Bratislava: Pedagogical Publishers, 1968), pp. 20–6.

[50] Jack C. Fisher, *Yugoslavia – A Multinational State* (San Francisco: Chandler Publishing Co., 1966), p. 9.

[51] Fisher, 'Planning the . . .', *op. cit.*, p. 252.

[52] Janusz Ziółkowski, 'Sociological Implications of Urban Planning', in J. C. Fisher (ed.), *City and Regional Planning in Poland* (Ithaca: Cornell University Press, 1966), pp. 199-200.

[53] Fisher, 'Yugoslavia . . .', *op. cit.*, p. 144.

[54] F. E. Ian Hamilton, *Yugoslavia – Patterns of Economic Activity* (New York: Frederick Praeger, 1968), pp. 111-13.

[55] Hamilton, *op. cit.*, p. 113.

[56] Schröder, *op. cit.*, p. 309.

Comments
LESZEK A. KOSIŃSKI

There is no doubt that the landscape reflects different cultural and political systems. Professor Rugg wants to utilize the landscape as a key to the understanding of the area and of the forces that shape the geography of Eastern Europe. I would suggest that there is another more direct way to understand these forces. Consequently I would be very sceptical about the diagnostic value of landscape analysis in any area unless there is no other information available.

Professor Rugg distinguished between three basic forces that shape the landscape of the area studied. National forces deeply rooted in historical tradition contribute to diversity. Forces of modernization represent the trend visible all over the world irrespective of political systems. Professor Rugg identifies them with the Western influence which seems to be rather questionable. Although the process of modernization began with the Industrial Revolution in Western Europe, at the present time the source area for the process of modernization is not necessarily limited to the Western world. It seems to me that a distinction should be made between the aspect of modernization that can be clearly linked with Western influence and those which are not identifiable as to their regional or geographic origin. Thirdly, he identifies the communist forces of industrialization. Here again, some of their aspects reflect simply a policy towards modernization of a society and one might argue that they are sponsored by the Communist government and by definition are communist. On the other hand only in some cases are the governments deliberately trying to alter the landscape in such a way that it would reflect the political beliefs and values.

An example, underestimated in Professor Rugg's analysis, seems to be the large-scale urban projects that are possible by tight control of the land and massive concentration of investment. The existence of these projects and their domination in the urban scene seems to be the most characteristic feature by which a socialist city differs from its capitalist counterpart.

Finally, I would strongly argue against the notion of the 'communist culture area'. It seems to me that although the political forces reflecting the prevailing ideology do play an important role in shaping the culture any definition of the term would require taking into consideration many other aspects in addition to purely political ones.

Remarks

In the discussion following Professor Rugg's paper, Dr Romanowski suggested that we view the problem in another way; that we introduce the concept of frontier and evolutionary landscapes – posing the question 'is the new element in the landscape caused by new ideas from within the culture, or is it an introduction of alien ideas?' In Eastern European landscapes, Dr Romanowski said, they are Soviet introductions as opposed to evolutionary. The characteristics of the landscape are developed over time, he said, and he did not think that the East European landscape has gone through the evolutionary process yet.

Dr Kristof then commented that the collective farm is one of the most important features of a socialist landscape; however, one must consider these changes in the light of what existed previous to the socialization of the landscape. Changes in the landscape were coming anyway and how much of the change can we say is due to socialism.

Dr Kristof commented that the present economic development in Eastern Europe results in the landscape being more varied, that there is incomparably more uniformity in the landscape of the several countries of Eastern Europe taken as a region when 70 to 90% of its population were peasants and agriculture practically the only economic activity. He pointed out that now there is a growing diversity in land use, type of settlement, economic activity, etc., and all of this diversity is recognizable in the landscape.

Dr Hamilton indicated that an ideology is involved; that socialism is not just national but international in scope and would lead to greater homogeneity in all of Eastern Europe.

The key to socialist landscape, according to Dr Hamilton, is that socialism aims at more equal distribution of wealth in a state. He indicated that the expression of this socialist philosophy will be shown in time on the landscape, and also pointed out that rural poverty is not just socialist but characteristic of underdeveloped countries.

Professor Ivanička then pointed out that the same results in landscapes do not necessarily evolve in the same processes. We must look at the past, he said, stating that there is no capitalist or socialist landscape, that Yugoslavia and Romania are different and that landscape is more complex than just an ideology.

Dr Fisher-Galati took exception to this remark saying that Romania is the wrong case, that it does not have a socialist landscape or even a Romanian landscape.

Dr Ivanička then suggested that there are three creators of landscape: (1) technology which is common to East and West, (2) ideology which is common to socialist countries, and (3) the tradition of the individual countries.

Dr Kromm then offered the view that many of the distinct characteristics of the socialist landscape can probably be revealed on a macroscale through the use of remote-sensing devices, while the more specific expression may best be recognized by ground level observation.

4

Geographic Research and Methodology on East-Central and Southeast European Agriculture*

JACEK I. ROMANOWSKI

This contribution is essentially a discussion of research done in East-Central and Southeast Europe in the field of agricultural geography. It does not attempt to describe the characteristics of the agriculture of that region, but is concerned with the methodology of and progress in arriving at a geographic synthesis of such characteristics. Agriculture is defined as the process of obtaining a livelihood from land tillage and/or animal husbandry. Forestry and fisheries as principal occupations are excluded. The pattern of life, when tied up in the agricultural process, is considered to be a necessary correlate of agriculture itself.

East-Central and Southeast Europe is still frequently referred to as an 'agricultural region' in contrast to Western Europe or even the U.S.S.R. This view no doubt stems from the traditional economic backwardness, low level of industrialization and high proportion of population engaged in agriculture in the region. The adjective 'agricultural' is, however, unfortunate because it singles out a sector of the economy which is, more often than not, very backward and neglected. In fact, as the share of industry in the national economies continues to grow, the adjective 'agricultural' can be applied only to selected regions within the individual countries. Overall, the area leads neither in agricultural productivity nor in agricultural growth. It is insignificant in the world's agricultural exports and is in fact a net importer.

Prior to World War II, the region may have been agricultural

---

* The author wishes to thank Professors Barbara Zakrzewska and Paul B. Alexander for their valuable suggestions which have been incorporated into this paper.

in terms of the percentage of the population employed, but this is not true today. The rise to power of Communist governments has resulted in a sharp change in economic policy. Industrial development and collectivization became the primary goals of the new states. Collectivization, industrialization and urbanization have disrupted the traditional agricultural pattern and have taxed the output of agriculture beyond capacity.

Only towards the end of the nineteenth century did geographers begin seriously to concern themselves with economic life in general and agriculture in particular. It is worth remembering that it was not until recently that the work of Johann Heinrich von Thünen was given the due attention by agricultural geographers.[1] Von Thünen was not a geographer, but was interested both in the spatial patterns of agriculture and in the reasons for the success or failure of various agricultural systems. The most encompassing definition of the term 'agricultural system' is 'the characteristics of obtaining a livelihood from agriculture as described for a specific situation or generalized for a specific area and period.' The agricultural system is thus not in itself a region but is rather a specific form of agricultural enterprise. It may, however, exist in a specific region.[2]

A complete understanding of success or failure and spatial variation of agriculture requires an extensive analysis of all the factors which affect agricultural operation and efficiency. These factors can be grouped into (*a*) physical environment (climate, soils, terrain, etc.); (*b*) economic environment (markets, profitability, security); (*c*) means of production (land, labour, capital); as well as (*d*) the efficiency of production, adaptability to changing conditions and chances for increased efficiency. It is understandable that, faced with such a broad task, geographers should have adopted a variety of approaches, each beginning with the study of selected aspects which in turn led to more comprehensive research.

This paper will survey the various schools of research in agricultural geography which have played a significant role in East-Central and Southeast Europe, the quality of information obtained and the research methods dominating in the individual countries of the region. The concern is with the comparability of the research accomplished in the various countries and schools of research. The measure of success of each of these schools will be their contribution to the understanding of the region's agriculture and their ability to synthesize that understanding into discrete, well-defined systems of agricultural operation.

SETTLEMENT AND LANDSCAPE GEOGRAPHY

As mentioned in the introduction, the spatial economic perspective is rather recent in geography. The earliest agricultural geographic work done in East-Central and Southeast Europe was concerned with the cultural landscape or with settlements or villages. This research orientation came from France and Germany where much work has been done either on the form and structure of the rural landscape (German) or on comprehensive syntheses of village life.[3] The study of the morphology of the rural landscape developed particularly strongly and continues as a branch of geography until this day.

Studies in settlement and cultural landscape geography were never meant to lead to agricultural typology, though the effort put into typologizing settlements has been appreciable and useful.[4] The immediate value of the approach has been the development of methods of inference from the visible landscape of many forces which themselves are not readily recognizable. Thus correlating house or barn type with cultural traits, village form with past conditions of production, and so forth, though not part of agricultural geography, has expanded the geographer's array of tools for the analysis of agriculture.

In the clarity of analysis and arrangement of the cultural landscape elements, German scholarship has been unequalled. The French on the other hand have produced descriptions of individual farm life situations which promote more understanding than could any mechanistic approach. The difficulty of the French studies, beautifully descriptive though they are, is that they do not lend themselves to comparative analysis and thus to structured typology. This dichotomy of approach, the mechanistic concern with structure and the comprehensive description of a site has been transferred to the countries of East-Central and Southeast Europe. Within each country the development of the subject has depended upon whether that country had closer ties with the German or with the French school. To the subject as a whole, the greatest contribution within the study area has been made by Polish geographers.

GERMAN RESEARCH IN EAST-CENTRAL AND
SOUTHEAST EUROPE

Landscape and settlement geography in East Germany and Poland has been largely influenced by the German school of geography, though French influence should not be completely denied. Since a fair portion of what now is part of Poland was part of Germany prior to World War II, German scholarship was directly responsible for much research in the area though one may argue about some of its conclusions and preconceptions. Silesia particularly had attracted the attention of the German geographer, resulting in voluminous literature on its agriculture, land organization and settlement.[5] Landscape and settlement geography were by no means the only fields of geography relating to agriculture, but they were certainly the best developed fields.

Since World War II, the geographers of the German Democratic Republic have largely abandoned research in this field, and the only valuable publication is the *Agricultural Atlas for the Territory of the G.D.R.* although it compares poorly with Otremba's publication in West Germany.[6] The East German atlas, though containing 67 sheets, is largely devoted to the natural conditions of agriculture. Other research in landscape geography is also mainly concerned with the physical environment of agriculture, an orientation that contributes little to the comprehensive understanding of the agricultural landscape. A very narrow but useful study is the book by Bochnig on the relationship of landscape to agricultural melioration.[7]

While field work on the territory of present-day Poland is understandably absent in West German scholarship, there are excellent general syntheses concerned with, or including, Poland which continue a long standing tradition.[8] Not part of our area of interest but significant from the methodological point of view in their influence on the geographers of Poland is a series of studies on the economic typology of settlements which appeared in two Berlin journals. These studies are important not only from the point of view of settlement typology but also with respect to planned modifications. Finally, a study of field systems in the western Alps which was published in East Germany, has become well known to Polish geographers.[9]

SETTLEMENT AND LANDSCAPE STUDIES IN POLAND

The most important geographic centre for settlement and landscape studies in pre-war Poland was the University of Lwów (now L'vov in the Ukraine). It was marked by greater reliance on quantitative and cartographic methods of deduction as opposed to the generally visual interpretations of the West European scholars of that day.[10] Geography at the Jagiellonian University in Kraków had a longer history than that of Lwów but in settlement and landscape studies it merits credit not so much for its research prior to World War II but for the training of some geographers who have contributed appreciably since the war. Particularly worthy of mention among them are, Zaborski, Dobrowolska, Leszczycki and Wrzosek, all of whom, prior to the war, published some articles pertaining to settlement.[11] Finally the University of Poznań produced one of the most productive settlement geographers of Poland, Kiełczewska-Zaleska, who studies the rural settlement and is active to this date.[12]

Landscape and settlement geography has continued in importance in Poland since World War II. The trend is towards greater concern with causal factors and relationships, with settlement studies clearly dominating those on the landscape. Perhaps the best work is carried out within the historical perspective, adding valuable information on the evolution of the present rural scene.[13] The studies on the current character of settlements and landscapes have an extremely broad range in Poland, and only a part of them contributes to agricultural geography. Similar to the trends in Western geography, Polish geography has moved from the study of the static structuring of spatial patterns to a more dynamic, functional analysis of spatial relationships of settlements.[14] The most recent trends have been the application of quantitative methods to geographic analysis. Quantitative geography is developing with great force at the Universities of Wrocław and Poznań and at the Institute of Geography of the Polish Academy of Sciences in Warsaw.[15] In this respect Polish geographers are keeping pace with Western quantitative geography and assimilate its methods, though most of the quantitative work is carried out outside the field of agricultural geography.

Living in a planned economy, it is only natural that Polish geographers should be drawn into the realm of planning, even if they are settlement or landscape geographers. This work is by no means a mere statistical inventory, as was the case of so many Soviet and East European studies. Rather, the research is in the framework of applied geography and is concerned with problems, the understanding and solution of which are of practical significance.[16] There are three principal directions of applied research in settlement geography in Poland. One is concerned with the actual or planned transformation of settlement networks. Of interest here is only the work on rural settlements.[17] The second is concerned with the actual or planned transformation of the village and its life, a long standing interest of the government which would like to socialize the 'private peasant'.[18] The last studies the relationship between the town and the village, with particular reference to the development of the peasant-worker class which has deep repercussions on the agriculture of the country.[19]

Though incapable of contributing much to agricultural typology, the Polish achievements in settlement and landscape geography have greatly deepened the understanding of the causality of settlement networks and significance of numerous social and economic factors within the oecumene. The Polish geographer is well acquainted with the complex relationships between the agriculturist, the character of his settlement, and the economic and natural environment. According to J. Kostrowicki, 'Settlement geography – the earliest separated and previously best developed branch of geography in the world as well as in Poland – has attained the greatest results. Though, however, before the war the subject of research in Poland was principally the rural settlement, after the war the geography of cities came to prevail ... In the realm of rural settlement studies, historical geographic studies prevail, as they did prior to the war, though the methodology of some studies came to be significantly altered and strengthened.'[20] It is a pity that despite this great output of research, there has been so little effort to synthesize the findings. There is to date no Polish Meitzen, not even an Otremba, who would attempt to order and classify the volume of information at hand.

CZECHOSLOVAKIA, HUNGARY AND ROMANIA

Compared to Polish geography, the other countries of East-Central and Southeast Europe have little achievement to claim in the settlement and landscape approach to agricultural geography. This is well illustrated for Czechoslovakia in an article devoted to Czech agricultural geography.[21] Unlike Poland, Czechoslovak geography, prior to World War II, gave little attention to settlements or human landscape. The only article worthy of note is Pohl's typology of rural settlements.[22] After the war, most of the work of settlement geography continued to be concerned with the town, and landscape geography continued to stress the physical landscape.

Of those few studies which could be classified as contributing to agricultural geography, only one belongs to the landscape school.[23] The others fall into settlement geography, mostly into the realm of functional classification.[24] There is one paper on the peasant-worker problem (cf. n. 19) and one which is concerned with description rather than classification.[25]

Hungarian geographers also have contributed little to the settlement and landscape approach to agricultural geography. What work there has been was largely influenced by the German school of *Siedlungsgeographie*. Research on settlements has, however, greatly increased in recent years. There are historical geographic studies of village and tenure patterns concerned with development of village systems and isolated settlements.[26] There are some detailed and, sometimes, comprehensive surveys of rural settlements,[27] as well as some studies of functional links between settlements and between the farm and field.[28]

In one aspect, Hungarian settlement geography probably equals that of the Poles in research output. This is in mostly contemporary studies of the present development of the village and the plans and predictions of future development as well as the future of settlement networks. This, as in the case of Poland (cf. n. 17 and 18), is connected with the transformation of the

Hungarian village from private to collective agricultural production and, no doubt, with the economic and technological changes in the country.[29] This research orientation can be recorded in each of the surveyed states.

Romanian geography was at first under the dominance of German geographers (Ritter, von Richthofen, Ratzel) but then fell completely under the influence of French geography and it is the French who have made a more marked impact. The French geographer most closely associated with Romania was de Martonne who, at the beginning of this century, published a number of significant works on that country.[30] An unusual study, comparing French and Romanian villages, was published by Ficheux.[31] Settlement and human geography was well developed. Perhaps an unequalled amount of work was devoted to the pastoral economy and transhumance found within the country.[32] Many of the studies on transhumance and pastoral life focus on the people involved and are more readily classified into the field of population or cultural geography.[33] A useful discussion of the state of Romanian geography before the war has been written by Popp.[34]

The Romanians have done much research on village life and structure, including historical studies, typology and relationships to the natural environment.[35] There are also a number of studies devoted to single villages. A study by Tufescu is an impressive monograph and a collective effort of the recent years is limited in size but superior in methodology.[36] Mihăilescu was concerned with the typology of rural settlements. He has produced a map of 'principal types of rural settlements in Romania' on which he differentiated three types of villages: compact, dispersed and isolated, the latter being the most common.[37] Mihăilescu also was the principal pre-war representative of Romanian settlement geography at international meetings. He presented the above-mentioned map at the 1931 Congress of the I.G.U. in Paris, and discussed his work on the history of settlement in Walachia (cf. n. 35) at the 1934 Congress of the I.G.U. in Warsaw.[38] He furthermore has contributed much to the establishment of the departments of geography at the universities in Romania. The total development in settlement and population geography does not, however, yield any comprehensive picture of Romania's countryside, and what there is is largely outdated due to the rapid changes in the economy after the war.

SOUTHEAST EUROPE

Little work is being done in the rural settlement and agricultural landscape of Bulgaria and Albania. The former has followed the German tradition, the latter cannot even be evaluated. Concerning Bulgaria, a somewhat synthetic study on settlements was written by Penkoff for a German journal and the best discussion of its settlement geography is presented in a recent Slovak publication.[39] In English there is a very good study by G. W. Hoffman, on rural settlement in Bulgaria, also published in a German version.[40] Concerning Albania, an early note appeared in the *Geographical Review*, and two very informative papers dealing with agricultural landscape and pastoralism were written by A. Blanc.[41]

Yugoslav agricultural geography leads in Southeast Europe. The French school is dominant and the studies within the settlement and landscape theme are more concerned with man than with structure. There is no work of note from before World War II but since the war publications have been numerous.

The leading scholars in the field are I. Crkvenčić and V. Klemenčič. They co-operated on a major article dealing with Yugoslav agricultural geography.[42] There is, in general, a healthy share of the analytical work in the overall output. I. Crkvenčić, for instance, has evaluated agricultural geography within the scope of social geography. One can note there the strong influences of West European geography on his approach to the field.[43] A somewhat more Marxist framework of the tasks and methodology of agrarian geography in the various republics was presented at the 1967 conference on agrarian geography at Maribor, not only surveying the current research but also relating the attainments of the past.[44]

As previously mentioned, the French school tends to prevail in Yugoslavia. This prevalence is by no means as strong as it was before the war but it is still evident. There are a number of more or less comprehensive studies of small areas, and A. Blanc (cf. n. 41) wrote a human geography of western Croatia.[45]

There is a long series of studies on the social side of human geography. These are concerned with the social (socio-economic) aspects of various forms of land use, with rural over- or de-population, and with the social effects of collectivization and industrial development; for instance, the effects of collectivization on the peasant character of the population.[46] In view of this social orientation, it is surprising that, except for an early article by Lefevre, there have been no geographic studies of the Zadruga.

From Yugoslavia there are not only the studies in the tradition of *geographic humaine* but also a fair amount of work on the cultural landscape. As elsewhere this type of research tends to be more structure oriented though we must remember that it is being carried out by the same scholars who are protagonists of social geography.[47] But even in landscape geography, there are a fair number of studies whose focus is as much on the *milieu humaine* as it is on the *Kulturlandschaft*. Much of the research is on the changing landscape, or better on the changing conditions of life and the resultant landscape.[48]

Finally, much recent work is concerned with aspects of applied geography. In the area of landscape and settlement studies, one applied aspect is the analysis of spatial patterns of economic life. Studies of land holding patterns or land fragmentation fall into this category.[49] Another area of applied geography concerns itself with regional planning. Thus the attractiveness of cultural landscapes to tourism and potential development of that industry, or the utility of settlement networks in the improvement of agriculture, are subjects dealt with in some studies. There are also some studies on the changing nature of the village.[50]

LAND USE AND OTHER APPROACHES

Settlement and landscape geography is the most established field of geography in East-Central and Southeast Europe, but it is by no means the only field concerned with agriculture. In fact it is rather marginal to agricultural geography. Other, more directly agricultural concerns are land use geography, socio-economic geography and social geography and the crop and commodity approach.

THE POLISH LAND USE SURVEY

Land use studies began in the United States in the 1920's but attained world recognition under the leadership of Professor L. D. Stamp of Great Britain.[51] There were a number of attempts at land use surveying in East-Central and Southeast Europe including those by the Bulgarians, Romanians and the Poles. It was J. Kostrowicki who adapted the British land use survey to the native conditions of Poland, modified it through field experience in other countries of the region, and produced a phenomenal volume of work in a very short period of time.[52]

The Polish Land Use Survey, as the Kostrowicki method is called, is much superior to its British predecessor in analytical power. Beside the land use map of a scale of 1:10,000 or 1:25,000, the survey considers the state of land improvement, field fragmentation, ownership fragmentation and land tenure conditions.[53] By the time of the 19th I.G.U. Congress in 1960, the Polish Land Use Survey was well on its road to success within the Bloc. This does not mean that there were no criticisms or objections. At the 1964 Budapest Conference on land use, many differences in opinion were voiced. This was particularly true of the expenses involved in the Polish survey, expenses which some geographers thought were unwarranted.[54] The Hungarians felt that they could attain better results by means of simpler land use surveying and deeper economic analysis.

The Polish Land Use Survey is no doubt a significant contribution judging by the volume of its output, despite the measured opposition to it, but its further utility to the field of agricultural geography is very limited.[55] This is so because, having developed a method of data collection, the business of collection can be relegated to technical personnel. Furthermore, the Polish Land Use Survey is much too time consuming to provide complete regional coverage in a reasonable span of years. This is also true of the simplified method while the so-called 'macroscale studies' are largely products based on statistical material rather than field research.[56]

But there is an even more serious criticism of the Land Use Survey. It measures essentially at one point in time the effects of agricultural life. The ephemeral nature of the survey (e.g. land use of the commune of Kruszwica as of 1957, the land use of the Czersk commune in 1962, etc.) reduces its utility in time, and any attempt at comparable coverage becomes an endless effort. The concern of land use surveys with the effects (land use) rather than the causes (factors of decision making) almost precludes any real understanding of agriculture.

Reflecting the above difficulties, the Kostrowicki group has attempted to introduce modifications to the Land Use Survey. There is no longer any serious plan of comprehensive coverage of Poland with the initial very detailed survey. One speaks instead of microscale sample surveys designed to verify more general studies.[57] The development of the macroscale surveys has already been mentioned above. At the same time the very costly detailed, multi-colour, land use maps are yielding to the newly developed, simplified land use map to be published at a scale of 1:200,000 or 1:300,000 instead of the original 1:10,000 or 1:25,000.[58]

The Polish Land Use Survey is also deepening the study of factors other than land use. One such study thus concerns itself with the relationship between land use and land ownership (state, collective, private) in Kujawy, a region northwest of Warsaw where the boundary between the Russian and Prussian Empires resulted in marked agricultural variation.[59] Another study relates the physical-geographic conditions of the Vistula delta to its land use. This study is very valuable as the selected region is not only strongly diversified in its physical conditions but also has an historical derived variance (The Free City of Danzig, East Prussia and Poland all had parts of this territory in the past), as well as the economic effect of its suburban (Gdańsk) location.[60] No doubt such correlative studies are designed to develop the Land Use Survey into a source of other agricultural information as well.

FROM LAND USE TO TYPOLOGY

The concern with a scientific typology of the world's agricultural systems is not new and has involved American, French, German, as well as Russian geographers.[61] The task has however never succeeded, reflecting no doubt the forbidding complexity of world agriculture. The problem has been to devise an agricultural typology which would be transferable in time and space. It is perhaps logical that J. Kostrowicki, after concerning himself with land use surveying, should turn to the problem of developing a typology.

One approach to typology has been by means of regionalization. Regionalization is a necessary part of any typology but cannot precede it. For this reason the definitions of the various agricultural regions are not definitions of types of agriculture but of agricultural regions, regions usually delimited with a great deal of intuition and 'common sense'.[62] At times the geographer performed phenomenal mental gymnastics in trying to 'scientify' his regional approach to reality. On the other end of the scale there were studies of individual farms or of very small areas, studies which were capable of fully describing the farm operations but were incapable of correlation into valid syntheses of larger regions.[63]

It was thus a powerful notion, the basing of a world agricultural typology on a world land use survey, which led Kostrowicki to extend his field of interest. But one might wonder, is not a land use based typology, a typology of forms of land use rather than of agriculture? After all, agriculture is not the business of using the land but rather that of obtaining a livelihood from the land. The difference in stress is critical, the agricultural type is conditioned by the socio-economic forces acting upon an environment and not by a form of land use. The Polish Land Use Survey does contain some economic indicators but they form in it only suffixes to the land use rather than its foundation.[64] Even if the recent modifications to the survey are fully adopted, the expenditure of effort on land use surveying and the stress placed on the land use itself seem unwarranted.

THE I.G.U. TYPOLOGY

As chairman of the Commission for Agricultural Typology of the I.G.U., Kostrowicki has striven towards the establishment of a world-wide agricultural typology. The procedure for agreeing upon a method has been quite democratic, but land use, nevertheless, seems to form the basis of that effort.[65] Thus the preliminary conclusions arrived at by Kostrowicki and Helburn include such socio-economic factors as ownership, management, labour supply, size of holdings and share of agriculture in the total employment. Among the agricultural factors, the system considers land fragmentation and patterns of land use, quality of land, water and climate management, biological control, cropping systems, husbandry techniques and the livestock to land ratio. The system also considers labour intensity and aggregate intensity of agriculture. Production characteristics form a whole sector of factors considered in this I.G.U. typology. These characteristics are land productivity, capital productivity, degree of commercialization, level of communication, agricultural orientation and specialization. But despite this wide array of factors, the typology itself still seems to be based on land use and the land use survey.

The I.G.U. agricultural typology is not just another wishful venture. It is based on extensive experience and is a direct outgrowth of the Polish Land Use Survey. The framework of the typology was laid in the criteria of the land use survey elaborated for Poland. That land use survey foresaw, from its conception, a typology of Polish agriculture. Many of the factors discussed in the I.G.U. papers already have been considered within the scope of the Land Use Survey.[66]

This Poland-based survey and potential typology was then expanded and modified to encompass all of East-Central and Southeast Europe. This tested the survey under far greater variations of physical environment, economic development and social conditions. On the basis of such work, Kostrowicki came up with two (or three) types of agriculture for the area; the Central European type, the Southeast European type and (perhaps) the Mediterranean type. Though based on still insufficient data, this typology seems to be far from satisfactory particularly considering the volume of work involved.[67]

THE SOCIO-ECONOMIC APPROACH

An approach to agricultural geography and typology which has been less productive thus far but is perhaps more promising for the future is the so-called socio-economic orientation advocated most strongly by Enyedi.[68] This is evident in Enyedi's reply to the I.G.U. questionnaire: 'I would like to underline that social aspects play a fundamental role in agricultural typology and serve as a starting point for it.'[69] Nor was he the only geographer placing much stress on socio-economic aspects.

In the same discussion on the I.G.U. questionnaire, the Russian geographer, Mukomel', makes an almost identical plea.

In his book on the agriculture of the world,[70] Enyedi points out that land use alone will not lead to understanding and he proceeds then deeper into the analysis of the social background and economic conditions. He calls his types of agriculture 'social types' and bases them on 'socio-economic motives'. The fact that his approach has a touch of the Marxist concept of economic evolution would not be a problem, but Enyedi does seem to be somewhat too deeply involved in the 'social' aspect at the expense of the 'economic'. The economic factor is characteristically relegated to the position of being merely a social policy, reflecting social conditions and not any economic forces.

Apart from the theoretical introductions, the book itself looks like an old-fashioned college text, with the old descriptive divisions of world agriculture. Presumably the most important of his types is market-oriented large-scale socialist agriculture. It is divided into three regional sub-types, (a) Asia and Cuba, (b) the U.S.S.R., and (c) East-Central Europe. The last sub-type, contiguous with our area of concern, is again divided into three groups. Albania, Bulgaria and Romania are considered to resemble the Soviet sub-type since collectivization occurred in them before any extensive industrial development. East Germany, Czechoslovakia and Hungary are the second group. Here collectivization occurred when the economy was already well developed. Agriculture is modern not only in social (collective) but also in economic and technological terms. Poland and Yugoslavia, where private farming still prevails and socialism takes on peculiar forms, are placed in the third group.

The East-Central European sub-type is then regrouped according to the orientation of farm production. East Germany and Czechoslovakia are called livestock economies of developed countries, Poland and Hungary are termed transitional of a mixed farming character. Romania's agriculture is dominated by plant crops though livestock is important, Yugoslavia is dominated by plant crops, mainly grains, and Bulgaria by vegetables and Mediterranean crops.

While one might agree with Enyedi's plea for more recognition of the social element in agriculture, his actual study of world types is no advance over traditional scholastic description. The closest that Enyedi comes to putting his ideas into action are small studies which largely reflect the Polish Land Use Survey.[71]

THE CROP OR COMMODITY APPROACH

If the settlement and landscape approach can be traced to German and French influences, the land use approach to British and then Polish efforts, then the crop or commodity approach can be credited to the Russians. In general, this approach to agricultural geography and typology reminds one of the crop or commercial geography of the 1920's in the West. Actually the degree of scholarship is much higher and the quality of the delimitation of agricultural regions or types in the U.S.S.R. does not trail behind any such efforts made in the West.[72] In their typology, the Russians rely mainly on the recognition of *specialization*, for example what are the principal crops, and of *concentration*, and on what crops are most of the inputs concentrated. The measurement of specialization is either through the structure of total production or that of market production; the unit of measure is either the state commodity price or a conventional (grain or livestock) unit equivalent. Very much work along this commodity line was done in the fifties throughout East-Central and Southeast Europe, directly influenced by the U.S.S.R.[73]

The commodity approach is widely used in East Germany in a greatly modified measure. In fact the Germans have been trying to measure and regionalize agricultural specialization and intensiveness by means of crop analysis. The system considers both the natural (land quality, climate, terrain) and the economic (transport, land use) factors but relies on crop analysis for its typology leaving the other factors as background information.

A critical study in this direction was made by E. Rübensam who divided East Germany into 17 zones and further established transitional regions.[74] Much work using the commodity-intensiveness approach has been done in West Germany by B. Andreae.[75] At first, Andreae stressed the structure of plant production but he did not succeed in obtaining a meaningful measure of intensiveness. It is interesting to note that Kostrowicki accounts for intensiveness of plant production simply by establishing a number of plant categories of intensiveness; e.g. structure forming, intensifying, extractive.[76] That, however, has little in common with economic intensiveness, that is with the intensity of inputs or production.

Unable to arrive at a typology on the basis of plant production, Andreae now relates the income of the farm to its form of specialization. This is a rather involved method but has unusual promise in its recognition of the structure of farm operation, a rather forgotten element in the settlement, land use and commodity approaches. But it is precisely the intricate linkage of the inputs of a farm, the operation with these inputs in order to produce the output, that seems to lie at the core of farm life, a farm establishment, and ultimately a farm type. Land use, type of building, or even the final crop *need not* be the central determinant. What differentiates Manchurian wheat from French wheat farming? What makes a Corn Belt farm more akin to a Kansas wheat farm than to a Romanian corn farm? What makes an *ante-bellum* plantation retain some of its elements of regeneration throughout the years of sharecropping? And, for that matter, are some collectives that much different from the farms of pre-socialist days as ideological literature would have us believe? If farming is a way of life, a business enterprise, then perhaps its definition can be best found in studying that way of life, the flow charts of that business operation.

CONCLUSIONS

Agricultural geography in East-Central and Southeast Europe is a well developed and well represented branch of the geographic sciences. Second only to the economists in volume of publication pertaining to agriculture, the geographers have already achieved marked practical results of their work. But while development in each Republic seems to be quite healthy (especially in Poland, the G.D.R., Hungary and Yugoslavia), one cannot say the same for the development of the region as a whole. Each country seems to have its own cherished traditions. This can lead to benefits from multiple forms of research and development but does pose problems of correlation and transferability.

The landscape and settlement studies were never intended to be uniform and there is no chance of developing any manuals for directed research without severely damaging the output of many scholars. The land use studies on the other hand lend themselves well to systematization. After all there are only so many possible agricultural land uses. But even here we find significant divergence despite the present dominance of the Polish survey. The socio-economic approach remains undeveloped but may provide some useful corrections to the Polish survey. The commodity approach is, in its simpler form, of use only to the planner and statistician. In the more involved elaborations it can be quite successful in describing agriculture.

The only serious effort to arrive at a typology of East-Central and Southeast European agriculture is being made by Kostrowicki and his associates. One may wonder about the wisdom of relying on land use to such an extent. In field work carried out at the University of Kansas, it soon became obvious that land use was not always essential to an understanding of a farm type. Typology after all need not be concerned with inventory. A greater concern with the structure of the operation and of the invisible factors of farm-life (values, opinions, attitudes) seems to be needed. Meticulous detail seems frequently a hindrance to analysis rather than an aid.

Finally, one might suggest that typology need not go hand in hand with regionalization. This seems to be the opinion of Kostrowicki as well but generally has been ignored in the past. It is one thing to arrive at a regional division of a larger unit and another thing to formulate specific types of agriculture. The type, ultimately, need not even exist in a region. The region may rather contain an array of deviations from the model type – thus, for instance, 65 to 80% of a fully self-sufficient, mixed farm type. The criteria for arriving at a farm type need not even be geographic, i.e. they need not have a spatial context and may be more suited for treatment in graph or tabular form than on a map.

Once a series of types is determined, the geographer then may have the task of developing methods of assigning the spatial reality to regions of one or another type of agriculture. This is still far removed, since no well-defined types have as yet been established. It is possible, however, that due to the sheer bulk of information accumulated in East-Central and Southeast Europe, the area is capable of providing the base for the first systematic, discrete and structured typology of agriculture.

The scope of possibilities of research by Western scholars on agricultural typology and geography in East-Central and Southeast Europe is great, always of course subject to political developments. But the native scholars are usually more than willing to co-operate in exchanges, visits and the transmission of information. There is particularly much potential in applying the scholarly attainment of the area to America and vice versa. American quantitative geography is rapidly absorbed in East-Central and Southeast Europe, particularly in Poland. There are other attainments in American geography which could be tested within the area. And there is much work from that area that could be tried in the United States. Certainly the world land use survey could occupy us to a much greater degree than the handful of responses given to the I.G.U. questionnaires.

NOTES

[1] Johann Heinrich von Thünen, *Der isolierte Staat in Beziehung auf Landwirtschaft und National-ökonomie* (The Isolated State With Respect to Agriculture and National Economy), (Rostock: Private edition, 1826); Andreas Grotewohl, 'Von Thünen in Retrospect', *Economic Geography*, Vol. 35 (1959), pp. 346–55; and Michael Chisholm, *Rural Settlement and Land Use* (London: Hutchinson University Library, 1962).

[2] Jerzy Kostrowicki and Nicholas Helburn, *Agricultural Typology, Principles and Methods, Preliminary Conclusions* (Boulder: I.G.U., 1967), 37 pp. and 12 pp. appendix.

[3] Friedrich Ratzel, *Anthropogeographie* (Anthropogeography), 2nd ed. (Stuttgart: J. Engelhorn, 1912); August Meitzen, *Siedlung und Agrarwesen der Westgermanen und Ostgermanen, Kelten, Römer, Finnen, und Slawen* (Settlement and Agriculture of the West Germans, East Germans, Celts, Romans, Finns and Slavs), (Berlin: N. Hertz, 1895); Albert Demangeon, 'L'habitation rurale en France. Essai de classification des principaux types' (The Rural Settlement in France. An Essay on Classification of Its Principal Types), *Annales de Géographie*, Vol. 29 (1920), pp. 353–75; and Roger Dion, *Essai sur la formation du paysage rural francais* (An Essay on the Formation of the French Rural Landscape), (Tours: Arrault et Cie, 1934).

[4] Albert Demangeon, 'The Origins and Causes of Settlement Types', edited by Phillip Wagner and Marvin Mikesell, *Readings in Cultural Geography* (Chicago: University of Chicago Press, 1962), pp. 506–16; Glenn T. Trewartha, 'Types of Rural Settlement in Colonial America', in *ibid.*, pp. 517–38; and Bogdan Zaborski, 'Criteria for Comparing Settlement Types', edited by F. E. Ian Hamilton, *Abstracts of Papers of the 20th International Geographical Congress* (London; Thomas Nelson and Sons, 1964), pp. 316–17.

[5] Josef Partsch, *Schlesien, eine Landeskunde für das deutsche Volk* (Silesia, A Geography for the German Nation), (Breslau: F. Hirt, 1896); Heinrich Loesch, 'Die Fränkische Hufe' (The Frankish Hide), *Zeitschrift des Vereins für Geschichte Schlesiens*, Vol. 61 (1927), pp. 81–107; and Herbert Knothe, *Vom deutschen Osten* (From the German East), (Breslau: Verlag von M. und H. Marcus, 1934).

[6] Rudolf Mätz, *Agraratlas über das Gebiet der DDR* (Agricultural Atlas for the Territory of the G.D.R.) Vol. 1 (Gotha: H. Haack, 1956); and Erich Otremba (ed.), *Atlas der deutschen Agrarlandschaft* (Atlas of the German Agricultural Landscape), (Wiesbaden: F. Steiner, 1965).

[7] Joachim Schultze, *Die Naturbedingten Landschaften der DDR* (The Nature affected Landscapes of the G.D.R.), Beiträge zur *Petermanns Geographische Mitteilungen*, No. 257 (1955); Reinhold Lingner and Carl Frank, *Landschaftsdiagnose der DDR. Ergebnisse einer zur Ermittlung von Landschaftschäden in den Jahren 1950 und 1952 durchgeführten Forschungsarbeit* (Landscape Diagnosis of the G.D.R. Results of Research on Determining the Landscape Damages of the years 1950 and 1952), (Berlin: Technik Verlag, 1957); and Erhard Bochnig, *Grundriss der Landschaftsgestaltung in der Landwirtschaftlichen Melioration* (Outline of Landscape Morphology in Agricultural Melioration), (Berlin; Deutscher Landwirtschaftsverlag, 1962).

[8] Meitzen, *Siedlung und Agrarwesen, op. cit.*; Wolfgang Ebert, *Ländliche Siedelformen im Deutschen Osten* (Rural Settlement Forms in the German East), (Berlin: E. S. Mittler und Sohn, 1937); Erich Otremba, *Allgemeine Agrar- und Industriegeographie* (General Agricultural and Industrial Geography), (Stuttgart: Franckhscher Verlag, 1953); and Gabriele Schwartz, *Allgemeine Siedlungsgeographie* (General Settlement Geography), Vol. 6 of *Lehrbuch der allgemeinen Geographie* (Textbook of General Geography), (Berlin: De Gruyter, 1959).

[9] H. Hüfner, 'Wirtschaftliche Gemeindetypen' (Economic Types of Communities), *Raum und Wirtschaft*, Vol. 3 (1952), pp. 43–57; Hans Linde, 'Grundfragen der Gemeindetypisierung' (Fundamental

Problems of Community Typology), *Raum und Wirtschaft*, Vol. 3 (1952); Helmut Lehmann, *Die Gemeindetypen. Beiträge zur Siedlungskundlichen Grundlegung von Stadt und Dorfplanung* (Community Types. Contributions to the Settlement-Geographic Foundation of City and Village Planning), *Schriften des Forschungsinstituts für Städtebau und Siedlungswesen* (Berlin: Verlag Technik, 1956), 67 pp.; Käthe Mittelhäusser, 'Funktionale Typen landischer Siedlungen auf statistischer Basis' (Functional types of Rural Settlement derived on a statistical basis), *Berichte zur deutschen Landeskunde*, Vol. 24 (1960); and Felix Monheim, *Agrargeographie der westlichen Hochalpen mit besonderer Berücksichtigung der Feldsysteme* (Agricultural Geography of the Western High Alps with particular reference to the Field Systems), *Beiträge zur Petermanns Geographische Mitteilungen*, No. 252 (1954), 136 pp.

[10] J. Albert, 'Z geografii osiedli wiejskich w dorzeczach Sanu' (On the Geography of Rural Settlements in the Regions of the San Tributaries), *Prace geograficzne*, Vol. 16 (1934), pp. 1–34; Józef Wąsowicz, 'Z geografii osadnictwa wiejskiego na Wołyniu' (On the Geography of Rural Settlements in Volhynia), *Czasopismo geograficzne*, Vol. 12 (1934), pp. 288–93; and Jan Ernst, 'Regiony geograficzno-rolnicze Polski' (The Geographic-Agricultural Regions of Poland), *Czasopismo geograficzne*, Vol. 10 (1932).

[11] Bogdan Zaborski, 'O krztałtach wsi w Polsce i ich rozmieszczeniu' (Concerning the Plan of Villages in Poland and Their Distribution), *Prace Komisji Etnograficznej P.A.U.*, Vol. 1 (1926); Maria Dobrowolska, 'Studia nad osadnictwem w dorzeczu Wisłoki i Białej' (Settlement Studies in the Wisloka and Biala Confluence), *Wiadomości geograficzne*, Vol. 6–7 (1931), pp. 103–8; Stanisław Leszczycki, 'Osadnictwo zachodnich Karpat Polskich' (Settlements of the Western Polish Carpathians), *Wiadomości geograficzne*, Vol. 9 (1934); and Antoni Wrzosek, *Własność ziemska na Pomorzu według narodowości* (Land Ownership in Pomerania according to Nationality), *Pamiętniki Instytutu Bałtyckiego*, Vol. 23 (1935).

[12] Maria Kiełczewska, 'Osadnictwo wicjskie Pomorza' (Rural Settlement of Pomerania), *Badania geograficzne nad Polską Północnozachodnią*, Vol. 14 (1934), pp. 1–41; 'Osadnictwo wiejskie wielkopolski' (Rural Settlement of Great Poland), *ibid.*, Vol. 15 (1935); and 'Osadnictwo wiejskie i miejskie Pomorza i Prus Wschodnich' (Rural and Urban Settlement of Pomerania and East Prussia), *Słownik geograficzny Państwa Polskiego*, Vol. 1 (1937), pp. 243–81.

[13] Maria Kiełczewska-Zaleska, 'O powstaniu i przeobrażeniu krztałtów wsi Pomorza Gdańskiego' (On the Origin and Transformation of Village Shapes of Gdansk Pomerania), *Prace geograficzne*

I. G. PAN, No. 5 (1956); Stanisława Zajchowska, 'Developpement de l'habitat en Posnanie' (Development of Settlement in the Poznań Region), *Przegląd geograficzny*, Vol. 32 (1960) supplement, pp. 227–234; Maria Dobrowolska, *Przemiany środowiska geograficznego Polski do XV wieku* (Changes in the Geographic Environment of Poland to the 15th Century). Warszawa: PWN, 1961); and Kazimierz Dziewoński, 'Geografia osadnictwa i zaludnienia – dorobek, podstawy teoretyczne i problemy badawcze' (Geography of Settlement and Population – Attainments, Theoretical Principles, and Problems of Research), *Przegląd geograficzny*, Vol. 28 (1956), pp. 721–62.

[14] Karol Bromek, 'Układ przestrzenny ośrodków usługowych w Polsce ze szczególnym uwzględnieniem woj. krakowskiego' (The Spatial Structure of Service Centres in Poland with Particular Stress on the Kraków Voivodship), *Przegląd geograficzny*, Vol. 21 (1948), pp. 286–91; and Michał Chilczuk, *Sieć ośrodków więzi społeczno-gospodarczej wsi w Polsce* (The Network of Centres of Socio-Economic Linkage of the Village in Poland), *Prace geogrficzne I. G. PAN*, No. 45 (1963).

[15] Wiesław Maik, 'Niektóre problemy badań nad układami osadniczymi' (Some Problems of Research on Settlement Patterns), *Czasopismo geograficzne*, Vol. 39 (1968), pp. 157–71; Zbyszko Chojnicki, 'Modele matematyczne w geografii ekonomicznej' (Quantitative Models in Economic Geography), *Przegląd geograficzny*, Vol. 37 (1965), pp. 115–34; and Ryszard Domański, 'Metody badania zbieżności układów przestrzennych' (Methods of Studying the Convergence of Spatial Patterns), *Przegląd geograficzny*, Vol. 41 (1969), pp. 79–92.

[16] Stanisław Leszczycki, 'Geografia stosowana, czy zastosowanie badań geograficznych dla celów praktycznych' (Applied Geography, Or The Application of Geographic Research to Practical Goals), *Przeglad geograficzny*, Vol. 34 (1962), pp. 3–24; (ed.), *'Problems of Applied Geography, II'*, *Geographia Polonica*, Vol. 3 (1964); and 'Perspektywy rozwoju badań geograficznych w Polsce' (Perspectives of Development of Geographic Research in Poland), *Przeglad geograficzny*, Vol. 36 (1964), pp. 411–26.

[17] Maria Dobrowolska, 'Przemiany struktury społeczno-gospodarczej wsi Małopolskiej' (Changes in the Socio-economic Structure of the Village in Little Poland), *Przegląd geograficzny*, Vol. 31 (1959), pp. 3–32; Maria Kiełczewska-Zaleska, 'O typach osiedli wiejskich w Polsce i planie ich przebudowy' (On the Rural Settlement Types in Poland and the Plans for Their Transformation), *Przegląd geograficzny*, Vol. 37 (1965), pp. 457–80; and Marian Benko, 'Uwagi o sprawie kierunku przemian wiejskiej sieci osadniczej w Polsce'

(Remarks on the Direction of Change of the Rural Settlement Network in Poland), *Miasto* (1967), No. 7, pp. 7-13.

[18] Jan Turowski, *Przemiany wsi pod wpływem zakładu przemysłowego. Studium rejonu Milejówa* (Transformation of the Village under the Influence of an Industrial Plant. A Study of the Milejów Region). *Studia K.P.Z.K. PAN*, Vol. 8 (1964).

[19] Maria Dziewicka, 'Zarobkowanie małorolnych chłopów poza gospodarstwem rolnym' (Wage Work of Small Farm Peasants Outside Agriculture), *Zeszyty Naukowe SGPiS*, No. 10 (1959); and Teofila Jarowiecka, 'Struktura zatrudnienia mieszkańców woj. krakowskiego' (Employment Structure of the Inhabitants of the Kraków Voivodship), *Rocznik naukowo-dydaktyczny WSP w Krakowie. Prace geograficzne*, No. 10 (1962).

[20] Jerzy Kostrowicki, 'Geografia polska w ostatnim XX-leciu' (Polish Geography in the Last 20 Years), *Przegląd geograficzny*, Vol. 36 (1964), pp. 427-50.

[21] Zdenek Hoffmann, 'Arbeitsrichtungen und -ergebnisse der Agrargeographie in der ČSSR' (Research Directions and Attainments of Agricultural Geography in Czechoslovakia), *Wissenschaftliche Zeitschrift der Martin Luther Universität zu Halle-Wittenberg*, Vol. 16 (1967).

[22] J. Pohl, 'Typy vesnických sidel v čechach' (Types of Rural Settlements in Bohemia), *Národopisný véstnik Československa*, Vol. 27 (1935), pp. 5-55.

[23] Koloman Ivanička, 'Prispevok k niektorým zmenám geografického prostredia na hornej Nitre' (A Note on Some Changes of the Geographic Environment in the Upper Nitra), *Geografický Časopis*, Vol. 11 (1959), pp. 207-21.

[24] Koloman Ivanička, A. Zelenská, and J. Mládek, 'Funkcjonalne typy vesnických sidel na Slovensku' (Functional Types of Rural Settlements in Slovakia), *Acta geologica et geographica Universitatis Comenianae, Geographica*, No. 6 (1968); Zdenek Láznička, 'Prispevok k klasifikacji funkcjonalnej ukladuv vesnických sidel v Jižnomoravským Okresu' (The Functional Classification of Rural Settlement Patterns of the South Moravian District), *Správy Geografického Ústavu ČSAV* (1966), No. 10, pp. 1-6; and Zdenek Láznička, 'Klasifikacje funkcjonalna vesnických sidel v Jižnomoravským Okresu' (Functional Classification of Rural Settlements in the South Moravian District), *Správy a vědecké činnosti ČSAV* (1967), No. 6, pp. 47-71.

[25] Oliver Bašovský, 'Dojezd do pracy jako element ekonomickogeografickej rajonizacii teritorii (na priklade Oravy)' (Commuting to Work as an Element of the Economic-Geographic Regionalization of

a Territory (Based on the Orava Region), *Acta geologica et geographica Universitatis Comenianae, Geographica,* No. 8 (1968); and J. Mládek, 'Vesnicke sidla v Vyhodnoslovenskym Železarnym Rejone' (Rural Settlements in the East Slovak Ironworks Region), *Acta geologica et geographica Universitatis Comenianae, Geographica,* No. 4 (1964), pp. 339–52.

[26] István Szabó, *A falurendszer kialakulása Magyarországon. IX–XV szazad* (The Development of the Village Systems in Hungary. Ninth to Fifteenth Centuries). (Budapest: Akademiai Kiadó, 1966); and István Balogh, *Tanyák és majorok Békés megyében a XVIII–XIX században* (Homesteads and Farmsteads in Békés County in the Eighteenth and Nineteenth Centuries), (Gyula: A Gyulai Kiadványai, 1961).

[27] Gyula Bédi, József Buda, and Jenö Tenyi, 'Falusi Települések komplex vizsgálata' (Investigations of the Village Settlement Complex), *Varosépités* (1965), No. 4, pp. 20–2; and Dénes Fülöp, Lajos Rendes, and Imre Tiba, 'A mezögazdasági jellegü településekkel kapcsolatos szociológiai kutatásokról' (On the Sociological Research Linked With Agrarian Type Settlements), *Településtudományi Közlemények* (1963), Vol. 15, pp. 45–55.

[28] Imre Kathy, 'Korszerü mezögazdasági település táji kapcsolatai' (Local Linkages of the Contemporary Agrarian Settlements), *Magyar Épitömüvészet,* 1961, No. 3, pp. 55–6; Mária Porpáczy, 'A gazdálkodás módjának és a település rendszerének kapcsolata az Örségben' (Relationships Between Systems of Production and Types of Settlement in the Orsegben), *Vasi Szemle,* Vol. 1 (1963), pp. 55–9; and József Becsei, 'A tanyai település néhány kérdéséről' (On Some Questions Concerning Homestead Settlement), *Földrajzi Ertesitö,* Vol. 15 (1966), pp. 385–406.

[29] Kálmán Faragó, 'A magyar falvak fejlödésének kérdéseiher' (Some Remarks on the Problem of the Development of Hungarian Villages), *Magyar Épitömüvészet,* 1961, No. 4, pp. 3–6; Imre Hegyi, 'A jövö falujáról' (On the Village of the Future), *Borsodi Szemle,* Vol. 6 (1962), No. 6, pp. 63–7; Károly Perczel, 'Milyenek lesznek a szocialista falvak Magyarországon?' (What will the Socialist Villages in Hungary be Like?), *Épitésügyi szemle,* Vol. 5 (1961), No. 7; Imre Kathy, 'A falu jövöje' (The Future of the Village), *Természettudományi Közlöny,* Vol. 7 (1963), pp. 97–101, 154–7, and 350–4; and Pál Beluszky, 'Mezögazdasági településkálozatunk jövöjéről' (On the Future of Our Agrarian Settlement Network), *Valóság,* Vol. 9 (1966), No. 6, pp. 9–15.

[30] Emmanuel de Martonne, 'La Valachie' (Walachia), *Buletinul*

Societății Regale Române de Geografie, Vol. 24 (1902); *La vie pasterale et la transhumance dans les Karpathes meridionales* (Pastoral Life and Transhumance in the Eastern Carpathians), (Leipzig, 1904); and *La Dobroudja* (The Dobroja), (Paris, 1918).

[31] Robert Ficheux, 'Villages de France et de Roumanie' (The Villages of France and Romania), *Buletinul Soc. Reg. Rom. de Geografie*, Vol. 48 (1929).

[32] Sabin Opreanu, 'Contribuțiuni la transhumanța diu Carpații Orientali (Remarks on Transhumance in the Eastern Carpathians), *Lucrarile Instituli de Geografie al Univ. diu Cluj*, Vol. 4 (1931); Laurian Someșan, 'Vieața pastorale in Muntii Calimanului' (Pastoral Life in the Calimani Mountains), *Buletinul Soc. Reg. Rom. de Geografie*, Vol. 52 (1933); and Victor Tufescu, 'Indrumător pentru studiul migrațiunilor temporare cu caracter economic' (Introduction to the Study of Temporary Migration and Its Economic Character), *Lucrarile Seminarului de Geografie Economica, Academie Comerciala*, 1946.

[33] Stefan Meteș, 'Les pasteurs transylvaines dans les Principautés Roumains' (The Transylvanian Shepherds in the Romanian Principalities), *Annales. Institut National diu Cluj*, 1925; N. Dragomir, 'La vie pastorale de Saliște' (The Pastoral Life of Saliște), *Traveaux Inst. Géogr. diu Cluj*, Vol. 2 (1925); D. I. Oancea, 'Contribuții la geografia pâstoritului din Masivul Bucegi' (Contributions to the Geography of Pastoralism in the Bucegi Massif), *Studii și Cercetari de Geologie, Geofizică, Geografie. Seria Geografie*, Vol. 1 (1966), pp. 91–97.

[34] Nicolae M. Popp, 'Geografia w Rumunii' (Geography in Romania), *Przegląd geograficzny*, Vol. 18 (1938), pp. 169–80.

[35] Vintilă Mihăilescu, 'L'evolution de l'habitat rural dans les collines de la Valachie entre 1790–1900' (The Evolution of the Rural Settlement on the Plains of Walachia in 1790–1900), *Buletinul Soc. Reg. Rom. de Geografie*, Vol. 56 (1936); Romulus Vuia, 'Chronologie des types des villages dans le Banat et la Transylvanie' (Chronology of Village Types in the Banat and Transylvania), *Revue de Transylvanie*, Vol. 3 (1937); and Laurian Someșan, 'Rolul factorilor geografiei in așezarea și in vieața economica a satelor din depresiunea Sibiului' (The Role of Geography in the Settlement and Economic Life of the Settlements of the Sibiu Depression), *Revue de Géographie Roumaine*, Vol. 5 (1942).

[36] Victor Tufescu, 'Nerej; un village d'une region archaique' (Nerej; A Village of an Archaic Region), *Institut des Recherches Sociologiques de Roumanie*, Vol. 1 (1940) and Ioan Șandru, Constantin

Martiniuc, et al., 'Contribuții geografice la studiul orașalui Huși' (Geographic Contributions to the Study of the Town of Husi), *Probleme de Geografie*, Vol. 8 (1957), pp. 231–324.

[37] Vintilă Mihăilescu, 'O harta a așezarilor rurale din România' (A Map of Rural Settlements in Romania), *Buletinul Soc. Reg. Rom. de Geografie*, Vol. 43 (1928).

[38] Vintilă Mihăilescu, 'Une carte de l'habitat rural en Roumanie' (A Map of Rural Occupance in Romania), *Comptes Rendus du Congrés International de Géographie, Paris, 1931*. (Paris: Colin, 1934), Vol. 3, pp. 33–5 and Vintilă Mihăilescu, 'L'evolution de l'habitat rural dans les collines de la Valachie entre 1790–1900' (Evolution of Rural Occupance in the Plains of Walachia, 1790–1900), *Comptes Rendus du Congrés International de Géographie, Varsovie, 1934*. (Paris: Colin, 1937), Vol. 3, pp. 478–81.

[39] Ignat Penkoff, 'Die Siedlungen Bulgariens, ihre Entwicklung, Veraenderungen und Klassifizierung' (The Settlements of Bulgaria, Their Evolution, Changes and Classification), *Geographische Berichte*, Vol. 5 (1960), pp. 211–27 and Štefan Očovský, 'Prehlad literatúry a náčrt vývoja geografie sidel v Bulharsku' (Survey of Literature and outline of the Development of Settlement Geography in Bulgaria), *Geografický časopis*, Vol. 21 (1969), pp. 159–68.

[40] George W. Hoffman, 'Transformation of Rural Settlement in Bulgaria', *Geographical Review*, Vol. 54 (1964), pp. 45–64, and 'Die Umwandlung der landwirtschaftlichen Siedlungen und der Landwirtschaft in Bulgarien' (Transformation of Agricultural Settlements and of Agriculture in Bulgaria), *Geographische Rundschau*, Vol. 7 (1965), pp. 352–61.

[41] Edna Scofield, 'Settlement Forms in Southern Albania', *Geographical Review*, Vol. 29 (1939), p. 677; André Blanc, 'Naissance et évolution des paysages agraires en Albanie' (Origin and Evolution of Rural Landscapes in Albania), *Geografiska Annaler*, Vol. 43 (1961), pp. 8–16; and André Blanc, 'L'évolution contemporaine de la vie pastorale en Albanie méridionale' (Current Development of Pastoral Life in Southern Albania), *Revue de Géographie Alpine*, Vol. 51 (1963), pp. 429–61.

[42] Ivan Crkvenčić and Vladimir Klemenčič, 'Razvoj i resultati agrarne geografije v Jugoslaviji' (Development and Achievements of Agrarian Geography in Yugoslavia), *Zbornik radova prvog Jugoslavenskog simpozija o agrarnoj geografiji* (Collection of Papers of the First Yugoslav Conference on Agrarian Geography), (Ljubljana: Kom. za Agr. Str. i Agr. Pejs., SGD. SFRJ, 1967), pp. 27–43, and Ivan Crkvenčić and Vladimir Klemenčič, 'Arbeitsrichtungen und -ergebnisse der Agrargeographie in Jugoslawien' (Research Direc-

tions and Results of Agrarian Geography in Yugoslavia), *Wissenschaftliche Zeitschrift der Universität zu Halle. Mathematisch-Naturwissenschaftliche Reihe*, Vol. 16 (1967), No. 2, *Mitteilungen für Agrargeographie, landwirtschaftliche Regionalplanung und ausländische Landwirtschaft*, No. 17, pp. 201-11.

[43] Ivan Crkvenčić, 'Naučno-istraživački rad agrarne geografije u svijetu suvremene socijalne geografije' (Scientific Research in Agrarian Geography in the Light of Contemporary Social Geography), *Zbornik* . . . (cf. note 42), pp. 52-64.

[44] Avguštin Lah, 'Neki aspekti programskog i metodološkog pravca agrarne geografije u Jugoslaviji' (Some Aspects of the Programmatic and Methodological Orientation of Agrarian Geography in Yugoslavia), *Zbornik* . . . (cf. n. 42), pp. 151-5; Miroslav Popović, 'Začeci, razvoj i stanje agrarne geografije u Srbiji' (The Origin, Development and State of Agrarian Geography in Serbia), *Zbornik* . . . (cf. n. 42), pp. 134-7; Tvrtko Kanaet, 'Rezultati naučno-istraživačkog rada na području agrarne geografije S.R. B.i H.' (Results of scientific research in Agrarian Geography in the People's Republic of Bosnia Herzegovina), *Zbornik* . . . (cf. n. 42), pp. 125-7; and Vladimir Klemenčič, 'Rezultati agrarno-geografiskih proučavanja Instituta za Geografiju Univerze u Ljubljani u godinama 1962-64' (Results of Agrarian Geographical Research of the Institute of Geography of the University of Ljubljana in the Years 1962-64), *Zbornik* . . . (cf. n. 42), pp. 73-83.

[45] Borut Belec, 'Antropogeografija vasi na Spodnjem Murskem polju' (Anthropogeography of a Section of the Lower Mursko Polje), *Geografski Vestnik*, Vol. 27-8 (1955-6), pp. 132-75; Vladimir Djurić, 'Pančevački Rit. Antropogeografska istraživanja' (Pančevački Rit. An Anthropogeographical Study), *Srpska A. N. Posebna Izdanja. Etnografiski Institut*, Vol. 5 (1953); and André Blanc, *La Croatie Occidentale. Étude de géographie humaine* (Western Croatia. A Study in Human Geography). *Travaux publies par l'Institut d'Études Slaves*, de l'Université de Paris, Vol. 25 (1957).

[46] Borivoje Ž. Milojević, 'Types of Villages and Village-Houses in Yugoslavia', *The Professional Geographer*, Vol. 5 (1953), No. 11, pp. 13-17; Ivo Baučič, Władysław Biegajło, and Ivan Crkvenčić, 'Socjalno-geografska obilježja sela Jesenice' (The Socio-Geographic Characteristics of the Village of Jesenice), *Geografski Glasnik*, Vol. 28 (1966), pp. 93-114; Vladimir Bračić, 'Socijalni i demogeografski problemi vinogradarske pokrajine' (Social and Demogeographic Problems of a Viticultural Region), *Zbornik* . . . (cf. note 42), pp. 111-13; and Borut Belec, 'Neke demogeografske i socijalne pojave

u procesu poljoprivredne proizvodnje u Ljutomersko-Ormoškim Goricama' (Some Demogeographic and Social Phenomena in the Process of Socialization of Agricultural Production in the Ljutomer and Ormoz Mountains), *Zbornik* . . . (cf. n. 42), pp. 114-18.

[47] Svetozar Ilešič, 'Die Flurformen Sloveniens im Lichte der europäischen Forschung' (The Field Patterns of Slovenia in Light of European Research), *Münchner Geographische Hefte*, Vol. 16 (1959); Ivan Crkvenčić, *Prigorje planinskog niza Ivančice* (The Foothills of the Ivancice Mountain Chain), *Radovi Geografskog Instituta Sveučilišta u Zagrebu*, Vol. 1 (1968), 113 pp.; and Ivan Crkvenčić, 'Kulturlandschaftliche Veränderungen in Hrvatsko Zagorje, Jugoslawien' (Changes in the Cultural Landscape of Croatian Zagorje, Yugoslavia), *Erdkunde*, Vol. 16 (1962).

[48] William B. Johnston and Ivan Crkvenčić, 'Examples of Changing Peasant Agriculture in Croatia, Yugoslavia', *Economic Geography*, Vol. 33 (1957), pp. 50-71; Ivan Gams, 'O vplivu agrarnog iskorištavanja zemljišta na karstna svojstva i procese' (On the Influence of Agricultural Land Use on Karst Features and Processes), *Zbornik* . . . (cf. n. 42), pp. 121-4; and Ivan Gams, *Pohorsko Podravje. Razvoj kulturne pokrajine* (The Pohorje Drava Region. Development of the Cultural Landscape), *Slovenska Akademja Znanosti i Umetnosti. Dela*, Vol. 9, *Inštitut za geografijo*, Vol. 5 (1959).

[49] Petar Marković and Darinka Kostić, 'Strukturalne promene na Jugoslavenskom selu u posleratnom periodu (1945-1962)' (Structural Changes in the Yugoslav Village in the Post War Period [1945-1962]), *Sociologija*, Vol. 6 (1964), pp. 170-95; Cene Malovrh, 'Značaj i važnost privredno-prostorne strukture individualnih poljoprivrednih gazdinstva' (Character and Significance of the Spatial Economic Structure of Individual Agricultural Farms), *Zbornik* . . . (cf. n. 42), pp. 100-10; and Sven Kalušić, 'Neke karakteristike agrarne strukture i prostorna diferenciranost posjeda stanovnika naselja Murtera i Betine' (Some Traits of Agrarian Structure and the Spatial Differentiation of Land Holdings of the Settlements of Murter and Betina), *Zbornik* . . . (cf. n. 42), pp. 128-30.

[50] Vladimir Kokole, 'Neki aspekti na studij agrarnih područja u smislu aplikacije u regionalnom planiranju' (Some Aspects of the Study of Rural Areas for Application in Regional Planning), *Zbornik* . . . (cf. n. 42), pp. 148-51; Milisav Lutovac, 'Uticaj industrije na poljoprivredu i seoska naselja' (The Influence of Industry on Agriculture and Rural Settlements), *Zbornik* . . . (cf. n. 42), pp. 155-7; and Stevan Vujadinovič, 'Preobražaj geografskog lika Poreča' (The Transformation of the Geographic Appearance of Poreče), *Glasnik Srpskog Geografskog Drustva*, Vol. 41 (1961), pp. 45-51.

[51] Preston E. James, Clarence F. Jones et al. (eds.), *American Geography. Inventory and Prospect* (Syracuse: Syracuse University Press, 1954); L. Dudley Stamp, *The Land of Britain, Its Use and Misuse* (London: Longmans, Green and Co., 1948); and L. Dudley Stamp, 'Commission on a World Land Use Survey', *The I.G.U. Newsletter*, Vol. 13 (1962), No. 1, pp. 28–31.

[52] Jerzy Kostrowicki (ed.), *Land Utilization. Methods and Problems of Research, Prace geograficzne I.G. PAN*, No. 31 (1962), 250 pp.; Jerzy Kostrowicki (ed.), *Land Utilization in East-Central Europe. Case Studies, Geographia Polonica*, Vol. 5 (1965), 512 pp.; Jerzy Kostrowicki and Wiesława Tyszkiewicz, *Land Use Studies in East-Central Europe. Dokumentacja geograficzna*, 1968, No. 3, 91 pp.; and Jerzy Kostrowicki (ed.), *Problematyka i metody geografii rolnictwa w pracach Zakładu Geografii Rolnictwa I.G. PAN* (Problems and Methods of Agricultural Geography in the Work of the Department of Agricultural Geography of the Institute of Geography of the Polish Academy of Sciences). *Dokumentacja geograficzna* (1968), No. 4, 113 pp.

[53] Jerzy Kostrowicki, 'Zdjęcie użytkowania ziemi i jego przydatność praktyczna' (The Land Use Survey and Its Practical Utility), *Biuletyn Komitetu Przestrzennego Zagospodarowania Kraju P.A.N.*, No. 42 (1966), pp. 211–15; 'Badania nad użytkowaniem ziemi w Polsce' (Studies on Land Use in Poland), *Przegląd geograficzny*, Vol. 31 (1959), pp. 517–33; and 'Problematyka geograficzno-rolnicza zdjęcia użytkowania ziemi Polski' (The Agricultural-Geographic Problems of the Land Use Survey of Poland), *Przegląd geograficzny*, Vol. 32 (1960), pp. 227–79.

[54] Maria A. Glazovskaya, 'Obzor raboty komisii po uchetu mirovogo ispol 'zovaniya zemel' ' (Review of the Work of the Commission on the World Land Use Survey), in Innokenti P. Gerasimov (ed.), *XIX Mezhdunarodnyi geograficheskiy kongress v Stokgol'me* (The 19th International Geographical Congress in Stockholm), (Moskva: A.N. S.S.S.R., 1961), pp. 382–7; Béla Sárfalvi (ed.), *Land Utilization in Eastern Europe. Studies in Geography in Hungary*, Vol. 4 (1967); and György Enyedi, *A mezögazdaság földrajzi tipusai Magyarországon* (The Geographical Types of Agriculture in Hungary), (Budapest: Akademiai Kiádo, 1965).

[55] David L. Armand, 'Budapeshtskaya konferentsiya po ispol 'zovaniyu zemel'. 6–10 maya 1964' (The Budapest Conference on Land Use. May 6–10, 1964), *Izvestiya Akademii Nauk S.S.S.R. Seriya Geograficheskaya*, 1964, No. 5, pp. 98–101; Walter Roubitschek, 'II Bodennutzungskonferenz der europäischen sozialistischen Staaten in Budapest, 1964' (The Second Land Use Conference of the European Socialist States in Budapest, 1964), *Petermanns Geographische Mitteilungen*, 1965; and L. Dudley Stamp, 'A Report on Land

Use Studies in East-Central and Eastern Europe, 1960-1964', *The I.G.U. Newsletter*, Vol. 15 (1964), No. 1-2, pp. 37-40.

[56] Wiesława Tyszkiewicz, 'Badania użytkowania ziemi w Bułgarii' (Land Use Studies in Bulgaria), *Przegląd geograficzny*, Vol. 38 (1966), pp. 327-8; Władysław Biegajło, 'Badania użytkowania ziemi w Rumunii' (Land Use Studies in Romania), *Przegląd geograficzny*, Vol. 39 (1967), pp. 635-9; and Roman Szczęsny, 'Problematyka i metody makroskalowych badań geograficzno-rolniczych' (Problems and Methods of Agricultural-Geographic Large Scale Studies), *Dokumentacja geograficzna*, 1968, No. 4, pp. 56-67.

[57] Władysław Biegajło, 'Problematyka i metody mikroskalowych sondażowych badań geograficzno-rolniczych' (Problems and Methods of Agricultural-Geographic Small Scale Sample Studies), *Dokumentacja geograficzna*, 1968, No. 4, pp. 34-55.

[58] Stefan Hauzer, 'Przeglądowe zdjęcie użytkowania ziemi. Założenia i metoda' (The Simplified Land Use Survey. Assumptions and Method), *Dokumentacja geograficzna*, 1968, No. 4, pp. 68-74.

[59] Wiesława Tyszkiewicz, 'Stosunki własnościowe a użytkowanie ziemi na przykładzie Kujaw' (Land Tenure Conditions and Land Use on the Example of Kujavia), *Dokumentacja geograficzna*, 1968, No. 4, pp. 83-92.

[60] Marian Matusik, 'Warunki przyrodnicze a użytkowanie ziemi na przykładzie dolnego Powiśla' (Natural Conditions and Land Use on the example of the lower Vistula Region), *Dokumentacja geograficzna*, 1968, No. 4, pp. 83-92.

[61] Derwent S. Whittlesey, 'Major Agricultural Regions of the Earth', *Annals, Association of American Geographers*, Vol. 26 (1936), pp. 199-240; Daniel Faucher, *Géographie agraire. Types de cultures* (Agricultural Geography. Types of Land Use). (Paris: Libr. de Médicis, 1949); and Innokenti P. Gerasimov, 'The Geographical Study of Agricultural Land', *Geographical Journal*, Vol. 124 (1958).

[62] Derwent S. Whittlesey, 'The Regional Concept and the Regional Method', in Preston E. James, Clarence F. Jones, *et al.* (eds), *American Geography. Inventory and Prospect*, (Syracuse: Syracuse University Press, 1954); David Grigg, 'The Logic of Regional Systems', *Annals, Association of American Geographers*, Vol. 55 (1965), pp. 465-91; and Michael Chisholm, 'Problems in the Classification and Use of Farming-Type Regions', *Institute of British Geographers. Transactions and Papers*, Vol. 35 (1964), pp. 91-103.

[63] Brian J. L. Berry, 'A Note Concerning Methods of Classification', *Annals, Association of American Geographers*, Vol. 48 (1958),

pp. 300–3; Ora S. Morgan, *Agricultural Systems of Middle Europe. A Symposium*, (New York; Macmillan, 1933); and Jacques Dubourg, 'La vie des paysans Mossi; Le Village de Toghella' (Life of the Mossi Peasants: The Village of Toghella), *Les Cahiers d'Outre-Mer* (1957), pp. 285–324.

[64] Jerzy Kostrowicki, 'Land Utilization Survey as a Basis for Geographical Typology of Agriculture', *Przegląd geograficzny*, Vol. 32 (1960), Supplement, pp. 169–83; 'Geographical Typology of Agriculture. Principles and Methods. An Invitation to Discussion', *Geographia Polonica*, Vol. 2 (1964), pp. 159–67; and Roman Szczęny, 'The Orientation in Agricultural Production of Poland', *Geographia Polonica*, Vol. 2 (1964), pp. 169–77.

[65] I.G.U. Commission for Agricultural Typology, *Questionnaire No. 2 on Methods and Techniques of Agricultural Typology*, March 1966; I.G.U. Commission for Agricultural Typology, *Discussion on the Commission Questionnaire No. 2* (Boulder: I.G.U., 1967); and Jerzy Kostrowicki and Nicholas Helburn, *Agricultural Typology. Principles and Methods. Preliminary Conclusions* (Boulder: I.G.U., 1967).

[66] Jerzy Kostrowicki, 'The Agricultural Problems Involved in the Polish Land Utilization Survey', *Prace geograficzne I.G. PAN*, No. 31 (1962), pp. 59–128; 'Geographical Typology of Agriculture in Poland. Methods and Problems', *Geographia Polonica*, Vol. 1 (1964), pp. 111–46; and Władysława Stola and Wiesława Tyszkiewicz, 'Znaczenie badań użytkowania ziemi w planowaniu przestrzennym' (The Significance of Land Use Surveying in Areal Planning), *Budownictwo wiejskie*, Vol. 11 (1964), pp. 11–13.

[67] Jerzy Kostrowicki, 'An Attempt to Determine the Geographical Types of Agriculture in East-Central Europe', *Geographia Polonica*, Vol. 2 (1964); 'An Attempt to Determine the Geographical Types of Agriculture in East-Central Europe on the Basis of the Case Studies on Land Utilization', *Geographia Polonica*, Vol. 5 (1965), pp. 453–98; and Jerzy Kostrowicki and Wiesława Tyszkiewicz, *Land Use Studies in East-Central Europe*, *Dokumentacja geograficzna* (1968), No. 3.

[68] György Enyedi, 'Eine Methode für die Abgrenzung von landwirtschaftlichen Rayons' (A Method for Delimiting Agricultural Regions), *Prace geograficzne I.G. PAN*, No. 27 (1961), pp. 285–93; György Enyedi, *A Föld Mezőgazdasága. Agrárföldrajzi tanulmány* (The Agriculture of the World. A Study in Agricultural Geography). (Budapest: Mezőgazdasági Kiado, 1965), 298 pp.; and György Enyedi, *A Mezőgazdaság földrajzi tipusai Magyarországon* (The Geographical Types of Agriculture in Hungary). (Budapest: Akadémiai Kiado, 1965), 71 pp.

[69] I.G.U. Commission for Agricultural Typology, *Discussion on the Commission Questionnaire No. 2* (Boulder: I.G.U., 1967), p. 14.

[70] György Enyedi, *A föld mezögazdasága. Agrárföldrajzi tanulmány,* op. cit.; and *The Agriculture of the World. A Study in Agricultural Geography*, Hungarian Academy of Sciences. Institute of Geography. Abstracts, No. 9 (1967).

[71] György Enyedi, 'Probleme der Bodennutzung in den Gebirgs- und Hügellandschaften Ungarns' (Land Use Problems in the Mountain and Hill Environs of Hungary), *Wissenschaftliche Zeitschrift der Universität Halle. Mathematisch-Naturwissenschaftliche Reihe,* Vol. 17 (1968), No. 2, *Mitteilungen für Agrargeographie, Landwirtschaftliche Regionalplanung und ausländische Landwirtschaft,* No. 31; György Enyedi and Gyula Szabo, 'A délkelet-Alföld mezögazdasági földrajzának alapvonásai' (The Characteristic Traits of Economic Geography in the Southeast Alföld), *Földrajzi Ertesitö,* Vol. 5 (1956), pp. 445–64, and Vol. 6 (1957), pp. 207–16; and Irene Enyedi, 'The "Kossuth" Collective Farm of Békécsaba in the Southern Part of the Great Hungarian Plain', *Geographia Polonica,* Vol. 5 (1965), pp. 407–20.

[72] Innokenti P. Gerasimov, 'The Geographical Study of Agricultural Land', *Geographical Journal,* Vol. 124 (1958); Sergey I. Sil'vestrov, 'Formy ispol'zovaniya zemel' kak osnova sistemy ikh ucheta i otsenki v SSSR' (Forms of Land Use as a Basis of a System for their Surveying and Evaluation in the USSR), *Prace geograficzne I.G. PAN,* No. 31 (1962), pp. 215–25; Andrey N. Rakitnikov, 'Izucheniye mestnykh osobennostey sel'skogo khozyaystva' (The Study of Local Peculiarities of Agriculture), *Prace geograficzne I.G. PAN,* No. 31 (1962), pp. 245–50; Valentina S. Mikheyeva, 'Ekonomiko-matematicheskaya model' razmeshcheniya sel'skokhozyaystvennogo proizvodstva po rayonam strany' (An Economic-Mathematical Model of Distribution of Agricultural Production In the Regions of the Country), *Vestnik Moskovskogo Universiteta. Seriya V, Geografiya,* 1962, No. 5, pp. 12–18; T. L. Bas'uk, *Organizatsiya sotsialisticheskogo sel'skokhozyaystvennogo proizvodstva* (The Organization of Socialist Agricultural Production), (Moskva, 1962), pp. 98–124; and T. L. Bas'uk, 'Rayonirovanie sel'skokhozyaystvennogo proizvodstva pri ispol'zovanii lineynogo planirovaniya' (Regionalization of Agricultural Production by Applying Linear Programming), (Moskva: unpublished, 1963).

[73] T. Jordanov, 'Razvitie i printsipi na ikonomicheskoto rayonirane v SSSR i znachenito im za ikonomicheskoto rayonirane na Blgariya' (The Development and Principles of Economic Regionalization in the USSR and its Significance for Economic Regionaliza-

tion in Bulgaria), *Trudovoye na Visshiya Ikonomicheski Institut 'K. Marks'*, Sofiya, 1955, No. 1, pp. 99–157; T. Chernokolev, *Rayonirane na selskostopanskite kulturi i zhivotni* (The Regionalization of Agricultural Cultivation and Husbandry), (Sofiya, 1956); F. Hamernik, 'Rayonisace zémědělské výroby v ČSR' (The Regionalization of Agricultural Production in the Czechoslovak Republic), *Vědecké Prace Vyzkumného Ústavu Zémědéské Ekonomiky v Praze*, Vol. 2 (1957), pp. 21–46; L. Sobotka, 'Základy teorie a metodiky rayonisace zémědělské výroby' (Principles of Theory and Methodology of Regionalization of Agricultural Production), *Sbornik Československe Akademie Zémědélských Ved. Zémědělska Ekonomika*, Vol. 6 (1957), pp. 441–54; V. Vrbenský, F. Hamernik, et al., 'Prirodni podminki zémědělské výroby. Rayonisace rostlinne výroby. Rayonisace živočišne výroby' (The Natural Bases of Agricultural Production. Regionalization of Plant Production. Regionalization of Animal Production), *Zbornik* ... (cf. above), pp. 455–6; Walter Roubitschek, 'Die regionale Differenzierung der agraren Bodennutzung 1935 im heutigen gebiet der Deutschen Demokratischen Republik' (Regional Differentiation of Agricultural Land Use in 1935 Within the Present Area of the German Democratic Republic), *Petermanns Geographische Mitteilungen*, Vol. 103 (1959), pp. 190–7; and R. Luntre, 'Geografia agriculturii R.P.R.' (Agricultural Geography of the Romanian People's Republic), *Natura* (1955).

[74] E. Rübensam, 'Die landwirtschaftlichen Produktionsgebiete und Produktionstypen in der D.D.R.' (The Agricultural Production Regions and Production Types in the G.D.R.), *Internationale Zeitschrift der Landwirtschaft*, 1957, No. 2.

[75] Bernd Andreae, 'Betriebsvereinfachung in der Landwirtschaft' (Operation Streamlining in Agriculture), *Berichte über Landwirtschaft*, Sonderheft 169 (1958); and Bernd Andreae, *Betriebsformen in der Landwirtschaft*, (Systems of Operation in Agriculture), (Stuttgart, 1964).

[76] Jerzy Kostrowicki, 'The Polish Detailed Survey of Land Utilization. Methods and Techniques of Research', *Dokumentacja geograficzna* (1964), No. 2, pp. 17–18.

Comments
BARBARA ZAKRZEWSKA

Professor Romanowski makes three contributions to methodology in agricultural geography. First, is the review of selected literature on the approaches to agricultural studies with special emphasis on contributions made by East European researchers. In this review he identifies the four major approaches used by most scholars.

The second contribution of this paper is the identification and analysis of the major methodological trends in agricultural studies in the individual countries of Eastern Europe. Here, Professor Romanowski identifies Poland as the major research centre in agricultural geography and singles out the work of Professor Kostrowicki, who is not only responsible for a great number of systematically conducted case studies in agricultural land use in Eastern Europe, but is a leading researcher in agricultural typology based on land use and applicable on the world scale.

In Hungary, Romanowski singles out the work of Professor Enyedi who stresses the still underdeveloped 'socio-economic' approach to the study of agriculture. On the basis of this approach Enyedi proposed three types and three sub-types of socialist agriculture.

Dr Romanowski also calls attention to the work of Professor Andreae in East Germany, who emphasizes a 'commodity-intensiveness' approach by relating the income of the farm to its form of production. This method is evaluated by Professor Romanowski as one that has 'unusual promise in its recognition of the structure of farm operation, a rather forgotten element in the settlement, land-use, and commodity approaches.'

Dr Romanowski's third contribution is made in a rather negative way and stems from the fact that he fails, purposely or unconsciously, really to define the major concepts with which his methodological paper is concerned. Thus, he offers no single definition of the words 'agriculture', 'agricultural region', 'agricultural geography', or even 'agricultural type', though for the latter he offers Kostrowicki's elaborate set of criteria (pp. 139-40). By avoiding precise definitions of these terms or concepts, Dr Romanowski forces the reader to search for such definitions and provides excellent material for discussion. He also provides a point for discussion by suggesting in the introduction to his paper that East-Central and Southeast Europe is perhaps not an agricultural region. We may begin our discussion with this last point.

Dr Romanowski implies, in the few words devoted to this topic, that agriculture in Eastern Europe is so undeveloped and neglected that the region cannot any longer be considered to be 'agricultural'. This suggests that the degree of development of a given economy, rather than the orientation of the economy and the way of life of a large segment of the population, determines whether a region is 'agricultural' or not. This is a debatable point.

Even though, in terms of contribution to the Gross National Product and to exports, agriculture is no longer the major branch of the economy, large parts of Eastern Europe are still predominantly agricultural and, for large segments of the population, agriculture is still the way of life.

With the rapid industrialization of East-Central and Southeast Europe after World War II, much agricultural labour has been transferred to industry and this has affected the rural/urban characteristics of the area. Nevertheless, in the middle 1960's, over half of the population of the four southern countries (Romania, Yugoslavia, Bulgaria and Albania) still made its living from agriculture, and large parts of these and the remaining East European countries were still almost entirely agricultural. In Bulgaria, foodstuffs are still the major export commodity.

Thus, while the post-war industrialization, coupled with the slow progress in agriculture, appears to have de-emphasized the agricultural characteristics of Eastern Europe as a whole, it must be kept in mind that large parts of this area, though not coinciding with national boundaries, are still very much agricultural. The evaluation of East-Central and Southeast Europe as an agricultural region depends, therefore, on the scale at which it is examined. If we use statistical data of the internal subdivisions rather than national boundaries, large parts of Eastern Europe would still be classified as strongly agricultural.

This discussion raises the question of what is 'agriculture' and what is 'an agricultural region'. This brings to mind the interesting paper by J. E. Spencer and R. J. Horvath concerned with the processes which create an agricultural region. ('How Does An Agricultural Region Originate?', *A.A.G. Annals*, Vol. 53, No. 1, March 1963, pp. 74–92), and a very recent paper by David Grigg on 'The Agricultural Regions of the World: Review and Reflections' (*Economic Geography*, Vol. 45, No. 2, April 1969, pp. 95–132).

Professor Romanowski does not specifically define the words 'agriculture' or 'agricultural region' but gives some insights into his thinking on these concepts. He is not alone in finding it difficult to crystallize the concept of 'agriculture'. Grigg devotes considerable space to the discussion of such terms as 'farming', 'agriculture', 'agricultural region', 'agrarian landscape', 'type of farming', etc. He suggests that, traditionally, the term 'agriculture' refers to the cultivation of soil and raising of livestock, but some authors have included in it the simple gathering economies and/or forestry. The problem remains that in different parts of the world different aspects of economy are included in agriculture.

Concerning agricultural regions, Grigg puts his finger on the problem in the following paragraph:

> Most systems of agricultural regions devised by geographers are based largely upon crop and livestock combinations, farming methods, and sometimes institutional features such as farm size or land tenure. But it can be convincingly argued that agriculture penetrates so many aspects of life in much of the world that to base a system of regions solely upon these largely economic attributes is to miss the essence of the regional differentiation of agriculture.
> (pp. 98–9)

Grigg further suggests that at present the term 'type of farming region' is more often used than the term 'agricultural region', and that the emphasis now seems to be on '... the farm as a business unit and the combination of enterprises which the farmer carries out on his farm' (p. 99). This agrees with one of Dr Romanowski's definitions of farming.

The major concept with which Dr Romanowski is concerned

in his paper is 'agricultural typology' and the criteria on which it should be based to be applicable through time and space. He appears to favour some form of the 'socio-economic' approach stressed by Enyedi, but realistically points out that, at present, Kostrowicki's approach based on 'land use' has the greatest chance for development. Dr Romanowski also separates the regionalization of agriculture from the typology of agriculture.

It is difficult to argue with this point of view, but the statement brings to mind the concept of 'agricultural geography' which Dr Romanowski uses in his paper several times but does not define. To most geographers the spatial context of any study is the most geographic one. If that concept is removed from agricultural typology, the question arises as to whether a geographer should occupy himself with the task of formulating the typology, or whether this should be left to a statistician concerned with methods of classification. Dr Romanowski suggests that the task of the geographer comes after the typology has been formulated. 'Once a series of types is determined, the geographer then may have the task of developing methods of assigning the spatial reality to regions of one or another type of agriculture' (p. 20).

It is interesting to note that Grigg, who is specifically interested in regionalization of agriculture, also recognizes the need for the separation of these two tasks. He recognizes two stages in the development of a system of agricultural regions for the whole world:

> First, types must be recognized and their criteria established. Second, different parts of the world must be assigned to each type. (p. 131)

Grigg recognizes the difficulties involved:

> First, the methodology of typology is not precise; ... Secondly, both the construction of a typology and assignment of areas to each type would require a vast amount of travel, understanding, and reading by one individual. (p. 131)

He also points out that one of the major problems in agricultural typology is the lack of precise, comparable data for all parts of the world, especially outside North America and Europe (p. 105). To me this seems to be the crux of the matter.

If comparable statistical data on world agriculture were available, several typologies based on different criteria could be devised, according to need.

Concerning agricultural regions, Grigg concludes that:

> ... one way to further the understanding of the agricultural regions of the world is to establish a largely deductive and genetic system, which can then be checked against reality. In short, on a world scale there seems little alternative to devising a series of models of agricultural types and their distribution, and arriving at a series of approximations that can be checked and revised by other workers. (p. 132)

As a final point of this discussion, I would like to return to the work of Dr Kostrowicki and his approach to agricultural typology. In connection with the 21st Congress of I.G.U. held in New Delhi, India, in 1968, Kostrowicki published a paper which summarizes the progress made by the I.G.U. Commission for Agricultural Typology of which he is the head ('Agricultural Typology. Agricultural Regionalization. Agricultural Development', *Geographia Polonica*, No. 14, 1968, pp. 265–74).

The Commission has, as of 1968, accepted the following principles and basic notions concerning agricultural typology:

> – 'type of agriculture' ... should be accepted as the supreme notion in agricultural typology,
> – type of agriculture should be understood in a broad meaning including all forms of crop growing and livestock breeding,
> – type of agriculture should be understood as a hierarchical notion encompassing types of the lowest order, several intermediate orders of types, up to the highest ones – world types of agriculture,
> – type of agriculture should be understood as a dynamic notion which changes either evolutionarily or revolutionarily along with the transformation of its basic characteristics,
> – type of agriculture should be understood as a complex notion combining several aspects or characteristics of agriculture. (p. 266)

In addition the Commission decided to consider, wherever possible, an *agricultural holding* to be a basic unit in agricultural typology. In macro-studies, other units (such as administrative) will be used with the awareness that in so doing one is dealing with aggregate indices and averages. Detailed sample surveys are recommended in the latter cases to determine the degree of divergence between the averages and the individual holdings (p. 267).

The Commission also decided that the definition of the type of agriculture should be based on *inherent* characteristics of agriculture which are grouped under three headings: (1) the social characteristics (who is the producer); (2) the organizational and technical characteristics (how the product is obtained); and (3) production characteristics (how much, what, and for what it is produced) (pp. 268–9). These factors can be given different weight in different types of agriculture.

External characteristics of agriculture, or conditions in which it develops, i.e. natural environment and especially land, though considered important in answering the question 'why' about some aspects of agriculture, are not considered the proper bases for determining types of agriculture. The rationale for this is that:

> Nature does not create or develop by itself any form of agriculture but it creates conditions which . . . only limit to some extent technical or economical possibilities of agricultural development. (p. 268)

In spite of the above separation of internal and external characteristics, the Commission realizes that:

> . . . each particular type of agriculture is the result of a combined action of a complex of social, technical, economic and cultural processes developing in defined natural conditions so that no type of agriculture develops in isolation . . .
> (p. 267)

This explains to some extent the great number of approaches to agriculture which Dr Romanowski had to review in his paper.

The Commission had also considered the problem of data availability on world scale and accepted the inevitability of using estimates in areas where data are not available or are unreliable. Kostrowicki points out, however, that in agricultural types of higher order only a few generally available indices will be needed. More detailed data will be needed to define lower order agricultural types or sub-types.

The Commission logically decided that the inherent characteristics of agriculture proposed above are subject to change pending development of case studies which are to be initiated by the Commission in various countries. Once the list of inherent characteristics is finalized, a method of arriving at types of agriculture will need to be worked out and tested for its applicability.

Kostrowicki also discusses the spatial concept of an agricultural region and points out that: 'Similarly to agricultural types, the definition of agricultural regions should be based rather on inherent characteristics of agriculture itself than on the conditions in which it develops' (p. 270).

Once the agricultural typology has been formulated, agricultural regions could be formed 'on the basis of dominance, co-dominance or co-existence of particular agricultural types in a given territory' (p. 271). Thus, the major task is still the formulation of agricultural typology, a topic extensively discussed in Dr Romanowski's paper.

Comments
PAUL B. ALEXANDER

In his papers Professor Romanowski clearly states the central problem in his conclusions, 'But while development in each Republic seems to be quite healthy (especially in Poland, the German Democratic Republic, Hungary and Yugoslavia), one cannot state the same for the development of the region as a whole. Each country seems to have its own cherished traditions.' The geographers involved in East European agriculture have some very pertinent questions to ask themselves. The particular questions which I have in mind are: to what extent is the current research useful and are we doing any more than gathering assorted detailed information? As Dr Romanowski has pointed out in his conclusions 'Meticulous detail seems frequently a hindrance to analysis rather than an aid.' Perhaps the most important question we could ask is what do we want to achieve in the future and what is the most efficient way of achieving this? In answering this question, it seems that most of us would agree that the ultimate objective in agricultural geography is a broad basic understanding of what makes agriculture function, what trends have developed, are developing, and what future trends will develop in world agriculture. It is in the last part, that is, the methodology of achieving these answers, that opinions differ. As my part in this conference, I would like to go beyond commenting on Dr Romanowski's paper and indicate some possibilities that I feel the geographers specializing in East-Central and Southeast European countries continue to overlook. On July 29, 1958, the United States Public Law 85–568 created the National Aeronautic Space Administration known to most of us as NASA. This is a civilian agency and one of its fundamental objectives is 'to cooperate with other nations in the effective use of aeronautical and space activities for peaceful and scientific purposes'. I would remind you that next year the first of the Civilian Earth Resources Technical Satellites will be in orbit, and recently the Military launched a 26 foot polar-orbited, sun-synchronized observation satellite. Agricultural and land use programmes are high on the list of NASA spacecraft priorities.

It is my opinion that by 1975, or at the latest 1980, all the countries of the world, including the countries involved in this conference, will be able to obtain panchromatic, colour, and colour infra-red photography of all the areas within their borders. Further, infra-red scanning imagery and radar imagery will likewise be available. Let us consider for a minute what implications this has for the Polish Land Use System.

The Polish scheme categorizes its information into six main areas of interest. These six areas of interest are obviously closely interrelated and as would be expected it is often difficult to categorize neatly all of the relative information on land use. From the great time and expense spent on the complicated colour maps, it would seem that the land utilization map itself is the most important single aspect of this survey. The six areas of interest are: (1) The physical basis of land use. This is a presentation of the natural environment and includes information relative to geology, landforms, soils, climate and vegetation. As a field investigator who has spent several hot Eastern European summers in the field recording the details needed in the Polish system, I find the prospects of having imagery available exceedingly exciting. Certainly a great deal of data needed for the physical basis of land use will be available in massive quantity. (2) The second category of the Polish Land Use System can best be translated as the cultural basis of land use. This is essentially the social background to land use. It centres upon the ownership categories, size of holdings, and the history of land fragmentation. Certainly here, imagery would be a great help. Size of holdings and land fragmentation will be easily identified on space imagery. (3) Primary categories of land use, concerned with the actual use of the land according to the classification into arable land, meadow, pasture, forest or non-productive land. There seems little doubt that computers will be able to assemble a map of primary categories of land use. Some problems may exist in discerning meadow from pasture, but in a sun-synchronized orbit, with imagery received every seventeen days, these difficulties would soon be resolved. (4) This is the technical or organizational aspect of land use such as crop rotation, fertilization and mechanization. These may be real problems for space imagery. (5) The fifth considers the secondary categories of land use, the specific types of crops planted. It is also concerned with the position of livestock on the farm. Livestock undoubtedly will present a problem. However, the satellites will be very efficient in computing a farm inventory map, including not only the types of crops but the vigour of the vegetation. (6) The final category deals with a summary of the agricultural production of the area and includes a consideration of crop yields and animal production. These again will be difficult targets for the satellite.

With the advent of space technology, the thought of field researchers diligently and painstakingly recording information about types of crops, soils, land fragmentation, etc., seems almost like ancient history. Five years ago, I spent two field seasons collecting data for the Polish Land Utilization System in northwestern Yugoslavia. When I think of the work that I and other field investigators went through to obtain information that will be generally inferior to what the satellite collects, I really wonder if we are thinking along the right lines in East European land use systems. This is being thrown out as food for thought but it seems to me that any evaluation of agricultural methodology in Eastern Europe should include the consideration that perhaps now is the time to gear ourselves for information retrieval systems of the future. Perhaps systems should be developed and field work carried on which would best enable the countries of East-Central and Southeast Europe and the world to make full use of space observation potentials.

Remarks

Dr Romanowski replied to the comment by Dr Alexander by stating that he agreed with the opinion that the development of NASA's satellite programme will be an extremely valuable method of ascertaining land use in the future. However, to deal with the future would encompass another paper. Concerning Dr Zakrzewska's suggestion that certain terms such as 'agriculture' be defined in some detail, Dr Romanowski indicated that he had not yet reached that stage, but that he was ready to discuss those factors which should be considered in a definition of agriculture.

A second point made by Dr Zakrzewska dealt with typology, and Dr Romanowski said that geographers need not be worried with the subject; they are just as qualified as anyone to study the problem, and that work on typology should be an interdisciplinary concern.

Professor Kansky offered the opinion that classification literature is very abundant and that the U.S.A. has a classification society in which its members argue constantly over the proper techniques in classification. He pointed out that the purpose of the system is the crux of classification, indicating that many geographers feel that regionalization is classification.

Dr Karcz commented that anthropological work in the area of labour supply in East-Central and Southeast Europe has failed to classify the role of the specialization of the family as its income rises. He suggested that the effects of income change as a system for agricultural analysis should be considered.

Professor Rugg suggested that we must understand that the attitude of the farmer towards his work has changed; that the farmer today looks upon himself as a businessman. Since the Eastern European collective farm is a business operation, this needs to be considered in the classification of land use.

Dr Romanowski agreed with Dr Rugg with some reservations presenting as an example the chicken farms in the northern Caucasus where the animals are never touched by humans, suggesting that in this respect, perhaps, we should classify such regions as food production areas and not agricultural areas.

On this point Dr Kristof indicated that farming as a way of life will die more quickly in Western Europe than in the Eastern part for the simple reason that the collective farms are supporting the poorer farms.

Professor Butler clarified Dr Kristof's remark by suggesting that in the West, social differences between the farmer and the industrial worker are decreasing while the economic differences are not.

Professor Karcz indicated that this is true in the East as well, stating that in some areas the government will now allow new farm villages to develop, but rather stresses the development of more small towns.

Dr Olsen at this point offered the interesting comment that this same governmental determination of the size of communities has developed in Oklahoma where the legislature, by establishing the minimum size of a public school district, has in effect established the size of communities. He suggested also that geographers should work towards an 'international land potential system' instead of a land use system.

Whereupon Professor Romanowski pointed out that we must realize that land potential and land use are not the same and that both systems should be used.

Mr Wilson asked if areal photography had been used in land use studies, especially in micro-studies and how this method compares to the NASA method.

Dr Alexander answered that we are dealing with a time factor and that the NASA method could cover the entire U.S.A. in seventeen days.

5
The Location of Industry in East-Central and Southeast Europe F. E. IAN HAMILTON

> *Cynics – are the dead weight in the ship of mankind,*
> *But idealists – are the helm of the sails.*
>
> *... Even a great man is forced, for the time being,*
> *to hide his face under make-up,*
> *... The October wind howls over him,*
> *He is the embodiment of decision,*
> *but he no longer can decide,*
> *and therein lies his greatness.*
>   Yevtushenku, *The Bratsk Station*

The task of this paper is not to make an exhaustive analysis of industrial location in this region. Rather it seeks to focus attention upon three basic questions: (1) the problems involved in studying the subject in this area, (2) the principles of socialist location and their apparent effects, and (3) certain tentatively formulated aspects of behaviour which are related to industrial location decision-making in planned economies. In some measure it attempts to answer Fleron's challenge that 'while ... some area specialists are more concerned with discovering regularities and formulating generalizations than with engaging in exhaustive description ... the large proportion of social scientists are nomothetic whereas the large proportion of area specialists are idiographic.'[1]

PROBLEMS OF RESEARCH

Work which has been published hitherto in any Western language upon the location of industry in the European socialist republics is restricted in volume and generally deficient in depth. Geography texts on the area[2] provide little more than an introductory description of the major industrial activities by place at a given point in time. However, recently, Pounds[3] has paid some attention to *explanation* of *trends* in industrial patterns, drawing, in part, upon his earlier research.[4] In related sciences, scholars have been concerned primarily with the pros and cons of post-1945 industrialization policies in order to assess the success of development strategies through time. Comments on industrial location in such works,[5] are scarce, largely descriptive and demonstrate consistency 'with the weak spatial perspective of orthodox economic thinking'.[6]

Yet the study of industrial location is fundamental to any understanding of the post-war geography of Eastern Europe. Since industry receives the highest priority in investment planning, it is the prime vector of changes in spatial economic structure. Moreover, industrial location decisions are conceived as an integral part of state and Party policy-making. As such, they are shaped towards the fulfilment of definitive planning goals. Those goals are rooted in a philosophy – socialism – which, inasmuch as it seeks to build a new society, also seeks to create a new geography of human endeavour. Socialist planning thus *aims* at the achievement of long-term spatial economic structure which is fundamentally different from that formed hitherto by capitalism. Attention in any modern geography of East-Central and Southeast Europe should be clearly focused, therefore, upon the principles, processes and purposes of changing spatially and through time the inherited capitalist pattern of activities. Industrial location analysis can also provide insight into the impact of the processes of decision-making in a planned socialist society upon the emerging spatial pattern of production and services.

Why, then, has research in this field been so restricted?[7] Four answers are obvious: the greater attractiveness of studying the problems in the U.S.S.R.; the greater difficulty of doing so for Eastern Europe on account of diversity in language, history and geography; a belief that such research is of limited value, assuming that location decisions are identical to those in the Soviet Union; and the assumption by some scholars that location choices are the outcome of authoritarian decisions which are rationally inexplicable.

There are, however, more basic reasons. The question of

industrial location has greater strategic significance than the location of any other productive activity. Secrecy was especially common before 1955 with respect to actual plant sizes and locations and to the motivation of given location decisions. Most medium-term plans yield some information on plant location,[8] though more data has become available retrospectively as the Cold War eased and as encyclopaedic sources of one kind or another were published.[9] Short of an intensive survey of daily national and provincial newspaper articles, of party conference reports, of a search in less obvious journals[10] and of a series of interviews with responsible officials, however, it is frequently hard to establish the precise motives for past plant location decisions.

Although the reliability of industrial statistics has improved substantially over the years, their general coverage within, and compatibility between, the socialist republics leave much to be desired. Frequently it is difficult to establish with adequate accuracy the spatial structure of industry for the base years 1938–40, representative of the 'culmination' of capitalist enterprise, and for 1945–46, expressing the 'war-modified' industrial base upon which socialist planning was built directly. The international boundary shifts of Poland, in 1945 present their own special problems, though this is partly compensated by the availability of excellent statistics for local and provincial administrative areas in the post-war period.[11] Sometimes, even where detailed data are on hand, however, alterations have been made to the numbers and sizes of local administrative and statistical units, rendering long-term comparisons hazardous. Substantial changes in the boundaries of Yugoslav communes and districts in 1958 and of Bulgarian *okrugs* in 1959 provide examples.[12]

CHANGING INDUSTRIAL DISTRIBUTION

Before World War II East-Central and Southeast Europe was called, in large measure justifiably, the 'agricultural belt of Europe'. As far as can be ascertained, mining and manufacturing employed fewer than 8 million (15%) of the 55 million working people in the area in 1939. The three northern states – East Germany, Poland and Czechoslovakia – then localized almost 85% of the region's industry (Table 5.1). More precisely, 75% of all industry was concentrated within a triangular area (Fig. 5.1) encompassing Łódź (Poland), Erfurt (Thuringia) and Budapest (Hungary). The triangle localized more than 90% of all Czechoslovak industry, 75% of East German and Polish industry and 65% of Hungarian industry. The explanation is

TABLE 5.1 – *East-Central and Southeast Europe: Employment in Industry 1938–39, 1950 and 1968* (millions and percentages)

	1938–39 Employment	%	1950 Employment	%	1968 Employment	%
German Democratic Republic	2·3(3·1)[13]	29	2·0	25	2·7	19
Poland	2·0(1·2)[14]	26	1·9	24	3·6	24·5
Czechoslovakia	2·3	29	1·7	21	2·6	18·5
Hungary	0·7	9	0·7	9	1·6	11
Romania	0·3	3	0·8	10	1·7	11·5
Yugoslavia	0·2	2·25	0·6	8	1·4	9·5
Bulgaria	0·15	1·75	0·35	4	1·1	6·0
Total	7·95	100	8·05	100	14·7	100

Sources: *U.N. Statistical Yearbooks; Rocznik Statystyczny Polskie Republike Ludowej*, Warsaw, 1968, pp. 629–30, 636.

that this was the one area that had possessed *all* the conditions necessary for nineteenth-century industrialization: relative political stability; the benefit of favourable economic policies from the Prussian, Austrian and Hungarian imperial governments; capital; coal and iron resources; and good transport.

Several well-defined sub-regional agglomerations figured prominently within the triangle: mining and basic industries in Upper Silesia, the Saxon Bay (in the south-eastern part of East Germany), northern Bohemia and Moravia Ostrava; and centres of labour- and market-oriented industries, above all, Łódź, Dresden, Zlin (now Gottwaldow) and Budapest. A variety of primary processing locations interspersed these agglomerations.

Very little manufacturing existed elsewhere, except in Berlin, Warsaw and northern Hungary, locations easily accessible by rail to and from, and functionally interdependent with, the industrial triangle. Scattered and usually small factories in Southeast Europe could not compete for markets with large-scale Central European producers. Exceptions were plants which (a) processed local resources with a high scarcity value (e.g. oil in Romania, timber or metals in Yugoslavia); (b) served protected home markets (e.g. textiles in Romania and Yugoslavia); (c) were fostered as state enterprises by Balkan governments (e.g. steel, railway vehicles, matches in Yugoslavia); or (d) were established as branch factories by foreign companies to control competition in Southeast European markets (e.g. *Bata* shoes, *Siemens* electrical). Yet here, too, industry was relatively localized: 78% in Yugoslavia lay north of the Danube–Sava line, 55% in Romania lay in the 'core' Transylvania-Muntenian areas. Only tiny Bulgarian industry showed any marked tendency towards dispersion, the Sofia region accounting for only one-third of the nation's industry.

Trends in industrial activity (see Table 5.1) during and after World War II were divergent in the European socialist republics as between the periods 1938/39–1950 and 1950–67. Substantially greater international equality in industrial development resulted from spatial redistribution of productive activity between 1939 and 1950. Reduced employment in the three northern states is explicable by heavy war damage to plant in, or post-war reparations of plant to the U.S.S.R. from, East Germany and Poland and by greatly reduced population in Poland (33%) and Czechoslovakia (25%).[15] Expatriation of Germans from Czechoslovakia (25% of the population) left much of industrial northern Bohemia without workers, skills,[16] and, partially, markets. These losses more than offset any gain in new German-built wartime plants. By contrast, the Southeast European states rapidly increased industrial employment from well below the 1938/39 average of 600,000 to more than 1,735,000 by 1950. This astonishing growth was achieved by absorbing surplus rural labour into the expansion of key minerals and energy production (especially in Romania) and into existing manufacturing industries where shift systems were introduced and labour was often substituted for capital in simple processes. War destruction of the small industrial base here was more quickly repaired, even in Yugoslavia, so that output soon surpassed 1939 levels. Moreover, that base also became inadequate sooner and demanded that effort be turned to entirely new industrial expansion.[17]

Between 1950 and 1967, the Southeast European countries expanded their absolute industrial employment by almost as much as did the northern states (i.e. by 3·35 million jobs compared with 3·5 million new jobs in the north) and raised

TABLE 5.2 – *East-Central and Southeast Europe; Indices of Increased Industrial Production and Employment 1950–1967*
(1950 = 100)

	1967 A. Production	1967 B. Employment	Ratio B ; A
German Democratic Republic	445	136	1 : 3·3
Poland	587	189	1 : 3·1
Czechoslovakia	411	141	1 : 2·9
Hungary	449	229	1 : 1·9
Romania	838	212	1 : 3·9
Yugoslavia	450	233	1 : 1·9
Bulgaria	880	314	1 : 2·8

Source: *Rocznik Statystycznej P.R.L., op. cit.*

their share of industrial employment in the region from 16% in 1939 to nearly 39% in 1968. Conversely, East Germany and Czechoslovakia, with the slowest rates of post-war growth, declined in their relative importance, while Poland and Hungary seemingly maintained their positions with a rate of expansion which was average for East Europe.

A more balanced view of broad geographic shifts in industrial activity may be obtained by comparing indices of growth in employment *and* in productivity (Table 5·2). Three conclusions may be drawn from this table: (1) greater improvements in labour productivity in the three northern states have enabled those states to maintain more of their former industrial lead than changes in employment would seem to indicate; (2) increases in productivity in Hungary and Yugoslavia have been much less impressive; Hungary is probably less important industrially now in relation to other socialist republics than it was in 1938/39 and (3) Romania and Bulgaria have clearly raised their contributions to Eastern European output by increasing labour productivity by as much, or more than, the northern countries.[18]

How far have the national and regional industrial location patterns also been modified? Fig. 5.1 shows the distribution of the more important industrial projects constructed between 1945 and 1969. A basis for comparison is afforded by distinguishing (*a*) completely new plant from (*b*) reconstructed, automated or expanded existing plants.

Two conflicting trends are apparent in post-war planned industrial location in the region: *dispersion* and *concentration*. The 'industrial triangle' (see Fig. 5.1) now localizes only half of the region's industry (compared with 75% in 1939), although the number of industrial jobs in the triangle has been increased from 5.9 to 6.7 million. Significantly, nearly 6 million new mine and factory jobs have been provided since 1939 in areas outside the triangle. Yet the triangle remains the only really highly industrialized region.

The *concentration* of new capacity in existing industrial centres or areas confirms, strengthens and perpetuates the distributional pattern of industry inherited from the capitalist past. In East-Central and Southeast Europe, such concentration has taken three basic forms: (1) the reconstruction, modernization and expansion of pre-war, wartime or war-damaged factories which required either large investments or long gestation periods: metallurgy in Upper Silesia (e.g. 11 steelworks) and Moravia Ostrava; chemicals in these areas, in Saxony and Bohemia; and mechanical and electrical engineering in most major cities; (2) all large- or medium-sized industrial centres have received completely new industries, sometimes of a linked nature; e.g. Budapest, with mainly textile, food and metal industries in 1946, gained 25 large new plants between 1946 and 1966 to manufacture machinery, vehicles, electrical appliances, chemicals, building materials and cellulose and (3) larger-scale factories and city industrial nodes were created from the transfer, or *relocation*, of factory installations from many rural locations in order to derive scale economies and to exploit transport and skilled labour opportunities more effectively. Such relocation of industry provided ready-made

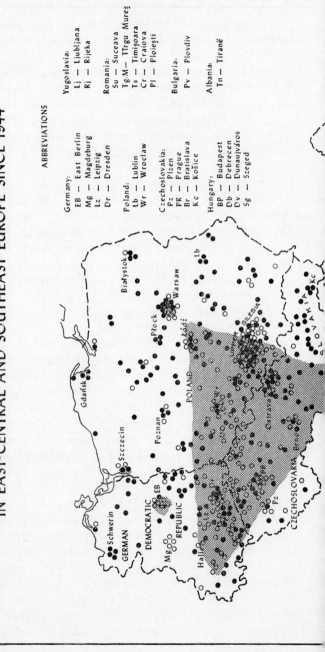

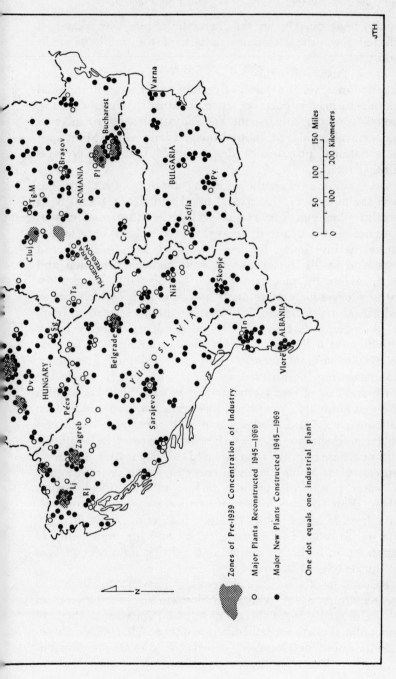

FIGURE 5.1 – DISTRIBUTION OF MAJOR INDUSTRIAL PLANTS CONSTRUCTED OR RECONSTRUCTED IN EAST-CENTRAL AND SOUTHEAST EUROPE, 1944–69. Data is based on field work in East-Central and Southeast Europe, 1963–66; unpublished Ph.D. thesis, The Location of Yugoslav Industry, University of London 1963: and on numerous sources published in the countries of East-Central and Southeast Europe.

employment quickly in the immediate post-war years; it affected most the locational pattern of the 'footloose' light industries.[19]

These forms of industrial concentration have combined either to increase the shares of certain major cities in national manufacturing importance (e.g. Warsaw, Budapest and Zagreb) or alternatively, for key industrial areas to increase (e.g. central Bosnia) or maintain (e.g. Transylvania-Muntenia) their national prominence despite rapid industrial progress in other regions.

Generally, however, the widening *dispersion* of industry has been the more pervasive spatial trend and clearly results from socialist planning. The controlled spatial transfer of capital from the more 'advanced' areas that accumulated most of that capital as investment for less developed areas has achieved dispersion at all levels, international, national, regional and local. Investment allocation in space has tended to bring industry to progressively more areas and settlements which lacked industry thirty years ago. Whole areas, usually termed *backward* or *underdeveloped* – because their dependence upon primary activities for a livelihood yields them a *per capita* GNP well below the national average – have received their first major manufacturing industries since 1945. Such areas include the north and east of East Germany and Poland, Slovakia, central and east Hungary, 'peripheral', western, northern and eastern Romania, upland Yugoslavia and most areas of Albania and Bulgaria. Dispersion has been generally greater in Southeast Europe where industries are oriented to fairly scattered mineral sources or to 'mobile' kinds of energy (electricity from lignite or water power, oil, natural gas) or to widespread surplus labour. East German industrial location patterns, by contrast, have been least subject to change on account of the diversity and localization of inherited capacity and the prominence of expansion in highly integrated and existing industries such as chemicals (e.g. at Bitterfeld).

Relocation of some 340 Bohemian factories contributed to greater dispersion of industry in Slovakia and was required to reduce unemployment there and to avoid scarcity of labour in depopulated northern Bohemia, including Sudetenland. Subsequently, new development raised the Slovakian contribution to total industrial output from 8% in 1939 to 24% in 1967. The tendency to allocate a wide variety of industrial among many medium- or small-sized towns has been strong throughout East-Central Europe.

These broad post-war trends are clearly reflected in *either* (1) the rising proportion of each nation's industry that is located in less developed areas: in central and eastern Yugoslavia from 22% (1945) to 45% (1968); in Poland's peripheral 'horseshoe' from 25% to 39%; *or* (2) at least a relative decline in the importance of the pre-war national 'industrial centre', as in Bulgaria where the share of the Sofia region in the nation's industry declined from 33% (1939) to 20% (1967).

At the regional level, industrial dispersal has been of two complementary types. These might be termed respectively, 'dispersed localization' and 'regionalized dispersion'. *Dispersed localization* consists in the increasing 'centrifugal' scatter of industrial growth in new plant locations near, but outside, existing agglomerations. Classic examples are provided by the development of new chemical, metallurgy and engineering industries in Nowa Huta, Oswiecim, Kety, Kedzierzyn and Częstochowa in the rural belt (Zone B) around the congested core (Zone A) of the Upper Silesian Industrial District. Another is the dispersal of new market- or labour-oriented industries outside, but within a radius of 25 miles of Budapest. 'Dispersed localization' also results from the relocation of production from congested urban areas to surrounding suburban, satellite or rural settlements, a phenomenon common in the Warsaw and Łódź metropolitan areas.

Regionalized dispersion is a form of 'centripetal' process in which new factories are allocated to, and grouped relatively loosely in space within, certain *selected* regions, largely to exploit their greater economic advantages. 'Regionalized dispersion' typifies the industrialization of such areas as central Bosnia or Transylvania-Muntenia. A very good Polish example is the recent development of major mining, power, metallurgy and petrochemical-based industries in several 'growth poles' (Konin, Koło, Kłodawa, Kutno, Płock, Wlocławek, Toruń and Inowrocław) which are grouped within an 8000 square-mile region bounded by Warsaw, Łódź, Poznań and Toruń. The concept implies that there are other areas (e.g. lying on several sides of the region defined) where very few industries have been located: dispersion is selective among regions.

PLANNING INDUSTRIAL LOCATION: THE PRINCIPLES AND THEIR IMPACT

State intervention in direct production in Eastern Europe is not new: it grew rapidly in the 1920's and 1930's to provide much needed industrial entrepreneurship in essentially backward or maladjusted economies. *Locational planning*, however, was restricted to a few large plants which were built to fulfil strategic or political needs, e.g. steel and engineering in Bosnia and Serbia. Only Poland could boast an integrated plan of industrial development and location for the Central Industrial District.[20] World War II quickly terminated the execution of the plan – as well as the lives of the few planners living in Eastern Europe.

Inevitably, the ideas on location conceived in the Soviet Union before 1940 shaped the locational decisions made in the first medium-term development plans of the European socialist republics. Moreover, after 1945 the U.S.S.R. constructed upwards of 300 industrial plants (as well as parts of many more) in these republics.[21] Locations for these plants were probably chosen, at least in consultation with, if not by, Soviet experts.[22] In this way the Soviet Union has exercised direct control over the location of one-half of all major industrial plants and power stations constructed in Albania, Hungary and Romania and one-third of those constructed in Bulgaria and Poland. By contrast, local *interpretations* of Soviet location principles have been more influential in Czechoslovakia, East Germany and Yugoslavia which received little Soviet assistance.

Emphasis should be placed upon the 'looseness' of applicable Soviet location principles, for neither were these formalized until 1954 in the writings of Livshits and Feigin[23] (i.e. after the first European development plans) nor have they been developed since into any strictly coherent theory.[24] None the less, with Soviet expertise on hand to advise (except in Yugoslavia after June, 1948), there is little doubt that planners in Eastern Europe were familiar with the location principles by 1948–49.

LOCATION PRINCIPLES

What are these location principles? How far have they been applied in practice? How far does this application explain the changed patterns of industrial distribution?

There are nine principles.

1. *Industries should be located near the natural resources they use,* especially, it is tacitly recognized, when those materials (or energy) lose weight in processing. The aim is to minimize 'the consumption of labour power', i.e. 'transport effort', or transport cost.[25] The translation, in 1925, of Weber's *Über den Standort der Industrien* (1909) into Russia stimulated this thought.[26] Lenin, however, had arrived earlier at a similar conclusion independently, although possibly he had familiarized himself with the content of Weber's original version while travelling and living in West and Central Europe prior to 1917.

This principle guided many location decisions after 1947, particularly for priority heavy industries directly engaging in processing materials and using large amounts of energy. It accounts for growth in the industrial triangle, especially Upper Silesia, where major natural resources are concentrated. Some industries were placed in new locations within the triangle to process newly discovered or newly worked deposits: brown coal near Cottbus (East Germany) and Konin (Poland), copper and sulphur in southern Poland. However, it is in underdeveloped and poorly surveyed areas where planners have attached most importance to deposits of metal ores, non-metallic minerals, timber, fuels and water power as locational determinants of processing capacities whence the dispersion of post-war manufacturing throughout Slovakia and in areas in, or bordering, the uplands of Bulgaria, Romania and Yugoslavia.

Between 1952 and 1964 an additional motive operated in Yugoslavia to favour the raw material-orientation of industry. Since natural resources constitute a major source of wealth in the underdeveloped regions, industries were built there to generate greater sources of local income which, under the decentralized economic system, could be reinvested locally in infrastructure and in getting bank loans for production to assist in achieving take-off towards self-sustaining growth. Such motivation was absent in the other, centrally planned, economies where, by the very definition of centralization, the location of income-generation was (and is) not necessarily important for the location of new investment.

2. *Industries, particularly the higher stages of manufacturing, should be located near the markets where their products are consumed.* Again, the approach is essentially Weberian – to reduce the transport costs of product marketing. Except in Yugoslavia, two circumstances have reinforced the Weberian approach: (a) planners have largely ignored demand functions and stressed production functions; and (b) fixed or manipulated prices, planned demand for capital goods, limited influence of consumer demand upon the production of consumer goods and the lack of competition encouraged planners to ignore the question of plant location to command the best spatial sales area as a means of maximizing profit (accumulation). No value is attached to a 'market potential' approach.[27] It is conceivable, though, that this may change when economic reforms towards more decentralized 'market-type' economies are extended.

Two further interrelated points need to be made. First, no Southeast European development plans stress the market as a location factor because in these underdeveloped countries the market area was, and is, essentially the whole state area. The choice of region was not necessarily important, or even dominant, for finishing industries as the market area (even in Yugoslavia) could be equally well served from many points. Second, planned expansion of, and structural change in, production in all states, especially the less developed, leads, in effect, to the creation of entirely new product market areas there (e.g. for vehicles, typewriters, electrical equipment, chemicals) and the expansion of existing market areas (e.g. for steel). Ultimately these changes modify the spatial patterns of consumer demand.

Broadly speaking, the market factor explains the location of industry in or near major metropolitan and industrial centres and in transport nodes throughout the region.

These principles point to two basic locational solutions in accordance with two environmental situations. In republics where resources required by and markets served by a given industry are dominantly coincident in space, then the industry will comprise vertically integrated plants (called by Lenin, *Kombinat*) manufacturing finished products from raw materials; e.g. the Plovdiv textile combine, East German chemicals at Bittefeld and Leuna; associated engineering and steel industries in Upper Silesia. In republics where resources required by and markets served by a given industry are spatially separate, then the industry will comprise a set of plants processing materials near the mines or quarries and distinct from another set of plants manufacturing finished products in respective market areas, e.g. Polish engineering, which uses Silesian steel, but which is dispersed among most medium- to large-sized cities to gain good transport access to market.

The issue is, however, oversimplified. The operation of other socialist principles may encourage industrial location outside areas of material supply and market demand.

3. *Industries should be allocated interregionally in such a manner as to develop regional specialization* according to regional natural resource endowment.

4. *Industries should be allocated interregionally to achieve greater regional self-sufficiency* according to regional market needs.

In effect, these are re-statements of principles 1 and 2, but cast within a regional framework for which the degrees of specialization and of self-sufficiency respectively may be expressed in tables of input-output and interindustry relations. Distance and transport are far less serious cost factors in East-Central Europe than they are in the U.S.S.R., for each European socialist republic is often smaller in size than an 'economic region' in the U.S.S.R. However, regional specialization *and* self-sufficiency have major implications for the East European countries within the context of COMECON (see below). Intra-nationally, regional specialization is more relevant and may be applied in terms of Kolossovsky-style 'energy-material-manufacturing' regional complexes to regions or industrial 'nodes'; e.g. in Poland it may be applied either to an area such as Upper Silesia or to an industrial complex like the emergent brown coal-electricity-alumina/aluminium refining-fabricated aluminium products complex centred on Konin.

5. *There should be an even distribution of industry throughout the country.* As now understood, the purpose of this principle is to achieve the fullest possible employment of human, and use of natural, resources within every region. It stems from the idea that idle resources (labour, materials, plant capacity) represent inefficiency. Yet it is rooted in Engel's argument that the late nineteenth-century technological advances – electricity, road vehicles – could free industry from locations near materials, fuel and power, and permit dispersion of industry in proportion to the distribution of population as a means of restricting unhealthy, large-city growth.[28] The principle implies the following:

(*a*) Inferior resources will be used in manufacturing to fulfil the socialist objective that everyone of working age has 'the right to work', i.e. that no one should be unemployed. Although this implies social motivation, which may well be related to the development of backward areas, there may be sound economic reasons for using low-grade materials, namely the greater expense of obtaining supplies of better grade materials from elsewhere, including abroad. The use of poor, high-cost Croatian coal in Zagreb industry, for example, is cheaper than good quality, low-cost Bosnian coal plus the cost of hauling that coal 250 miles to Zagreb.

(*b*) Industries should be distributed according to the spatial incidence of unemployed or underemployed, mainly agrarian, labour, i.e. to labour-supply zones. The prime argument here is that a socialist (Marxist) economy must minimize expenditure on 'non-productive investments', namely services and housing. Costs of infrastructure to support new industry may be reduced by locating industry (if necessary by dividing plants[29]) where labour is already housed within a short commuting radius. This factor was vitally important in Poland where so much city housing was destroyed in wartime and it explains much dispersal of finishing industries to smaller cities there. A related factor for all East-Central Europe was the need to utilize the scarce skill of handicraftsmen.

(c) 'Industrial location is also the means of creating bastions of socialism throughout the countryside.'[30] Marxist doctrine stresses the inseparability of institutions and economic results: a working proletariat sympathetic to Communism must be formed everywhere to undermine spatially the dominance of the individualistic peasant who threatens socialist production.

Few principles are stressed as much as this one in development plans. Undoubtedly greater industrial dispersion has resulted but it is difficult to discern which motive was decisive in each case. More important, the ambiguity and vagueness of the principle virtually gave planners a free hand in choosing locations, particularly for those industries not tied closely to sources of weight-losing materials. However, Secomski stressed that, by 1955, research in Poland had altered the concept from 'the equal distribution ...' to 'the rational distribution ... of productive forces' involving 'a choice among alternatives to develop each region towards interregional equality'.[31] This clearly overlaps with principle six.

6. *Industry should be located so as to advance the economic and cultural development of all regions inhabited by national minorities.* This principle originated in Lenin's belief that the economic dichotomy between 'metropolitan-industrial' European Russia and the 'colonial non-Russian periphery' was unhealthy to political, social and economic viability in the U.S.S.R. Backward regions inhabited by ethnic minorities had to be industrialized. In Europe, the principle is very relevant in Yugoslavia with its complex ethnography and highly differentiated regional economies[32] and in Czechoslovakia where in 1937 'Slovakia was 70 years behind Bohemia-Moravia in economic development'.[33] The problem in Romania is 'inverted' as the Hungarian Empire left the Hungarian- and German-inhabited areas of Transylvania more advanced than the Romanian-inhabited Carpathian ring and lowland periphery. Despite some observers' allegations that Romanian nationalism has discriminated against

'developed' Transylvania since 1955 in favour of 'underdeveloped' Romanian areas, Transylvania has maintained an industrial growth rate around the state average.

The regional ethno-economic problem does not exist in other East European countries; there, the equivalent idea is to develop the backward regions at a faster rate than the developed regions. In practice the absence of national minority interests (which are far more heavily vested in locational problems) in these states has encouraged the supremacy of 'homogeneous' national allocational aspects of economic choice to achieve most rapid over-all growth; deliberate development of backward areas in East Germany, Poland and Hungary has thus been of subordinate importance.[34]

What strategies are, or have been, applied in attempts to industrialize underdeveloped regions in the European socialist republics? Basically, there have been three:

(i) Where natural resources were available locally, priority investment went into the development of metallurgy, power and chemicals, as in the Hunedoara and Maramureş regions (Romania) and Bosnia–Herzegovina (Yugoslavia). (ii) Where natural resources were lacking, or not discovered at the time of the decision to initiate regional industrialization, heavy industries *were* constructed if transport access to imported materials was potentially good. It is significant that, in contradiction to principle one, all new iron and steel plants constructed in East-Central Europe (except the Zenica and Sisak plants, Yugoslavia)[35] were located in backward areas which lacked an adequate resource basis for them: plants at Eisenhüttenstadt (East Germany), Košice (Slovakia), Dunaújváros (Hungary), Nikšić and Skopje (Yugoslavia), and Galaţi (Romania). All except the Yugoslav plants depend on supplies of iron ore from Krivoi Rog and coke from the Donbas. (iii) Where neither raw materials for heavy industry nor satisfactory transport links with other regions existed, attempts have been made to introduce labour-intensive industries which manufacture high-value products, e.g. tools, plastic products in the Yugoslav Karst country.[36]

7. *Industrial location policy should eliminate economic and social differences between town and country*, chiefly by greater interdependence between agriculture and industry, the one supplying raw materials for processing, the other the capital goods for raising agricultural productivity. Clearly this could lead to the development of processing industries in many rural settlements and of agricultural producers' goods in more central nodes.[37]

8. *Industrial location planning must take heed of defence needs.* Already operative in the U.S.S.R. after the XVIII CPSU Congress meeting in 1937, in the Polish Central Industrial District, 1936, and in the location of Yugoslav metallurgy in 1938, the importance of this principle was greatly inflated by bitter wartime experience and subsequent cold-war tensions. Despite their small area, the European socialist republics could offer strategic locational advantages in their eastern regions: in eastern Poland, where implementation of the Central Industrial District plan contributed to the industrialization of Rzeszów *Województwa*; Slovakia; the Tisza Valley; Moldavia and eastern Bulgaria. By contrast, strategic locations in Yugoslavia lay in the mountains as far away as possible from borders with major Cominform neighbours.[38] In every country, until 1955, however, the Ministry of Defence enforced interregional dispersion of key metal-working, engineering and chemical industries, intra-regional dispersion of plants and the siting of plants at safe distances from urban agglomerations.

9. *Industrial location choices should facilitate international division of labour with COMECON,*[39] i.e. greater national industrial specialization in accordance with natural and human resource potentials and hence greater international economic interdependence. From 1949 to 1959, however, national autarky encouraged duplication of varied industries and plants between nations, with consequent losses in the economies of scale. Many industries were developed in each country irrespective of the availability of, or of the costs of, supplying the materials, fuel and power required. Metallurgical plants at Eisenhüttenstadt and Dunaújváros stand as testimonies.

Before Khruschev, international division of labour was

still considered taboo since it was equated by Marxists with bourgeois empires and colonial economic policy. Even so, some specialization among COMECON nations was planned to make inter-state aid more effective. The U.S.S.R. supplied upwards of 300 industrial (mainly steel and engineering) plants to the European socialist republics between 1945 and 1965 (99 to Albania, 79 to Bulgaria, 75 to Poland, 63 to Romania, 27 to Hungary and 5 to Czechoslovakia). Other countries supplied plants to those countries below them in the 'hierarchy of economic development': East Germany supplied all nations, whereas Bulgaria and Romania supplied only Albania with factory equipment. The burden of the transfer of wealth fell heavily on East Germany and Czechoslovakia, each of which supplied over 70 factories to their southern partners and to Poland. Whereas East Germany supplied mainly engineering and chemical plants, Czechoslovakia was responsible for most supplies of power station equipment.

By 1960 the need to obtain economies of scale from large specialized plants was recognized. The policy now became the location of large plants to supply international COMECON markets from nodal locations either near the cheapest or most accessible source of materials, e.g. sulphur at Tarnobrzeg, southeast Poland, or to permit the cheapest distribution of products to market areas, e.g. steel output at Košice or Galaţi; oil refining at Płock, Bratislava and Szashalombatta (south of Budapest). Czechoslovakia is actively investing in industrial expansion in regions which may supply her with deficient raw materials or products: brown coal, electricity, copper, sulphur and superphosphates in southern Poland, East German potash products, Hungarian bauxite-aluminium and the Volga oilfields. Attempts at specialization have been mainly confined to engineering (e.g. Polish shipping and railway vehicle production) and to the co-ordination of changing patterns of energy production and flows. The latter has involved: (*a*) increasing dependence upon Soviet coking coal and oil supplies; (*b*) co-ordinated exploitation of major energy resources which straddle international boundaries (Ostravian–Silesian coal, the upper Nysa Valley brown-coal basin, Danubian hydroelectricity);

(c) the substitution of oil and gas from the U.S.S.R. and Pannonia for coal in the growth of the chemical industry, a trend which has encouraged the location of new chemical capacities away from coalfields; and (d) the increasing importance of brown-coal and lignite in electricity and coal-gas production.

However, four factors operate to restrain the emergence of highly specialized industrial centres within COMECON. Less developed nations (especially Romania) resist specialization on primary processing of varied minerals and insist on developing manufacturing to reduce their historical legacy of backwardness. Conversely, more developed states (especially East Germany and Czechoslovakia) are reluctant to help eliminate international economic disparities by crediting modern factory construction in Southeast Europe since this may erode their long-term competitive ability. Moreover, fear of losing political sovereignty with supra-nationality overrides economic advantages of production in large-scale plants. Finally, no country is willing, either for economic or ideological reasons, to admit industrial obsolescence since this would undermine confidence in planning.

PLANNING LOCATION: SOME BEHAVIOURAL CONSIDERATIONS

In reality, location decisions emerge from a complex interaction of complementary and conflicting economic, social and political pressures. It is now more commonly appreciated that 'for many economists (and geographers) the "real" world is one of (minerals and manufactures) comparative advantage, supply and demand, transportation infrastructure, etc. They, and the theories they espouse, exist professionally only in an ideal world. The world of grubby politics, nationalism, prejudice, primordial loyalties... in which economic activity is embodied is not recognized as being of the same order of reality.'[40] Relatively little behavioural research has applied to industrial location under capitalism.[41] What hope, then, is there of finding comparable research for the socialist world? Wiles provides a terse answer: '... Marxism is a secularized religion of a Western kind, and therefore it must close itself to whole areas of fruitful scientific inquiry. We find, for instance, even in Poland, nothing about... the behaviour of socialist entrepreneurs.'[42] The reasons are not far to seek: scanty information and the intention of socialist economy to plan according to 'objective laws'.

Those 'laws' are the principles set out and analysed above as to their apparent relevance. Nevertheless, even the most totalitarian system is applied by men and groups of men who, as human beings, interact both with each other and with a system, using it as well as being conditioned by it. A tentative outline is made below, therefore, of industrial location choice in Eastern Europe within a behavioural framework. Attention focuses upon decision-makers' response to the economic system, the organization of decision-taking and the rationality of location decisions.

THE ECONOMIC SYSTEM AND HUMAN RESPONSE

That Marxist economic thought has evolved in isolation from mainstream economics[43] explains why the command economy, dominant in our area until 1956 and still pervasive, was unconcerned *per se* with the allocation of scarce resources among many competing alternatives. Such allocation implied choice, which necessitated the application of economic criteria. 'Attempts... to define investment criteria were rejected because acceptance of any one criterion would have restricted the freedom of investment (i.e. locational) choice.'[44] Such criteria were considered to threaten Party sovereignty (which may be termed *independent discretion*)[45] or the infalilbility of leaders. In the absence of economic criteria, location choices could be arbitrary, at Party will, but they were undoubtedly guided by the socialist location principles.

Decentralized decision-taking became more frequent following Stalin's assertion (1952) that 'economics could be an objective science' and the 'thaw' of 1953–6 (in Yugoslavia after 1950). It demanded the evolution of criteria by which unlike projects could be evaluated. Except in Yugoslavia, however, the 'liberalized' system continued to suffer from the defects of a command economy: ineffectiveness of demand and fixed or semi-fixed prices which conferred neither utility nor scarcity values to goods and services but which functioned primarily to accumulate capital (achieve profit).[46] Throughout Eastern Europe, profit or *rentability*[48] were adopted by 1958 (in Yugoslavia by 1954) as the key success indicators for allocating investment between projects (and hence locations).[48] Whenever criteria are applied to calculate rentability, the answers are distorted by the degree of irrationality of the pricing system.

Location decision-makers at all levels – planners, ministers, managers of industrial associations and factories – respond to the system as it is, without entering into its rationale or logistics. The economic system is outside their *authority spheres* although it determines decisions they make in their *activity spheres*.[49] The consequences may be profound. Arbitrarily high prices for (and thus profit from) certain products (e.g. plastics, artificial fibres, consumer goods) will call forth greater investment from the Planning Commission, regional authorities and factories in the expansion of production of those items, the more so the more funds are decentralized along with authority. Arbitrarily low prices for energy and raw materials invoke the opposite response.[50] Resultant structural changes in industry affect locational patterns since every production decision (even to have zero expansion in the present location) is a location decision. High prices, designed to achieve high profits (in the consumer sector), encourage planners to pay less attention to costs and more attention to the guiding principles in locating new or expanded plants. It would be interesting to examine the impact on location decision-makers of a pricing policy common throughout Eastern Europe (except Yugoslavia) which calculates average costs and profits for all producers within an industry and then simply transfers the extra profits from efficient plants as budget subsidies to inefficient plants.

The utility of investment efficiency indicators for selecting development projects has been restricted further by inadequate appraisal of *which* costs need to be calculated. Throughout Eastern Europe until 1960 (Yugoslavia until 1952) the rentabilities of alternative industrial projects were assessed on input costs calculations only. This favoured capital-intensive but low-per-unit output projects (e.g. hydroelectric rather than thermal power stations), an effect of locational significance for dependent manufacturing industries. Where input costs were computed to include social expenses – mainly workers' housing – the results similarly favoured capital-intensive industries in order to restrict the required labour force, and hence, 'non-productive' housing investment. Transport costs were often omitted from calculations because equalized pricing for the whole national product market encouraged decision-makers to believe that transport costs were inconsequential.[51] Clearly this encouraged arbitrary location choices, although planners argued that plant efficiency in the small European Socialist states could offset higher transport costs and permit location more frequently in underdeveloped areas. When, later, transport costs were included, all the idiosyncrasies of subsidized and 'deviational' tariffs evoked spatial responses by planners, ministers and entrepreneurs in the same manner as did pricing in general.[52] Labour costs are not considered because employment is simply a planned factory 'norm'. Wages are related to basic living needs, not to value of work or to value as expressed by an interplay of demand and supply. Labour, therefore, influences location decisions primarily through social investment requirements.

More realistic costing methods were applied after 1952 in Yugoslavia, 1956 in Poland and 1960 elsewhere, though their raw materials are still largely arbitrary prices. Calculations for alternative plants and locations now include investment costs per unit output in plant and social infrastructure;[53] production costs for different input-mixes per unit output; and transport costs on materials and products (though costs of supplying market areas were often omitted – a reflection of the production-function approach). This trend expresses a growing recognition in East-Central and Southeast Europe of the quintessence of modern economic theory (abstracted from any political system): scarcity dictated choice among alternatives and resource allocation to maximize given ends. It especially reflects the initial impact upon economic decision-making of linear programming and computer techniques innovated by Kantorovitch and Nemchinov in the U.S.S.R. and by Lange and Kalecki in Poland, which open the way to optimum locational solutions.

The innovation of economic criteria to select investment projects should lead to a rigorous ranking of location principles, but complete rejection of 'non-economic' principles is unacceptable given the aims of socialism. In the absence of criteria before *circa* 1960, this ranking was determined by the ideals of the political leadership. During the decade after 1948 most location principles were subordinated to the main goal of the Communist governments, 'the fastest possible economic growth rate as the best possible method of satisfying political and economic needs'.[54] This gave priority to the expansion of existing industrial plants at low infrastructural investment cost and restricted development in backward and minority areas (except those possessing resources crucial for heavy industrial growth). Priority for heavy industries, however, made the defence factor of disproportionate significance (especially for engineering, steel and chemicals). Decisions, therefore, involved compromise between the advantages of developed urban-industrial areas and the advantages of underdeveloped areas. Experience makes it impossible to conclude for the European socialist countries – as Koropeckyj does for the U.S.S.R.[55] – that defence was the overriding location principle while the lowest-ranked principles were the elimination of town-country differences and the even distribution of industry. For, with the sole exception of northern East Germany and Poland, there has been a broad spatial convergence in East-Central and Southeast Europe of scattered natural resources (and thus raw material-oriented industry) with regional resource-based specialization, of defence facilities, of over-populated backward regions, of national minority regions and of benefits from utilizing existing housing and social infrastructure more intensively. The degree to which planners are able to disperse industry may be investigated through the 'spatial elasticity of investment planning'.

THE SPATIAL ELASTICITY OF INVESTMENT PLANNING

Severe restraints upon industrial dispersion are imposed by the fact that, for any period, planners have to work at the margin with existing projects and capacities which absorb into ongoing production large proportions of available capital. The concept of the 'spatial elasticity of investment plans' is relevant here. A definition would run as follows: the degree of change in the spatial pattern of production that is possible, or results, from a planned change in the volume and branch allocation of investment in the planning period as compared with the previous period. If the spatial change in production is less than proportionate to the change in investment allocation, the plan is said to be spatially inelastic; if it is more than proportionate, the plan is elastic. Important variables in the concept are (*a*) the size of existing capacity; (*b*) its diversity and location; (*c*) the balance of investment allocation among heavy industries (i.e. few projects) and light industries (many projects) and between individual branches to establish the number of optimum-sized plants that the plan permits; (*d*) the allocation of investment between plant-expansion (i.e. spatially inelastic) and new plant construction (potentially spatially elastic); (*e*) the spatial patterns of natural and human resources.

Investment plans have been spatially inelastic in Poland (especially the Six-Year Plan 1950–5, extended to 1959) and East Germany. The Polish Plan allocated 90% of industrial investment to existing plants; much of the remaining 10% went into steel mills at Nowa Huta and Częstochowa. Moreover, most capital went into heavy industries – electricity, coal-processing, metallurgy chemicals – which were already located in, or were newly allocated to, the major natural resource region of Poland, the Upper Silesian coal basin. This region received 70% of all planned industrial investment. Little scope remained for industrial location elsewhere. East German plans similarly concentrated upon existing capacity, in this case rationalization of production between formerly competitive (and hence duplicate) plants in a wide variety of industries.[56] By contrast, a limited industrial base, many new industries and scattered resources permitted a high elasticity of spatial plans in Bulgaria.

The problems with this concept are its theoretical significance and its practical application. There is a problem of measurement. Numbers of optimal-sized plants vary between industries and with different levels of economic development. Potential numbers of locations vary and so do the complex relationships between technology, investment, employment, output and value-added. Clearly a standardized language is needed to turn the concept into a useful tool of analysis in relating plans to emerging spatial patterns.

THE MAKING OF DECISIONS

Until recently, with the major exception of the decentralized system in Yugoslavia, there has been in the European socialist republics a more or less pyramidic structure of industrial decision-making. At higher, more powerful levels, decision-makers may exercise *independent discretion* (free will). The first development plans were formulated during the 'cult of personality'. Location choices for major plants were subject to the approval, suggestion or edict of the supreme leader. For example, it is intimated that, in Poland, Beirut influenced the establishment of engineering and electrical industries in Lublin. The pitfalls in accepting this are all too apparent, for several motives for such a choice converged. Lublin was situated in a backward area with few industrial resources. People there needed work. The city had elected Beirut to parliament and had been the seat of the first Polish socialist government (1944). In addition, the location in Lublin of a plant for the assembly of trucks, manufactured in Minsk (Belorussia) and destined for the Polish market, had economic rationale *à la Weber*.

The independent discretion of political leaders ranges from the selection of locations for individual plants to the direction of planning officers in the application of socialist location principles. Highly respected planners may also have such discretion in working out detailed plans of location, e.g. Boris Kidrič in Yugoslavia; so may directors of industrial associations in rationalizing outputs of plants in their association.

However, 'the monopoly of decision-taking by one man was, to a large extent, an illusion'.[57] In the absence of scarcity prices, resource allocation is the outcome of the *bargaining force of participants in planning*.[58] Individuals or groups develop vested interests which are largely identified with their respective 'activity spheres', whether functionally they are near the top or near the bottom of the decision-making scale. The executive, chief planners, ministers of various industries, bank officials who allocate investment credit, regional or local political or economic spokesmen, national minority representatives, heads of industrial associations (groups of enterprises producing related items), or even influential factory managers – all represent different interests in bargaining.

Each of these individuals may co-operate with any or all of the others to produce industrial location projects for consideration by planners or bankers which would yield greater benefits than if there had not been co-operation: for example, a plan for a complex interrelated industries in a national minority region may emerge from cross-ministerial co-operation with regional representatives. Planning Commissions provide very crucial links in the chain of interaction between the political leadership and the proposals for projects from lower ranks. This interaction needs to be analysed spatially. Unfortunately, such *co-operative discretion* is not as common as it ought to be in a planned system, mainly because, within the Party framework, identity with an activity or area becomes identity with a vested interest and, eventually, personal integrity or ambition.

More commonly, location strategies or decisions are the outcome of *interdependent discretion* involving *conflict interdependence*; i.e. where decisions are made at the expense of an area, an individual, a group or an 'opportunity'. The Executive and Central Committee of the Party are 'always riven by deep divergences of outlook and sharp conflicts over policy'.[59] The outcome of latent or actual power struggles inevitably concerns major policy issues which ultimately affect, often substantially, short-run or long-run industrial location decisions at all spatial scales. The importance that is attached to investment in agriculture will condition the growth of food-processing and farm-serving (engineering and chemical) industries. The relative emphasis placed upon heavy *v.* light industries will affect the spatial elasticity of investment. The ascendance of representatives of sector planning – the ministers, heads of industrial associations or powerful factory managers – or of regional (in Yugoslavia, republic or commune) representatives, or of groups of planners who support one or more of these vested interests, will largely determine the degree to which planning either subordinates regional industrial development to national (or federal) needs or stresses a spatial approach which involves the assessment of the scale and character of the contributions that each region can make to national industrial progress. This particular conflict frequently resolves itself into expansion in existing industrial areas at lower short-term production cost (and probably imperceived higher long-term social costs) *v.* expansion in underdeveloped areas at lower long-term production *and* social costs.

Hungarian experience before 1960 testifies to the prevalence of 'sector' planners, who emphasized the production cost savings of location in the Budapest region, over regional planners, who stressed the savings from restricting congestion in that region that could accrue from substantial industrial development in 'counter-magnetic poles of growth' located in the peripheral areas of the state: after 1960 influence shifted in favour of regional planners who have incorporated into recent plans major industrial projects in such areas (e.g. at Győr, Kazincbarcika, Tiszapalkonya, Szeged, Pécs and Komlo).[60]

A shift of power between 'national Communists' and 'Muscovites' can have fundamental consequences for the international pattern of industrial growth among the European socialist republics – witness the Romanian 'industrial miracle' of the last decade.

Finally, one major source of group pressure outside the administration must be mentioned: the Church, particularly the Polish Catholic Church. In this case conflict involved *competition to avoid 'obnoxious' elements*: both State and Church seek to overcome the other's ideological and moral challenge. The State threw industrial location into the conflict according to principle 5 (*c*) above – the creation of socialist bastions – to support the other weapons in its armoury. Whence the Polish government sited its three major post-war steelworks (all part of the Six-Year Plan) near the largest nuclei of Catholic endeavour, Kraków (i.e. Nowa Huta), Częstochowa and Warsaw. The same factor may have motivated major industrial developments around Lublin and Poznań, Pecs (Hungary), Zagreb and Sarajevo (Yugoslavia).

RATIONALITY AND LOCATION DECISIONS

Potentially, a planned socialist system can achieve highly rational and efficient organizational structures for data input and processing and for data output embodying locational decisions. Data may be collected for standardized territorial statistical units for very comprehensive sets of intimately interlocking economic, social and spatial variables. Such a system also has the ability to establish, and act upon, social benefits and costs. There need not be divided and conflicting responsibilities over economic and social affairs nor secrecy about production costs which riddle non-Communist societies. In the last analysis, linear programming and computer methods can preform the mathematical solution of complex planning problems. However, in reality, optimal location decisions are not achieved for five major reasons.

First, information is very imperfect. Inevitably, as scientific knowledge and methodology progress, there is at any given point in time either insufficient data on or insufficient awareness of relevant problems. Secret or manipulated statistics in the 1950's provided decision-makers with poor data. Ministers and factory managers inflated or deflated their productive capabilities in order to gain prestige (and income) by overfulfilling plans. Lines of communication for collecting data and applying planning decisions are still dominantly vertical (i.e. within Ministries for particular industries, from the Minister down to factory or area representatives) and the horizontal links (i.e. inter-industry) at proper spatial levels are usually weak. Information relating to the co-ordination of essentially linked activities at local or regional levels, therefore, has often been unavailable, with a resulting loss of external economies, e.g. in industries processing cattle-feed in Yugoslavia.[61] Basically, however, the problem is that information is of poor quality as long as prices, and hence costs, and profits, remain arbitrary: success indicators and investment criteria are of limited value.

Second, available information has often been poorly processed. The appointment of politically reliable people in administration, from the Planning Commissions down, has restricted the flow of necessary expertise into appropriate offices. At times decisions have been based upon superficial or naïve analysis, e.g. interregional growth rates in Yugoslavia.[62] The regime, however, cannot be blamed for this as there is a lack of even secondary school education among the first generation of planners, let alone any real planning experience. Universities are producing a new crop of experts now who are eager to apply modern computational methods to planning and locational problems.⁶³ But the innovation of efficient data processing is being restrained by politicians who fear that linear programmes and computers, in producing optimal planning solutions, undermine their own authority and credibility.

Third, even given perfectly rational pricing, data collection, data processing and decision-taking, there is still no guarantee that optimal location solutions could be produced. There are serious limitations upon time, cost and the comprehension of planners to devise exhaustive alternative plans and projects for objective socio-economic comparison. There are also serious limitations to the ability of planning bodies to compare competing projects satisfactorily. During the years of the open competition for investment in Yugoslavia, the investment bank was flooded with 250 to 750 page project proposals, most of which received little more than adequate analysis at the cost of two- to three-year delays in reaching decisions. Moreover, factory workers' councils, regional councils and industrial associations usually have insufficient expertise and knowledge to produce really sound economic projects. These people find it difficult to use properly the instruction books that are available on how to establish project costs, profitability or rentability.⁶⁴

Fourth, planners are not perfectly rational; they are only boundedly rational.⁶⁵ In the absence of rational economic criteria for decision-making, East European planners could be rational only in some social or political sense. None the less, economic rationality has been sought more frequently since 1956 to justify the apparent application of the rather vague principles of socialist location. The problem may also be seen in reverse: motivation for industrial location decisions – defence, proletarianization – may affect the ability of planners to use otherwise pertinent economic information.

The ideological conviction of the planners – or the ideological impact of the political leadership upon the planners – certainly influences their sequential attention to the principles of location and to any other overriding short- or long-term planning objective (e.g. full employment, defence). In assessing alternatives for any planned expansion in output, the planner's rationality depends upon his awareness of the relevant *spatial activity sphere:* i.e. the set of locations which should be studied (*a*) to gauge the availability of relevant information for those locations and subsequently (*b*) to compare their merits and demerits with respect to actual plant location choice. The planners' rationality also depends upon both the order and the thoroughness in which the environment is searched for possibly alternative locations. There is a strong tendency to choose the first or easily found satisfactory location, bearing in mind the relevance of the theory of least effort to any decision-making process – especially when decisions are made under the acute pressure that typifies planning offices.

The order in which the environment is searched depends upon the planner's knowledge and experience, which may or may not be inter-related with his own *spatial identity*, i.e. his home region (especially in highly provincialized states such as Yugoslavia or Romania), or with his own *spatial preferences*, i.e. regions he likes or knows well. Evidence from Yugoslavia seems to show that planning in republic capitals for respective republics involves a convergence of space identity, space activity sphere and space preference which tends to encourage in the planner's mind a centrifugal or centripetal view of the republic in relation to its capital. His search for suitable plant locations begins in the capital and usually ends in, or at relatively short distances from, the capital as soon as a satisfactory location has been found. In short, planners are liable to leave several or many alternative locations for plants 'unsearched' and therefore 'unfound', so that there is no guarantee that any plant is located optimally. Pressure from vested interests of any kind, or the system of open competition evolved in Yugoslavia, can correct the planners' search procedure by bringing a greater number of alternatives to his notice. Conversely, influential pressure may halt the search earlier, especially if that pressure comes from an appointee manager to a new plant who prefers to live in or near a major urban centre rather than in some more isolated, less attractive underdeveloped area, even though the location principles demand the latter choice.

Fifth, as economic efficiency and rationality become more important in decision-making in the European socialist republics, so do *adaptive* location decisions which are made in response to careful analysis of environmental opportunities. Under the command economy, however, the *adoptive* approach was more common: industries were 'planted' in locations or regions which did not necessarily possess the skilled labour force, the infrastructure or the right quality of raw materials or fuel to support them. Location decisions were made frequently in ignorance of, or as a result of, poor perception of the environment, resulting in high-cost operations or wasted investments, e.g. the plants developed at Schwarze Pumpe (East Germany) and Lukavac (Yugoslavia) for manufacturing high-grade coal products from poor brown coal. The location of many large or 'political' factories in underdeveloped areas was clearly of an adoptive nature, motivated by the desire to create a 'technical demonstration effect' of educative value[66] in demonstrating the ability of the planned system to radically alter regional economies at Party will. Yet, the adoptive approach could yield a project in an underdeveloped area which at first would be unprofitable, but which would render other industries profitable, and would react upon the 'location leader' plant and justify it economically *ex post facto*.

CONCLUSION

The location of industry in East-Central and Southeast Europe has undergone profound changes since 1945 under the impact of planning towards a new ideal. Nine broad principles are invoked to guide the spatial allocation of new capacity at the all-industry, the industrial branch and the individual plant levels. As yet there is no comprehensive, rigorous body of location theory incorporating these principles. Varied in origin and often loose in meaning, the principles essentially polarize location trends around two extremes: concentration (principles 1, 2, 3 and 9) and dispersion (principles 1, 2, 4, 5, 6, 7 and 8). In so far as natural resources are scattered, over-population widespread and access is to a relatively evenly distributed ultimate consumer market (e.g. Bulgaria), dispersion will prevail. Compromise between concentration and dispersion becomes a crucial planning and policy issue when there are highly localized natural resources and inherited industrial capacity (e.g. East Germany, Poland) or highly articulate national minorities and a historical legacy of regional differentials in infrastructure (e.g. Yugoslavia).

Within this framework industrial location choices have been subject to changing decision mechanisms: from a command economy built on Party sovereignty, free will, decisions from the centre and subordination of economic principles, to more democratic, more decentralized economies in which decisions are the result of the interaction of individual, group and area interests with each other and with increasingly rational criteria for assessing economic efficiency. Nevertheless, complete rationality is impossible without rational economic prices and objective decision-makers. Here there is a great contrast between Yugoslavia, where the system permits greater rationality, but where regional pressures cause irrational 'division of the federal investment cake' and the other European socialist republics where the reverse is more true.

The analysis attempts to open up many gaps in our knowledge, especially planner's decision-making strategies and the precise channels of contact and hence influence on decisions. We need to know far more about the scale and process of planned capital mobility within a socialist economy in order to appreciate the extent to which this phenomenon can contribute to the success of new locational patterns. Generally there is insufficient spatial economic data pertinent to investment, production, transport and social benefit-costs. We need more research on the lines of Lacko's railway transportation cost relief map[67] or of Malisz's correlations of city thresholds with costs of city and regional economic development.[68] One suspects that there are more attempts to back even political location decisions by economic rationale than appears superficially. It is also probable that many non-economic location choices, especially for industries not tied locationally by large losses of material weight in processing, are neither more nor less random than most in a capitalist society. One must carefully qualify, therefore, Wiles' contention that 'the most important Communist criterion by far is *local political pull*. Plants go to areas where local officials, especially Party *officials, are important or unscrupulous.*'[69]

NOTES

[1] Frederic J. Fleron, Jr., 'Soviet Area Studies and the Social Sciences: Some Methodological Problems in Communist Studies', *Soviet Studies*, Vol. 19 (1968), pp. 313-39.

[2] Examples are: George W. Hoffman, editor, *A Geography of Europe*, 3rd edition (New York: The Ronald Press, 1969); R. H. Osborne, *East-Central Europe: A Geographical Introduction to Seven Socialist States* (London: Chatto and Windus, and New York: Praeger, 1967); Norman J. G. Pounds, *Europe and the Soviet Union* (New York: McGraw-Hill, 1967); and Margaret R. Shackleton, *Europe: A Regional Geography*, 7th revised edition (London: Longmans, 1966, and New York: Praeger, 1969). To a certain extent this is also true of G. W. Hoffman and F. W. Neal, *Yugoslavia and the New Communism* (New York: Twentieth Century Fund, 1962).

[3] Norman J. G. Pounds, *East Europe* (London: Longmans, Green & Co., Ltd and Chicago: Aldine, 1969).

[4] Norman J. G. Pounds, 'The Industrial Geography of Modern Poland', *Economic Geography*, Vol. 36 (1960), pp. 231-53, and his *The Upper Silesian Industrial District* (Bloomington: University of Indiana Press, 1958), chapter X.

[5] For example, see: Robert Byrnes, editor, *Yugoslavia* (New York: Praeger, 1957); L. A. Dellin, editor, *Bulgaria* (New York: Praeger, 1957); Stephen Fischer-Galati, editor, *Romania* (New York: Praeger, 1957); Oskar Halecki, editor, *Poland* (New York: Praeger, 1957); Jan M. Michal, *Central Planning in Czechoslovakia* (Palo Alto: Stanford University Press, 1960); John M. Montias, *Central Planning in Poland* (New Haven: Yale University Press, 1962); Stavro Skenki, editor, *Albania* (New York: Praeger, 1957); Nicholas Spulber, *The Economics of Communist East Europe* (Cambridge, Mass.: Massachusetts Institute of Technology Press, 1957).

[6] W. Isard, *Location and the Space Economy* (Cambridge: Massachusetts Institute of Technology Press, 1956).

[7] The author's own research has attempted to fill some of the gaps in this respect. See: F. E. Ian Hamilton, *Recent Changes in Industrial Location in Yugoslavia* (Unpublished Doctoral Dissertation, University of London, 1963); 'The Changing Pattern of Yugoslav Industry', *Tijdschrift voor Economische en Sociale Geographie*, 54 (1963), pp. 96-106; 'The Skopje Disaster', *T.E.S.G.*, 55 (1964), pp. 76-9; 'Location Factors in the Yugoslav Iron and Steel Industry', *Economic Geography*, Vol. 40 (1964), pp. 46-64; 'Geological Research, Planning and Economic Development in Poland', *T.E.S.G.*, 55 (1964), pp. 391-3; 'Models of Industrial Location', Chapter 10, in R. J.

Chorlcy and P. Haggett, editors, *Models in Geography* (London: Methuen, and New York: Barnes and Noble, 1967), pp. 381-6; a large number of the issues raised in this paper are touched upon in the author's major work, *Yugoslavia: Patterns of Economic Activity* (London: Bell, and New York: Praeger, 1968). That book also focuses upon the principles, processes and purposes of changing spatially, and, through time, the inherited capitalist pattern of activities.

[8] For example, the first Yugoslav Five Year Plan 1947-51. See: *Petogodišnji Plan Razvitka F.N.R. Jugoslavije, 1947-51* (Belgrade, 1947).

[9] See, for example: *Katalog Privrede F.N.R. Jugoslavije* (Zagreb: Prosvjeta, 1959), 3 vols.; *Rocznik Gospodarczy i Polityczny Polskie Republike Ludowej* (Warsaw: P.W.N., annually); Emil Hoffmann, *COMECON: Der gemeinsame Markt in Osteuropa* (Opladen: C. W. Lerke Verlag, 1961).

[10] While on research in Yugoslavia, the author found that published discussions of the motives of plant location decisions were frequently contained in the monthly engineering journal, *Tehnika*.

[11] Głowny Urząd Statystyczny, *Statystyka Przemsłu; statystyka zakładów przemysłowych i rzemieśliczych* (Warsaw: G.U.S., 1958), mimeographed. Also see: *Rozwój Gospodarczy Powiatów 1950-65* and *Statystyka Powiatów 1966* (Warsaw: G.U.S., 1967).

[12] R. Petrovic, 'Determiniranje Nekih Prostornih Relacija u Komunalnom Sistemu F.N.R.J.', *Zbornik VI Kongres Geografov F.L.R.J. v L.R. Sloveniji*, 1961, pp. 303-14; and R. H. Osborne, 'Economic Regionalization in Bulgaria', *Geography*, 44 (1960), pp. 291-4. For a fuller analysis see: Thomas M. Poulsen's chapter in this book.

[13] The figure of 2·3 millions relates to the present territory of the German Democratic Republic; that in brackets refers to the area of the present G.D.R. plus Upper and Lower Silesia, Lubus, Pomerania and the Mazurian Lakes area which were administered by Germany until the end of World War II.

[14] The figure of 2·0 millions refers to the area of present-day Poland; the figure in brackets is the total of industrial employment within Poland's 1919-39 boundaries.

[15] The population of Poland in 1939 was about 35 million; in 1946 it was only 23 million.

[16] Josef Goldman and Joseph Fleck, *Planned Economy in Czechoslovakia* (Prague: Orbis, 1949).

[17] Hamilton, *Yugoslavia, op. cit.*, pp. 101-2.

[18] The reasons for these differences are complex and have been insufficiently studied. Explanations seem to lie, however, in different

industrial structures, different technological development, the age of industry, the efficiency of organizations and the impact of varied economic systems.

[19] Hamilton, *Yugoslavia, op. cit.*, pp. 234–5; also *Proizvodne Snage N.R. Srbije* (Belgrade: Economic Institute, 1953).

[20] This plan was designed to decentralize the growth of Polish production in strategically important industries – metallurgy, chemicals, engineering – by locating new capacity away from the Upper Silesian Industrial District. This district, being divided between Poland and Germany, was very vulnerable to German attack. A triangular region, lying between Kraków, the San-Vistula confluence, Lvov and the Polish-Czech boundary was thus delimited to receive new capacity. The original document in fact, provided a master-plan for the dispersion of manufacturing throughout south-eastern Poland after 1945.

[21] Emil Hoffmann, *op. cit.*, pp. 27–67.

[22] There is little direct evidence on this point, although it has been frequently heard by the author. For example, the alternatives that confronted planners in Poland in 1949 were to expand steel production in either Szczecin and Gliwice using Swedish ores, or at Nowa Huta using Ukrainian ore. The Russians are alleged to have insisted on the latter choice. However, other evidence would seem to confirm this observation. Namely, that the United Kingdom Board of Trade, acting on behalf of British exporting companies (e.g. I.C.I., Courtauld's, Davy Steel), has shown reluctance to sign agreements for the supply of factories to East-Central Europe unless the locations designated for them by local planning agencies appear to be viable economically. Certainly this has been true of terylene, steel and viscose-cellulose factories supplied to Poland, Romania and Yugoslavia.

[23] R. S. Livshits, *Ocherki po Razmeshcheniyu Promyshlennosti SSSR* (Moscow, 1954), and Y. G. Feigin, *Razmeshechenie Proizvodstva pri Kapitalizmei Sotsializme* (Moscow, 1958).

[24] Kazimierz Secomski, 'Z zagadnień teorii rozmieszczenie sił wytwórczych w gospodarcze socjalystycznej', *Ekonomista*, 2 (1956), p. 4.

[25] V. I. Lenin, *Collected Works*, Vol. 27 (1918), p. 320, and A. M. Ginzburg, *Ekonomija Promyshlennosti II* (Moscow, 1927), pp. 397–8.

[26] Stalin later rejected Weber as a 'bourgeois apologist for capitalism' and support for Weberian ideas became dangerous.

[27] Edgar S. Dunn, 'The Market Potential Concept and the Analysis of Location', *Papers and Proceedings, Regional Science Association*, 2 (1956), pp. 183–94.

[28] Friedrich Engels, *Anti-Dühring*, trans. Emile Burns (New York: International Publisher, n.d.), p. 71.
[29] For a model of this idea see: Chorley and Haggett, *op. cit.*, pp. 384–6.
[30] Hilary Minc, *Osiągnięcia i Plany Gospodarcze* (Warsaw: Ksiazka i Wiedza, 1949), p. 70.
[31] Secomski, *op. cit.*, pp. 5–6.
[32] Hamilton, *Yugoslavia, op. cit.*, Chapters 2, 6, 8, 16.
[33] Radoslav Seluchy, 'The Economic Equalization of Slovakia with the Czech Lands', *East Europian Economics*, 4 (1965), pp. 16–24.
[34] A. Kuklinski, 'Changes in the Regional Structure of Industry in People's Poland', *Geographia Polonica*, II (1967), p. 107.
[35] Hamilton, 'Location Factor in the Yugoslav Iron & Steel Industry', *Economic Geography, op. cit.*, pp. 46–64, and *Yugoslavia, op. cit.*, pp. 238, 257–9.
[36] Hamilton, *Yugoslavia, op. cit.*, pp. 341–2.
[37] This theme is taken up in a nomothetic context elsewhere. See Hamilton, *Yugoslavia*, (1967), *op. cit.*, pp. 407–8.
[38] Hamilton, *Yugoslavia, op. cit.*, pp. 236–9.
[39] For details on the development, growth and purposes of COMECON, see: Michael Kaser, *COMECON: Integration Problems of the Planned Economies* (London and New York: Oxford University Press, 1967), also a forthcoming Searchlight book by R. H. H. Mellor.
[40] James D. Clarkson, *Ecologic and Spatial Analysis: Towards Adaptive Research in Developing Countries* (Honolulu: Social Science Research Institute, Working Papers 7, University of Hawaii, Honolulu, 1967), p. 26.
[41] Examples of such work include: Melvin L. Greenhut, *Microeconomics and Space Economy* (Chapel Hill: University of North Carolina Press, 1964); Melvin L. Greenhut and M. R. Colberg, *Factors in the Location of Florida Industry* (Tallahasee: University of Florida Press, 1962); Walter Isard and Tony Smith, 'Location Games: With Applications to Classic Location Problems', *Papers, Regional Science Association*, 19 (1967), pp. 45–80; and Gunter Krumme, 'Towards a Geography of Enterprise', *Economic Geography*, 45 (1969), pp. 30–40.
[42] P. J. D. Wiles, *The Political Economy of Communism* (Oxford: Blackwell, 1962), p. 48. To my knowledge the only studies of an enterprise functioning in a European socialist country have been made by Pounds, but this is historical not behavioural, and Feiwel whose study is economic not locational. See: Norman J. G. Pounds, 'Fabryka Im. Juliana Marchlewskiego: A Textile Plant in Łódź,

Poland', in Richard S. Thoman and Donald J. Patton, editors, *Focus on Geographic Activity: A Collection of Original Studies* (New York: McGraw-Hill, 1964), pp. 154–8, and George R. Feiwel, *The Economics of a Socialist Enterprise* (New York: Praeger, 1966).

[43] P. J. D. Wiles, *op. cit.*, Chapter 3; and Vladimir Treml, 'The Revival of Soviet Economics and the New Generation of Soviet Economists', *Studies on the Soviet Union*, Munich, 5 (1965), pp. 1–22.

[44] Stefan J. Kurowski, *Szkice Optymistyczne* (Warsaw: Pax, 1957), pp. 292–3.

[45] For this term, and 'co-operative discretion' and 'interdependent discretion' the author is indebted to the research of Julian Wolpert.

[46] S. G. Strumilin, 'Concerning the Problem of Optimum Proportions', *Planovoe Khozyaistva*, 6 (1962) and Wiles, *op. cit.*, pp. 55–6.

[47] For a Yugoslav definition and discussion of 'rentability' see: Hamilton, *Yugoslavia, op. cit.*, pp. 111–15.

[48] The various criteria that are used in different countries are: (1) the index of profitability (profit/total costs of goods sold); (2) rate of profit, (profit/value of fixed and working capital); (3) 'rentability' which is defined as either total costs/total revenue or total income (i.e., total revenue − total costs)/fixed and variable assets. See: Hamilton, *Yugoslavia, op. cit.*, and Vaclav Holesovsky, 'Planning Reform in Czechoslovakia', *Soviet Studies*, 19 (1968), p. 545.

[49] For a definition of these concepts see: Janusz Zielinski, 'On the Theory of Success Indicators', *Economics of Planning*, 7 (1967), pp. 36–45.

[50] Industry in northwestern Yugoslavia has expanded almost more rapidly than elsewhere in the country, primarily because of the decentralized response to semi-controlled pricing systems which in turn have been subject to differential rates of inflation. See: Hamilton, *Yugoslavia, op. cit.*, p. 148; also, Eugeniusz Szyr, *Niektore Problemy Rozwoju Gospodarki Narodowej w Latach 1959–1965* (Warsaw: Ksiazka i Wiedza, 1959).

[51] Hamilton, *Yugoslavia, op. cit.*, Chapter 12.

[52] Hamilton, *Yugoslavia, op. cit.*, pp. 288–94. In the discussion of this paper Professor Earl Brown raised the point that there might be differences in transport rate structures and rates per ton-mile between the East-Central European States. This is a field which needs to be investigated. Professor Chauncy D. Harris, however, considered any analysis of transport costs in command economies as virtually futile because arbitrary tariffs are often used as accounting devices, not rational economic indicators of transport costs.

[53] This not only includes housing but also all extensions to public utilities, cultural facilities and, where necessary, commuter bus transport facilities especially where factories have to provide their own buses because of exceedingly scarce public transport facilities (e.g. in the Bitola region of southwestern Macedonia).

[54] Włodzimierz Brus, *Ogólne Problemy Funkcjonowania Gospodarki Socjalistycznej* (Warsaw: P.I.W., 1961), p. 117.

[55] Iwan S. Koropeckyj, 'The Development of Soviet Location Theory', *Soviet Studies*, 19 (1967), pp. 1–28 and 19 (1967), pp. 232–44.

[56] Henri Smotkine, 'Un type de complexe industriel: le district de Karl Marx Stadt en République Democratique Allemande', *Annales de Geographie*, 414 (1967), pp. 152–67.

[57] H. Gordon Skilling, *The Governments of Communist East Europe* (New York: Crowell, 1966), p. 86.

[58] W. Eucken, *Grundsätze der Wirtschaftspolitik* (Tubingen, 1952), p. 79.

[59] Skilling, *op. cit.*, p. 91.

[60] A lengthy discussion of these and related problems of vested interest and behaviour in Yugoslav locational planning is contained in Hamilton, *Yugoslavia, op. cit.*, pp. 138–53, 235–40, 251–2, 282–3, 292–3 and 350–62. In summary, it is useful to note that conflict between regional vested interests is best studied in Yugoslavia. However, the system there of open competition between factories and communes for limited investment funds from the investment bank, which operated between 1956 and 1965, is a fertile and largely virgin field of research into (1) the motivation of worker's councils and people's councils in seeking an expansion of industrial capacities, (2) the behaviour of bank experts in choosing projects for investment, (3) the ways in which factory and commune representatives brought pressure to bear upon bank officials and (4) the interaction of the same bank officials with federal republic planners and the political leaders in counteracting such pressures.

[61] For example, see: Hamilton, *Yugoslavia, op. cit.*, pp. 248–50.

[62] Hamilton, *Yugoslavia, op. cit.*, pp. 142–3.

[63] An example is provided by a programme for the location of growth and change in the cement industry in Poland for the period 1965–80. See: Jan Zurkowski, 'Programowanie Lokalizacji Produkcji', *Studia Komitetu Przestrzennej Zagospodarowania Kraju*, 23 (Warsaw, 1968), p. 169.

[64] This is a common weakness – as relevant in Poland with its high educational traditions as in Yugoslavia with its tradition of illiteracy. Examples of such guide books are: for Poland, Komisja Planowania Przy Radzie Ministrow, *Instrukcja Ogolna w Sprawie*

Investicja w Gospodarce (Warsaw, 1960); for Yugoslavia see: Momčilo Pejovic and Radoslav Nikotić, *Priručnik za Investitore* (Belgrade: Nolit, 1958), 443 pp.

[65] For a discussion see: Allan Pred, 'Behaviour and Location', *Lund Studies in Geography, B: Human Geography*, 27 (1967).

[66] Wiles, *op. cit.*, p. 319.

[67] Laszlo Lacko, 'Toward the Development of New Economic Maps', *Tijdschrift voor Economische en Sociale Geografie*, 60 (1969), pp. 24–31.

[68] B. Malisz, 'Ekonomika Ksałtowania Miast', *Studia KPZK* (Warsaw, 4), 1963.

[69] Wiles, *op. cit.*, p. 148.

Comments

S. EARL BROWN

Dr Hamilton has provided an excellent analysis of industrial location in Eastern Europe which is consistent not only with modern tendencies in our discipline, but also with his personal concern for such an approach. Such concern is clearly evident in his major contribution to *Models in Geography*. The attempted application of normative models of location theory modified by the behavioural aspects of decision making presently provides an exciting and challenging research frontier in location analysis. Such work is difficult at best and certainly is complicated by the political environment of the socialist societies with which we are concerned. In this valiant, if not completely successful, attempt to be 'nomothetic' rather than 'idiographic', he has provided a refreshing perspective on the problems of industrial location and presented a number of interesting ideas, several of which might well be expanded by coping with the following questions.

First, some elaboration on the high labour productivity in Romania and Bulgaria and the low productivity in Hungary revealed in Table 5.1 would seem to be desirable. What explanations exist for these characteristics? Does Romania's emphasis on petroleum and petro-chemical industries, highly capital-intensive rather than labour-intensive, suffice to explain the figure? Conversely, is Hungary's low position a reflection of a limited resource base for capital-intensive industry and thus a forced labour-intensive position? If so, how unfortunate then that Hungary, with its low rate of population growth, seems saddled with this type of industrial structure whereas Romania, which seems to need the labour-intensive type of structure, finds it lacking. What implications for the future of manufacturing in Hungary can be derived from this data? What changes in rank of individual countries would appear if the ratio were based on capital investment to production rather than employment to production?

Second, it might be useful to indicate what methods are used in computing transport costs in these individual political units.

To what extent are product differentiation, the tapering principle, and other aspects of capitalist transport economics treated differently in this area? More importantly, how does the application of the socialist transportation cost structure affect any attempted application of Weberian location theory? How will the introduction of such data into computer programs affect the comparability with Western studies using similar mathematical techniques?

Certain of the concepts presented provide fertile ground for further testing and development in the area of Eastern Europe and elsewhere. Thus, it is hoped that Dr Hamilton or others working in the field of industrial location, might cope with the problems posed in his discussion of the 'spatial elasticity of investment plans'. It is a concept which is fundamental to the discipline of geography and which lends itself very well to useful comparative studies which have long been a part of our field. Most assuredly it has important implications in its attempts to satisfy the demands in the socialist societies for greater industrial dispersion.

Similarly, the concepts of 'dispersed localization' and 'regionalized dispersion' could be applied with benefit to comparative studies within our area of concern as well as elsewhere. The problems of industrial concentration and dispersion are universal and such comparative study might well yield results which are useful both to socialist and capitalist societies.

It should not be surprising that the comments thus far are based predominantly on the first part of the paper dealing with the principles and effects of socialist location policy. Nor should it be surprising that the behavioural section of the paper, though of great interest, is the less satisfactory portion of the study. This area represents, at least for the industrial geographers and location theorists, a relatively new approach to our studies. Such excursions into virgin territory are always difficult and fraught with frustrations. One only has to read the two recent reviews of Alan Pred's pioneer work in *Behavior and Location* to be aware of the pitfalls involved in such new research. Yet Dr Hamilton has attempted to apply Pred's behavioural ideas, although not his behavioural matrix, to the emerging industrial pattern of the area. Thus, the ideas bounded rationality' and of the 'adaptive–adoptive' dichotomy are referred to in analysing locational decisions. Both concepts appear to have considerable relevance to seemingly irrational location decisions in Eastern Europe. The unsolvable problem of defining the bounds of rationality arises. What are the limits to the rational application of the nine principles of location theory used in the socialist economy? Which of the locational decisions reflect the attempted application of the principles and which reflect reaching the bounds of rationality? Where does one place the locational decision of the Košice steel works in this classification? Or would its classification depend on the time the decision was made, thus complicating the picture further with the temporal dimension which Pred promised in a future study? Since we cannot enter the minds of the decision-makers, it becomes impossible to answer such intriguing questions.

While analysis of the nine location principles may be objective, analysis of the behavioural interpretation of these principles by mere man – be he communist or capitalist – makes impossible any comprehensive, rigorous and generally applicable location theory in Eastern Europe. This Dr Hamilton readily admits. That he has made an attempt in this direction is indeed commendable. He has not only provided a useful review of the problems of industrial location in our area of concern, but he has also provided an important base for future study. In so doing he has taken a significant step in answering Fleron's challenge for a more nomothetic, social scientifically oriented area study and we are indebted to him for this contribution.

Comments
KAREL J. KANSKY

In this very welcome paper Hamilton properly insists on a nomothetic approach to the process of industrial location. If we accept the frequently mentioned definition that 'a nomothetic approach seeks to establish abstract general laws for indefinitely repeatable events and processes'[1] then Hamilton's paper has to be classified as more of an idiographic rather than nomothetic nature.

The paper seems to me to serve four purposes satisfactorily:

1. to draw attention to the problem of explaining changes in the spatial patterns of industry in East-Central and Southeast Europe;
2. to provide valuable information on past and present spatial distribution of industry in the European socialist republics;
3. to identify the locational strategies as utilized in the European socialist republics after World War II;
4. to propose a behavioural approach as the best-suited method to formulating generalizations about the process of industrial location in East-Central and Southeast Europe.

Less satisfactory is the fact that no attempt has been made in this paper to propose a well-defined theory of socialist industrial location. Geographers' capabilities in formulating well-defined theories are limited because of lack of experience; and because of lack of sufficient and reliable data it is almost impossible to construct a theory. Nevertheless, daring and innovative experiments would be highly welcome by a large portion of the profession.

1. There is no need at this time to discuss the proposition that area specialists tend to be more idiographic rather than nomothetic in their studies, or to argue whether or not the

[1] E. Nagel, *The Structure of Science* (New York; Harcourt, Brace and World, Inc., 1961), p. 547.

spatial distribution of industry of a geographic area can be explained in theoretical terms.

2. Nor is there a need here to discuss in detail the impressive collection of facts, or the reliability of data or the bibliographic sources. The past and present industrial distributions, as given by the author, seem to be complete. They should be recognized as a remarkable source of information and as such an important contribution to our knowledge of East-Central and Southeast Europe. The selection of sources of numerical information as well as the author's knowledge of the literature clearly indicate a high degree of reliability of the presented industrial distributions. The list of library sources, relevant to the topic, is much longer than any list I have ever seen before. In discussing the changing industrial distribution Hamilton interprets the spatial shifts correctly.

3. The main thrust of Hamilton's paper lies in his attempt to identify the 'socialist' locational principles and to estimate the degree to which the principles were applied in the European socialist republics. The author lists nine locational principles: 1. proximity to natural resources; 2. proximity to markets; 3. interregional proximity to natural resources; 4. interregional proximity to markets; 5. even distribution of industry; 6. socio-economic development of minority groups; 7. elimination of socio-economic differences between town and country; 8. defence needs, 9. interdependence within COMECON.

There is little to say about Hamilton's discussion of the applicability of the above principles. His knowledge of the subject matter and his verbal interpretations of the available information is satisfactory. The only thing that is missing is an organized summary of his analysis. More specifically, the interpretation could have resulted in a formal scheme that would integrate the widely dispersed conclusions on the principles' applicability. What I have in my mind is the following: since Hamilton is examining the role each locational principle has played in influencing the spatial shifts in industrial distributions in each of the European socialist republics, his method was that of a correlation analysis of qualitative data. More precisely, he has attempted to relate a set of locational decisions Y_i, where $Y_i .. (y_{1'}, y_{2'}, ..., y_n)$ to a set of variables X_j, where $X_j = (x_{1'}, x_{2'}, ..., x_m)$. That is to say, that each locational decision Y_j is viewed as a dependent variable that is functionally related to a set of independent variables X_j. Hence, a table or matrix M could have been constructed such that the entries of the matrix would express the importance of the x_j variable in deriving the y_i locational decision. The degree of importance of each variable x_j could have been expressed along a nominal, ordinal, or ratio scale. Even by expressing the degree of importance along a nominal scale (e.g., the score of 1, if the x variable has been considered in deriving the locational decision y_i and the score of 0 if the variable x_j has not been considered in deriving the locational decision y_i), the matrix M would make it possible, for instance, to establish a rank order among the locational principles. Similarly, the matrix could have been used to establish a rank order among the European socialist republics according to their locational strategies.

The matrix M would also facilitate an easy identification of missing data. Another possible mode of analysis would be a comparison between columns of the matrix M. The purpose of this examination would obviously be to group those variables that are correlated; or, in other words, those variables that exert the same kind of influence, that are complementary to each other. And similarly, by comparing the rows of the matrix M one could define groups of regions in which the same locational strategies were applied.

4. There is no doubt that the process of locating industry can be approached from a behavioural point of view. The industrial location choice is always made by 'man or groups of men', hence, man or a group of men can be accepted as the basic unit of research, and his or their behaviour described or analysed. Although the author is promising an outline of a 'behavioural framework' the outline is not included in this paper. The discussion that follows the title (heading) 'The Economic System and Human Response' is not of a behaviouristic nature. The actual behaviour of man or of a group of men is not described and analysed. Only several groups of decision-makers are identified (for example the Party, planners, ministers, managers of industrial associations, etc.) and some locational decisions hypothetically ascribed to them. The roles of the groups are not detailed. It seems to me that the 'behaviouristic' part is very similar to the 'non-behaviouristic'

part of Hamilton's paper. As far as the structure of the argument is considered, in both parts the locational decisions are viewed as dependent variables related to a set of independent variables. The only difference between these two parts is that some additional independent variables (locational principles) are suggested and discussed in the latter part of the paper (for example, conflict between state and church, conflict interdependence, etc.). The concept of 'The Spatial Elasticity of Investment Planning' is worth mentioning here. This seems to be an interesting notion to which Hamilton might address himself in the future. At the present this concept is not well-defined and its empirical applicability is not obvious to the discussant. I suspect that the idea is inspired by economics. Also, the entire 'behaviouristic' section of the paper, including the sections titled 'The Making of Decisions' and 'Rationality and Location Decisions', resembles an economist's approach. What I mean is, that both the concept of spatial elasticity and the behavioural section of the paper read more like a paper written by a regional economist rather than one written by a behavioural scientist.

Hamilton's paper does not terminate with a theory of socialist industrial location. But very few, if any, papers written by geographers attempt to reach that difficult goal. It is obvious that before geographers will be able to define specific or general theories they have to desire to do so. This paper illustrates that Hamilton has a desire to reach that difficult goal. It demonstrates also that the author has a capability to formulate a well-defined theory of socialist industrial location.

The discussant's opinion is that at the present time the behavioural approach to formulating a theory of socialist industrial location cannot be fruitful. The lack of data on decision-makers' behaviour and the low degree of reliability of most of the socio-economic data originating in European socialist republics are the principal but not only reasons. The paper itself supports this opinion. Hence, a purely spatial approach to the changing industrial distributions might facilitate the formulation of a theory. The nearest neighbour analysis, diffusion models, or surface fitting perhaps could identify some spatial regularities in industrial distributions that could be included as building stones of a theory. It is reasonable to believe that complex and possibly not entirely rational locational analysis of the socialist decision-makers could generate some spatial regularities in the industrial distributions. Usually the black box approach to formulating theories of complex phenomena is the method most fruitful at the beginning stage of theory formation.

Remarks

In Dr Hamilton's response to the comments following his paper, he stated that he realized the problem of the behavioural approach and that he was only a beginner in the field. He pointed out the lack of a socialist location theory, even in the socialist countries and that even though socialist planners had access to materials, the political environment and pressure of planning on time made such work difficult.

Dr Hamilton also pointed out that another difficulty in this work was the fluidity of the social and economic situation and agreed with Professor Kansky's suggestion that such methods as the diffusion of ideas and nearest neighbour analysis be used, but pointed out that there were problems in trying to 'marry the normative and behavioural approaches'. He considered Professor Brown's idea of computing transportation costs and the problems which would come up due to the difference between East and West.

Regarding the question of breaking the restraints in COMECON, Dr Hamilton said that the problem was difficult to solve since it raised the issue of supranationality.

Professor Harris commented on the excellent quality of the paper which covered new ground in difficult terrain. Due to the difficulty in distinguishing between the political exhortations of the Soviet aims and the diverse political methods of Eastern Europe, Harris thought Hamilton's emphasis on administrative decisions a sound one. He thought transportation cost analysis useless due to the arbitrary freight rate structure.

Professor Harris then indicated that a major question in Eastern European planning was whether or not to concentrate on backward areas, and that since this decision is *always* a political or administrative one then location analysis can illuminate but not pinpoint the exact decision. Even with these problems, Professor Harris continued, the difference between capitalist and socialist countries is not as strong as is often thought, i.e. the bureaucratic structure of the OEO in the

United States. Therefore, the paper, in his opinion, is 'good in recognizing the complexity of the problem and in not offering a clear solution as there isn't one'.

Dr Rugg then asked Dr Hamilton's statistical sources and also to what degree social attitudes towards underdeveloped areas was prevalent in Eastern Europe and the Soviet Union.

Dr Hamilton answered saying that his data was personal computation gleaned from statistical yearbooks; as for backward areas, it is up to the planners to mediate between the aim of social equality and the benefit/costs of investment in these areas. He felt that this is the question many socialist planners have not approached, and that this was especially true in 1957–64, but less so now. He pointed out that the backward areas of Eastern Europe and the U.S.S.R. differ in that those of Eastern Europe are heavily populated and that labour resources are an additional incentive to development whereas most in the U.S.S.R. are sparse or uninhabited. He also stated that the issue varies from country to country.

Dr Karger agreed with Professor Harris that it is difficult to extract individual location theories not only because of the variety of environments but because of rapidly changing conditions, not to mention principles.

Dr Hamilton commented that the important point was the shift in recent times from indefinite administrative approaches to approaches based on definite criteria and therefore interpretative principles. This denotes the gradual realization in Eastern Europe that economics is an objective science and socialist countries only differ from capitalist countries in their political framework.

Professor Hoffman then commented that much too often the location of specific industries is based on a single objective in order to be sure that individual districts can show that they, too, have an industry. Often this decision is made without considering such facts as an existing infrastructure, including labour supply and education. The literature of the socialist countries and even talks by political leaders is full of statements emphasizing the need for industrial dispersion, though some countries, lately, such as Romania and Bulgaria, have given more serious attention to locational factors.

6

Administration and Regional Structure in East-Central and Southeast Europe

THOMAS M. POULSEN

Although bureaucracy is one of the most characteristic phenomena of the complex world of modern society, few geographers have considered its relation to their field of interest. Most of the available literature on bureaucratic organization has been produced by political scientists, sociologists and scholars in the field of business administration. They have been concerned with such topics as the formal structure of bureaucratic decision-making, distortion of information due to an excess of organization levels, or the role of staff in influencing the policy of organizations.[1] Bureaucratic administration does have its specific geographic aspects, however. These include such problems as the delimitation of field service regions, the territorial co-ordination of agencies, and the impact of administrative organization upon patterns of circulation, manufacturing and location of settlements. The geography of bureaucratic organization is particularly pertinent to the study of communist states where such a large share of economic, social and political activities have been formally structured by their regimes. It also has relevance to the geography of the non-communist world, which is now in a period of increasing governmental participation in economic activities and steady combination of private enterprises on national and international scales. Decision-making is more and more being transferred from a multitude of entrepreneurs at the local level to the apexes of bureaucratic pyramids. This increasing centralization of authority has far-reaching implications for traditional geographic concepts and models of spatial form and interaction.

This paper examines the impact of administrative organization and decision-making upon the economic, social and political geography of the countries of East-Central and Southeast Europe and considers to what extent the geographic elements of these countries have had modifying effects upon structures of bureaucracy. It also suggests some implications of the weakening of highly centralized decision-making in the communist states to future patterns of spatial interaction.

The study is based upon observations made in Poland, Czechoslovakia, Hungary, Romania, Yugoslavia and Bulgaria during the 1966–7 academic year. A number of problems presented themselves, not the least of which was the difficulty of obtaining precise data and documents concerning the actual reaching of decisions. Moreover, there have been few studies in either the East or the West that specifically examine the nature of public administration in a socialist society. Western political scientists examining Eastern Europe have been concerned with questions involving the nature of power, politics within ruling parties, or the formulation of internal and external policy, and not with the day-to-day work of administering a socialist complex of economic, social and political functions.[2] Similarly, with the notable exceptions of Shabad, Fisher and Helin,[3] few geographers have examined the phenomenon of communist territorial administration. Because of the paucity of data and sparseness of convergent research, the present study is introductory and suggestive rather than definitive and exhaustive.

REGIONAL STRUCTURE BEFORE THE ESTABLISHMENT OF COMMUNIST REGIMES

Throughout the countries of East-Central and Southeast Europe the element of administration has long had a significant role in influencing patterns of spatial organization. With the general low level of industrial development in these outlying territories of the nineteenth-century Russian, German, Austro-Hungarian and Turkish empires, administration was a principal basic economic activity of a majority of provincial capitals. The judicial, fiscal and organizational functions of these cities, moreover, made them focal points of regional circulation. Most were historic centres that had come into prominence in the eighteenth and nineteenth centuries with the increasing administrative rationalization of the imperial governments.

When industry above the handicraft level began to develop after the middle of the last century, mainly through foreign investment, it tended to gravitate to those towns which were administrative centres, and thus reinforced their local pre-eminence. Such towns offered advantages of established amenities, routes of communication and other elements of what Wrobel terms 'economies of common location'.[4] Only a few industrial nodes developed independently of administrative centres, chiefly in Silesia (Katowice, Ostrava). At a lower level of spatial organization, patterns of circulation for marketing and social affairs were generally constricted within communes, the origins of which could often be traced to church parishes established in the Middle Ages.[5]

Following World War I the comparative advantages of towns designated as administrative centres increased. The capitals of the successor states became typical primate cities, while provincial and district centres benefited from the efforts of the new governments to stimulate lagging economies through direct investment and management of industrial plants and commercial enterprises.[6] This was particularly true in Czechoslovakia, where government agencies in provinces and districts controlled factories, savings banks and rural electrical networks, and local commune governments were proprietors of bus lines, breweries, apartment houses and dairies.[7] In Yugoslavia the central government invested in steel, machinery and paper plants, and it nationalized such industries as coal mining, railways and tobacco processing.[8] The greater part of governmental activity, however, was devoted to the conventional tasks of preserving public order, maintaining the security of the central authorities and collecting taxes. It should also be noted that there was increasing scope for private initiative. Several spatially significant developments occurred, including the transformation of the Czech town of Zlin into a major industrial centre by the Bat'a shoe manufacturing firm.

The spatial structure of administration in the interwar period was initially based upon the territorial units inherited from the past. A major problem was a contrast in forms and sizes of administrative regions, since several of the new states had inherited units from two, or in the cases of Poland and Romania three, different imperial organizations. The authorities of the successor states were also concerned about the regional antagonisms that they had inherited. Such differences were often related territorially to patterns of internal subdivisions, particularly when such subdivisions had boundaries that had once been international ones. In order to provide more orderly administration and reduce internal disunity, some reorganization of territory took place in the interwar years affecting various levels of administration in nearly all of the countries examined. Thus Romania changed its local government areas in 1920, 1925, 1929, 1931 and 1938.[9] In Yugoslavia, several internal reorganizations also occurred, culminating in nine *banovinas* that were established in 1931 and neutrally named after river basins in order to weaken the nationalisms of the major ethnic groups.[10]

The chief impact of territorial reorganizations was upon cities and towns that lost or gained governmental functions. Several centres that saw the elimination or reduction of their areas of administration began to decline as central places of economic and social activity. These included the towns of Siedlce in Poland, Vidin in Bulgaria, and Pozega in Yugoslavia. A number of cities, however, gained new centrality from being named to head administrative areas, and this had a tendency to reshape local patterns of spatial interaction to a significant extent. Among such examples were Ostrava in Czechoslovakia and Craiova in Romania.

THE COMMUNIST ERA

The political and economic organization of Eastern Europe was fundamentally changed following World War II. Throughout the area regimes dominated by communist parties and oriented to the U.S.S.R. were established. Although non-communist elements were not completely eliminated until 1948, governments were early modelled on the forms of the highly centralized public administration and economic management prevailing in the U.S.S.R.

This Soviet model of organization should not be viewed as a direct product of Marxian ideology nor as a management system carefully designed by social scientists. Rather, it was a collection of *ad hoc* practices that had emerged over the previous quarter century of Bolshevik rule. Many standard operating procedures could be traced to the tsarist period and even to patterns of organization of the Mongols and Byzantines.[11]

From its inception the Soviet system of government and economic management had been one of centralized bureaucracy. It involved a series of parallel line hierarchies allocating authority from centre to locality through offices in the centres of provinces and districts. Local government in the English or American sense really did not exist; local affairs were managed simply through the execution of Moscow-determined policy by the lowest tier of bureaucrats.[12]

The ramifying bureaucratic structures were co-ordinated by being fitted into identical administrative-region moulds, each of which was supervised by communist party officials controlled through the separate hierarchy of the party secretariat. Lines of responsibility of administrative officials to their superiors at higher levels and to party secretaries in their province and district were never clearly stated, and this pressure both from the top and from the side was one of the elements that kept the bureaucracy from becoming an independent, conservative force with a self-perpetuating momentum comparable to governmental machinery in France and other countries.

The establishment of Soviet-style regimes in the countries of East-Central and Southeast Europe had many immediate effects on their economic geography. A pronounced feature was the change in the 'rules of the game' for the location of economic activities. Differential competition for land and maximization of profits lost their determining roles. The initiatives of a multitude of decision-makers at the enterprise and consumer level were replaced by a structured chain of command in which political consideration, often of a personal and quite irrational nature, became dominant. Moreover, an evolutionary development of economies was replaced by a forced march towards industrialization. A host of long-range patterns of spatial interaction were set into motion in a relatively short time.

One profound effect of the bureaucratization of decision-making was the enhanced significance given to administrative regions. These areas became the spatial framework of decision-making for a far greater range of operations and activities than they had ever had before, particularly in the economic sphere.

A Yugoslav study recently pointed out that typical communes had more than two hundred and fifty separate governmental functions in the broad categories of general law, economy, finance, city planning and construction, education, culture, health, social policies, labour and labour relations, communal service, internal affairs and aspects of military administration.[13]

REDISTRICTING ON THE SOVIET MODEL

The inherited frameworks of administrative regions were certainly not ideally designed for the demands placed upon them. Substantial changes in provinces and districts were introduced in each country by 1951, most likely to make them more amenable as vehicles for co-ordinating bureaucracy. Table 6.1 summarizes this immediate post-war redistricting.

TABLE 6.1 – *East-Central and Southeast Europe: Summary of Changes in Administrative Regions, 1945–1951*

BULGARIA	*Pre-Communist Pattern:* 10 provinces (*oblasts*), 95 districts (*okolias*), c. 2,000 communes (*obshchinas*)
	December, 1947: Elimination of all provinces (*oblasts*), leaving country divided directly into 95 districts
	September, 1949: Creation of 14 provinces (*okrugi*)
	1951: Two provinces (Vidin and Yambol *okrugi*) eliminated, leaving 12
CZECHOSLOVAKIA	*Pre-Communist Pattern:* 3 provinces (*zeme*), 269 districts (*okresy*), 142,000 communes (*obce*)
	December, 1948: Replacement of *zeme* by 19 new provinces (*kraje*)
HUNGARY	*Pre-Communist Pattern*[2] 27 provinces (*megyes*), 113 districts (*jurases*), c. 3,000 communes (*közössegs*)
	December, 1949: Reduction of provinces to 19 by combining two or three units; several shifts of districts to adjacent provinces
POLAND	*Pre-Communist Pattern:* (in 1939 boundaries) 16 provinces (*wojewódstwos*), 264 districts (*powiats*), 14,609 communes (*gminas*)
	1945–6: 17 provinces (*wojewódstwos*) established, of which six were in annexed German territories; boundaries of ten pre-war provinces remaining to Poland substantially changed, and one new province (Bydgoszcz *wojewódstwo*) established from portions of four others; 317 districts (*powiats*) existed by 1949, subdivided into c. 3,000 communes (*gminas*) and 40,000 localities (*gromadas*)

ROMANIA	*Pre-Communist Pattern:* (after loss of Bukovina, Bessarabia and Southern Dobruja) 58 provinces (*judeteţ*), 421 districts (*plasis*), 6,248 communes (*communas*) *September 1950:* Establishment of new framework of 28 provinces (*regiunes*), 177 districts (*raionuls*); and 4,052 communes (*communas*)
YUGOSLAVIA	*Pre-Communist Pattern:* 9 provinces (*banovinas*), 378 districts (*srezovi* and *kotari*), 4,645 communes (*opštinaš* and *općine*) *1946–51:* frequent changes took place at all levels within framework of 8 provinces based on nationality groups: republics of Serbia, Croatia, Slovenia, Bosnia and Herzegovina, Macedonia, and Montenegro, plus the autonomous province of the Vojvodina and the autonomous region (*oblast*) of Kosovo and Metohija within the Serbian republic; end-of the year totals were: 1946 – 50 sub-provinces (2 *oblasts* and 48 *okrugi*), 407 districts (*srezovi* and *kotari*), 11,566 communes (*narodni odbors*); 1947 – 340 districts (2 *oblasts* and 338 *kotari* and *srezovi*), 7,866 communes (*narodni odbors*); 1948 – 340 districts (1 *oblast* and 339 *srezovi* and *kotari*), and 7,967 communes (*narodni odbors*); 1949 – 23 sub-provinces (*oblasts*), 344 districts (*srezovi* and *kotari*); 7,782 communes (*narodni odbors*); 1950 – 20 sub-provinces (*oblasts*), 360 districts (*kotari* and *srezovi*), 7,102 communes (*narodni odbors*); 1951 – 360 districts (*srezovi* and *kotari*), 7,104 communes (*narodni odbors*)

This table was compiled from a variety of sources. The standard statistical yearbooks of each country generally include a section describing the 'administrative-territorial regionalization' of the state, although the number of communes is often omitted. The Soviet geographer, S. F. Burenko, published detailed descriptions of the 1949–51 changes in Bulgaria, Czechoslovakia, Hungary and Romania in *Izvestiya Vsesoyuznogo Geograficheskogo Obshchestva*: 'Novoye Administrativnoye Deleniye Narodnoi Respubliki Bolgarii', Vol. 83, No. 4 (July–August, 1951), pp. 386–8; 'Novoye Administrativnoye Deleniye Chekhoslovakii', Vol. 83, No. 5 (September–October, 1951), pp. 522–7; 'Novoye Administrativnoye Deleniye Vengrii', Vol. 83, No. 4 (July–August, 1951), pp. 388–90; 'Novoye Administrativno-Ekonomicheskoye Raionirovaniye Narodnoi Respubliki Rumynii', Vol. 82, No. 4 (July–August, 1950), pp. 623–5. Polish changes are described in Yu. V. Ilinin, *Pol'sha* (Poland) (Moscow: Mysl, 1966), pp. 7–12. An exhaustive study of modifications of the administrative map of Yugoslavia is contained in Irina Perko-Separović, *Veličina Lokalnih Jedinica* (The Sizes of Local Units), (unpublished doctoral dissertation, University of Zagreb, 1965).

The reasons for specific changes in this period are difficult if not impossible to establish. Although discussions in the nominal parliaments were published by the countries concerned, they were generally in the nature of rationalizations or slogans concerning the need to 'harmonize economic and administrative regions' or 'bring government closer to the people'. So far as the author has been able to ascertain, the stenographic reports of discussions in higher party councils for the period in which policy was established have not been released.

However, it is instructive to compare the new patterns of provinces among themselves, and also with their counterparts in the U.S.S.R. The results of such comparison are summarized in Table 6.2.[14] Because of the great variations in the geography of the Soviet Union, especially its large area of non-oecumene, it was deemed useful to utilize data from administrative regions in the comparably situated and populated Ukraine, in addition to averages for the U.S.S.R. as a whole. The Ukraine has the advantage for comparison of being an ethnically defined unit with problems of post-war political control and rehabilitation comparable to that of its neighbouring 'people's democracies' to the west.

From the table it is clear that with the possible exception of Poland, the provincial-rank units were comparable neither in size nor in population with each other nor with *oblasts* in the U.S.S.R. Except for the Polish *voivodships*, the revised provinces had less than half the numbers of people and amount of area of their Ukrainian counterparts and were generally closer to one-third. On the other hand, at the district level the difference between countries was not as great. The population and area of Ukrainian *raions* in the U.S.S.R. fell approximately in the middle of the range among the East-Central and Southeast European countries. At the provincial level one noteworthy similarity concerned the total numbers of units. These ranged from 14 in the case of Bulgaria to 28 for Romania. In comparison, there were 26 provinces in the Ukraine. The average number of district units within a province was far less uniform, ranging from 6 to 19. This was substantially below the 28 districts (*raions*) per province average in the Ukraine. At the commune level the number found within a district approximated the Ukrainian commune (*selsoviet*) average of 23 per district with the exceptions of 50 in Czechoslovakia and 9 in Poland.[15]

The similarity with the U.S.S.R. in total numbers of provinces per country rather than in their sizes or populations reflects the purpose of provinces in the immediate post-war years. The communists had consolidated power by 1948, eliminating their

TABLE 6.2 – *East-Central and Southeast Europe: Characteristics of Provinces and Districts, 1948–1950*

Country	Provinces	Average km²	Average Pop.	Average Number of Districts Per Province	Total Districts		Average km²	Average Pop.	Average Number of Communes per District
Bulgaria	14 *Okrugi*	7,856	502,000	6	95	*Okolias*	1,157	72,000	21
Czechoslovakia	19 *Kraje*	6,736	649,000	14	269	*Okresy*	478	46,000	50
Hungary	19 *Megyes*	4,894	484,000	6	113	*Jarases*	823	81,000	25
Poland	17 *Voivodships*	18,300	1,465,000	19	317	*Powiats*	984	79,000	9
Romania	28 *Regiunes*	8,464	567,000	7	177	*Raionuls*	1,340	90,000	23
Yugoslavia	24 *Oblasts*	11,130	736,000	16	344	*Kotari/Srezovi*	744	46,000	24
U.S.S.R.	157 *Oblasts*, etc.	142,700	1,274,000	28	4,368	*Raions*	5,130	45,000	18
Ukrainian S.S.R.	26 *Oblasts*	23,100	1,562,000	28	726	*Raions*	828	55,900	23

For sources see Table 6.1.

last effective political opponents. They saw their main task as establishing and maintaining effective control over the countryside from the centre, particularly with the revolutionary task of agricultural collectivization on their agendas. They lacked large numbers of skilled cadres, however, and had to rely upon holdover officials for day-to-day government housekeeping, especially in the outlying areas. Strong central control over politically unreliable local officials demanded close supervision in the field. At the same time there was an upper limit to the number of supervisory officials who could be effectively co-ordinated from the centre.[16]

In American corporations an optimum span of control has been considered to be in the range of five to ten persons per supervisor.[17] However, Soviet experience had most likely indicated that the number could safely be expanded to 25 or 30, a figure representative not only for *oblasts* in the Ukraine, the most populous republic outside of the sprawling RSFSR, but also for *raions* within *oblasts*. Beyond this number there were probably too many instances of 'harassed supervisors and frustrated subordinates'.[18] That such a broad span itself could be tolerated reflects two special circumstances: the near-identical tasks of the party secretaries and the fact that there were few cross communications and accompanying possibilities for dispute between them. The task of a supervisor or top executive, after all, is not just to issue orders but to resolve or adjudicate conflicts among subordinates. Such conflicts are more likely to arise when subordinates represent different 'lines' of endeavour. In a corporation such different 'lines' could be functional differences such as sales, accounting, processing, etc., or different types of production such as synthetic textiles, tyres, and plastic kitchenware in a chemical firm. In such a situation each subordinate's position presents a unique set of problems. Problems are compounded by the possibility or need for contacts across the different line hierarchies at lower echelons. In the young socialist regimes, however, economic management was secondary to political imperatives based on one-party authority. Assignments to the party officials in the provinces from the central committee were identical. Moreover, disputes were few, because party secretaries were in geographically dispersed locations, with relatively simple and direct lines of communication to the capitals. Such a situation is condusive to broader spans.[19]

It should be noted also that a broad span of control was preferable to establishing an intermediate link in the territorial-administrative hierarchy that would permit closer supervision through narrower spans. The reorganization of economic and social activities and establishment of central planning required a high degree of operational autonomy at the province and district levels. The secretary had to have room to manœuvre, with freedom for his staff to co-operate and reach decisions on the site, rather than being subject to bureaucratic delay and distortion of information as essential communications with the centre filtered through intermediate steps.

When uncertainty prevails, potential relationships among the possible components of a task cannot be foreseen accurately. Hence the task cannot be divided into many parts assigned to specialists unless the specialists are in constant communication with each other and can continually redefine their relationships as they gain more knowledge. This requirement is best served by a flat hierarchy, since it provides greater authority to each official and allows greater emphasis upon direct horizontal relationships. These factors are essential because:
1. Each official must be free to co-ordinate directly with a great many others in unpredictable ways, so formal channels cannot be set up in advance.
2. The need for dialogues among officials and for constant redefining of tasks makes working through intermediaries inefficient.
3. Communications among officials who have about the same status are less likely to be inhibited than those among officials on different levels.
4. Coping with highly uncertain tasks requires very talented specialists who can be retained in the organization only if they are given relatively high status and responsible positions incompatible with a many-level hierarchy.
5. Talented specialists working under novel conditions often know much more than their supervisors about how to co-ordinate their activities.[20]

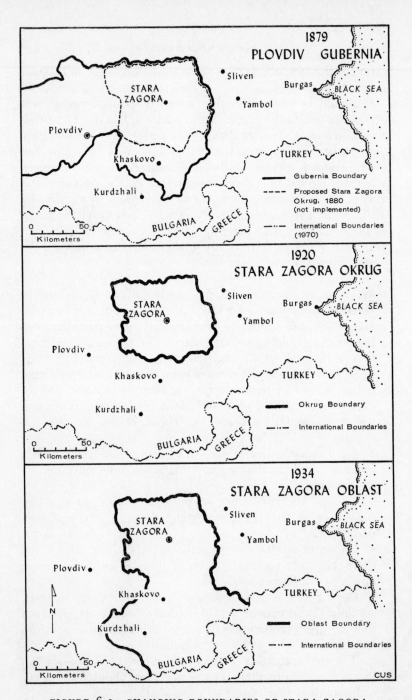

FIGURE 6.1 – CHANGING BOUNDARIES OF STARA ZAGORA
PROVINCE, BULGARIA
The area of administrative jurisdiction of the city of Stara Zagora has been

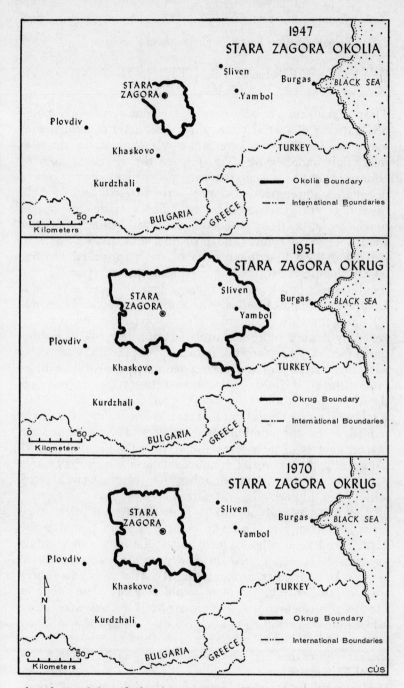

altered several times during the past century. The city's role as a focal point for the economic and social organization of territory has likewise waxed and waned in accordance with the changing pattern of the surrounding province.

EFFECTS OF REDISTRICTING ON REGIONAL STRUCTURE

The reorganization of administrative regions particularly modified patterns of spatial interaction. The zones of influence of many cities and towns were profoundly changed by the new internal boundaries (see Fig. 6.1 for the changing area of administration of a representative provincial centre). Newly designated administrative centres were invested with a wide range of central government branch offices in trade and services to which citizens from all communities within the given area were obliged to come. The provinces and districts became sharply defined functional areas of their appointed centres, which themselves formed a communist counterpart to the spontaneously developed hierarchy of central places in Western Europe and Anglo-America. The Polish geographer Dziewoński states that 'the economic role of voivodship towns is ... generally defined quite uniformly ... and in result of institutional factors their spheres of influence are practically identical with the territory of the corresponding administrative units (voivodships).'[21] Enhanced administrative status encouraged the development of non-governmental functions also. Thus Kosiński remarks on the fact that several towns that were newly named to be the administrative centres of districts became market centres of their surrounding areas for the first time.[22] Conversely, the territorial extent of influence of existing centres was abridged when new administrative regions were created within their hinterlands.

Such changes did not occur immediately, nor were they necessarily complete, however. A time lag was bound to be experienced while schools, medical facilities, and other public agencies were established in newly created administrative centres. In some cases, former patterns continued because there was no possibility or intent to establish a complete range of functions in new units. For instance, Kosiński notes that establishments providing repair service for machinery continued to serve a number of administrative districts from an intermediate centre that had lost functions.[23] Similarly, special vocational high schools were established in only a limited number of places and drew students from several neighbouring districts.

Generally, where people had a choice that was free of institutional limitations, traditional patterns would continue. This occurred in the functional zones for marketing, particularly when a town was located near the borders of its administrative district. Likewise employment in industry was not confined to people living within the administrative district where the factory was located. Patterns of commuting to work show major flows across administrative boundaries, particularly where rapid transit facilities were available. This was particularly true in the Upper Silesian region of southern Poland.[24] Much of such commuting reflected the housing shortage in industrial cities and tended to become confined to the area immediately adjacent to industrial centres as workers' housing was made available.

It should also be noted that not all government agencies were able to use (nor chose to use) the framework of provinces and districts. Presumably if regional co-ordination is not considered, each type of function would lend itself to its own unique optimum deconcentration of authority in field regions. In the United States, for instance, where little co-ordination regionally is evident, James Fesler has noted that more than 140 systems of federal government field service areas were in existence, more than half of which did not treat New England as an integrated whole.[25] Bora indicates that beginning in 1950 at least 30 central government agencies in Hungary did not use the provinces (*megyes*) in allocating authority downward.[26] Most used aspects of the system, but in areas of minor service demands would group two or more provinces together. These included bureaus in charge of bus systems, truck transport, brick-making and management of water resources. Some, however, used completely different systems of regionalization. Among these, Bora singles out the electric service regions, which he considers had great theoretical interest to geographers due to their similarities to hypothetical economic regions advanced by academic bodies.[27]

A factor in the nodality of both new and old administrative units was the tendency to locate not only the offices of service institutions and other agencies of local government in the administrative centres, but also new factories and communal facilities which made the centres even more attractive for future development.[28] The nodality was further strengthened by the selective improvement or establishment of completely new transportation facilities focused upon the newly named centres.[29] In his investigation of Bialystok *voivodship*, Dziewoński has noted what he terms an 'over-centralization' in the provincial centre.[30] The provincial capitals as a group have tended to show a rate of annual population growth greater than either the national capitals or the cities and towns without provincial functions.[31] As in the pre-war period, administrative employment provided a sizeable part of the total basic employment.

It is also quite likely that the rapid growth of provincial cities has been due to a tendency for each party secretary to press for a full range of economic and cultural functions in his province. Similar pressures existed at the district level. In each country it is common practice to talk about 'political factories' established with little consideration of economic location factors. An example of this can be seen in Bulgaria where each of the districts (*okolias*) sought to establish its own central milk-processing plant. Two new plants were built in Popovo and Tŭrgovishte which were only 37 km apart, even though plants existed already in the adjacent district centres of Preslav, Kolarovgrad, Razgrad and Ruse.[32] Pressures for such autarkial development are reflected also in the tendency to develop marginal resources of fuel, building materials and the like in preference to dependency upon outside sources.

Similar programmes for self-sufficiency in agricultural production are also evident. In the Bulgarian study cited above, it should be noted that in each of the major provinces (*okrugi*) the proportion of milk that is imported into the province in no case exceeds 5% of the total consumed (with the exception of the three largest cities in the country). In most cases, milk that is brought into a province or sent across boundaries to other provinces amounts to less than 2% of the total produced within the region.[33]

One of the 'sins of omission' resulting from the pattern of new territorial units was that problems overlapping more than one administrative region were more likely to be ignored due to lack of attention or to disputes over responsibility for improvement.[34] Sometimes such disputes were related to the hierarchical structure of administration. Thus in Yugoslavia, the main highway between Belgrade and Zagreb for many years was in a state of disrepair with a posted speed limit in sizeable sections of less than 50 km per hour. A major factor was the insistence by the Croatian Republic government that it was the responsibility of the federal Yugoslav government to repair it, with a reverse contention by central authorities.

In addition to setting the territorial pattern for operational control and management of the society, the administrative regions of the socialist states also became the frameworks for spatial planning. Such planning involved not only centrally co-ordinated areal programmes of economic and social investment developed in conjunction with financial planning, but also operational plans for agriculture, retailing, local industry, education, health and welfare elaborated by local people's councils. Jack Fisher terms the former 'regional' planning, and the latter 'territorial' planning.[35] In Poland plans of both types are usually developed in accordance with the boundaries of provinces, districts and communes, although provisions are made for drafting regional plans covering portions of two or more provinces or for groups of districts within a province.

It is difficult to assess the role of such plans in the formation of new patterns of geography, however. On the surface one would expect that they would tend to promote the highest degree of uniform development and coherence of administrative units. The critical question is the degree of implementation such plans have received after formulation. Like their counterparts in the West, socialist town and country planners investigate and prescribe, but they do not construct. They set limits and identify needs, but it is the responsibility of other agencies to carry out their recommendations.

Handicaps to implementation of spatial planning have included an inherent conflict between locally produced territorial plans based upon perceived needs, and centrally elaborated regional plans developed according to the cumbersome Soviet-created system of establishing economic 'balances'. Obviously, central plans take precedence and blunt local autarkial tendencies, if only because investment funds are disbursed from the centre. One scholar has pointed out that in fact in Czechoslovakia most decisions and requests of local government councils have been ignored or vetoed by central authorities, with less than 25% actually implemented.[36] Fisher cites the haphazard provision of commercial services in Yugoslav cities because of the absence of a continuously funded agency to implement plans, in contrast to the success in housing construction due to the existence of such a body.[37]

A further handicap is the imperfection of plans themselves. It has long been recognized in the West that planning must be viewed as a continuing process of decision-making rather than the preparation of blueprints of regional development. Too many decisions are necessarily made independently of the planners. As more and more authority is delegated to local governments and individual enterprises, the same confrontation with accomplishments by agencies outside planners' control is likely to prevail in Eastern Europe. Moreover, technology and basic wants and needs are subject to continual change. Simply bringing the technological level of transportation up to Western European standards would drastically alter the premises upon which many plans in East-Central and Southeast Europe have been constructed. The result is that existing plans can and do serve as guidelines for determining investment priorities, but they are subject to continual change and up-dating and cannot in themselves assure the ideal patterns of form and interaction sought by their designers. Finally, there is the problem of change in the administrative regions themselves. Each of the countries has seen revisions in its administrative map, and there is no assurance that it will not change in the future for political or economic reasons, voiding regional plans based on the goal of creating coherent administrative units.

Nevertheless, the planning process necessarily has major geographic significance. At the very least regional plans reinforce or justify managerial tendencies towards uniformity of regional development and closed circulatory systems within administrative regions of equivalent rank.

In forming provinces, consideration was no doubt given to creating units as nearly equivalent in size and population as possible within each country. This is understandable from the point of view of bureaucratic desires for order and uniformity particularly when partitioning was being done on the basis of a map, rather than as a result of field work.[38] The creation of several provinces in the otherwise coherent region of Polish Upper Silesia probably reflects this factor. However, the geographic structure of the new units suggests that areal uniformity was secondary to another consideration based on Soviet experience – the need to organize a potentially rebellious countryside around centres of proletarian strength. The Soviet experience had demonstrated that collectivization of agriculture was a task demanding the utmost attention of the party and its supporters, who almost entirely came from the cities. Provinces and districts were thus viewed as areas of organization and control by cities, and not simply as two-dimensional partitionings of space. The nodality of territorial-administrative regions was then a conscious consideration in their original design.

It should be noted that such a view was consistent with traditional organizational patterns in East-Central and Southeast Europe. The historic province (*zeme*) of the Czech lands and the interwar provinces (*banovinas*) of Yugoslavia appear to be the only exceptions to a general pattern of nodal administrative regions. The communist consolidation of small, predominantly agricultural provinces in Hungary quite likely took into consideration their absence of urban nuclei, as did the elimination of such essentially rural units as the district (*judeţ*) of Salaj in Romania.[39] Rude Petrović, citing a speech by Kardelj, has noted that a consideration in modifying the pattern of districts (*srezovi*) in Yugoslavia in the 1950's was to ensure having a non-agricultural majority in units wherever possible.[40] It is interesting to note, however, that tradition and nodal advantages in terms of a central position and intersection of transportation lines took precedence over proletarian concentrations when the semi-rural Romanian centre of Deva, with a population of 10,000, was named the administrative centre of Hunedoara province in preference to the industrial city of Hunedoara, which had a population in excess of 100,000.[41]

SUBSEQUENT REDISTRICTING

In considering the significance of post-war redistricting to alterations in patterns of circulation and changes in regional trends of economic development, one must note that such modifications were often of an ephemeral character since many of the initial provinces and districts were themselves short-lived. Soon after their introduction it became clear that in many cases the political–bureaucratic basis was counterproductive to economic aims of the regimes. The multitude of units created to ensure centralized party control throughout the territory of each state had resulted in a geographic framework of regions too small for effective planning, and too much at variance with established complexes of regional development. As industrialization progressed the regimes realized that their limited funds for investment were being dissipated in unnecessary duplication of facilities and development of marginal resources by autarkial tendencies in small units too numerous for close supervision. Obvious linkages between industrial plants were frustrated by arbitrarily established boundaries.[42]

To remedy such economic problems, and also for other motivations of a political or administrative nature, a number of subsequent readjustments of the maps of administrative

TABLE 6.3 – *East-Central and Southeast Europe: Summary of Post-1951 Changes in Administrative Regions*

BULGARIA	*July, 1959:* Introduction of new framework of 27 provinces (*okrugi*) subdivided into 867 districts (*obshchinas*), replacing former system which had grown by local changes to 12 provinces (*okrugi*), 117 districts (*okolias*), and 1,833 communes (*obshchinas*)
CZECHOSLOVAKIA	*April, 1960:* Provinces (*kraje*) reduced from 19 to 10 and districts (*okresy*) from 269 to 108; communes (*obce*) had previously declined gradually from 14,200 to c. 11,000
HUNGARY	Remained relatively unchanged, although districts (*jurases*) increased gradually from 113 to 128 by 1960
POLAND	*1954:* Replacement of 3,000 communes (*gromadas*) and 40,000 localities (*gminas*) by 8,787 new communes (*gminas*); gradual changes in subsequent years saw increase in districts (*powiats*) from 317 to 391 and reduction in communes (*gminas*) to 5,245 (1961)
ROMANIA	*1952:* Provinces (*regiunes*) reduced from 28 to 18, with one being designated as a 'Magyar autonomous region'
	1956: Provinces (*regiunes*) reduced to 16 with minor regional transfers of territory
	December, 1960: Districts (*raionuls*) reduced from 177 to 146; boundaries of provinces modified in several areas, Magyar autonomous region reduced in size and renamed 'Mures autonomous region', and seven of fifteen other provinces also renamed
	January, 1968: Establishment of new framework of 39 provinces (*judeteţe*), subdivided directly into communes, eliminating district (*raionul*) level

YUGOSLAVIA   A continuing steady decline in the numbers of districts (*srezovi* and *kotari*) from 360 in 1951, to 107 in 1955, to 24 in 1965; communes (renamed *opshtinas* or *općine* in 1952) similarly declined from 7,104 in 1951, to 3,811 in 1952, to 1,479 in 1955, to 548 in 1965; it appears that the nominal districts (*kotari* and *srezovi*) lost nearly all their functions except for the collection of statistics following a 1955 decision to shift the centre of gravity of decision-making to the commune level – thus, the *opshtina* and *općine* have been *de facto* districts since that time; the last *srezovi* were abolished by 1968

This table was compiled from a variety of sources, including the standard statistical yearbooks of each country. The following works contain detailed descriptions of changes in administrative regions for the respective countries: (Bulgaria) Nikola N. Muleshkov, 'Administrativno-teritorialno Ustroisvo na Narodna Republika Bulgariya' (Administrative-Territorial Structure in the People's Republic of Bulgaria), *Godishnik na Sofiiskiya Universitet, Yuridicheski Fakultet*, Vol. 4, Book 2 (Sofia: Dürzhavno izd. 'Nauka i Izkustvo', 1964, pp. 353–413; (Czechoslovakia) I. M. Maergoiz, *Chekhoslovatskaya Sotsialisticheskaya Respublika: Ekonomicheskaya Geografiya* (The Czechoslovak Socialist Republic: Economic Geography), (Moscow: Mysl, 1964), p. 99 ff.; (Hungary) Gy. Bora, 'A Rayonkutatás Jelentösége és Problémája a Magyar Gazdaságföldrajzban' (The Significance and Problems of Regionalization to Hungarian Economic Geography), *Földrajzi Értesitö*, Vol. 9, No. 2 (1960), pp. 192–215; (Poland) Waclaw Brzezinski, 'Ksztaltowanie Teritorialnego Podzialu Pahstwa' (The Formation of the Territorial Division of the State), *Państwo i Prawo*, No. 3 (1963); (Romania) Walter Roubitscheck, 'Die administrative Gliederung der Rumänischen Volksrepublik' (The Administrative Regionalization of the Romanian People's Republic), *Petermanns Geographische Mitteilungen*, Vol. 107, No. 1 (1963), pp. 45–8; (Yugoslavia) Irina Perko-Separović, *Veličina Lokalnih Jedinica* (The Sizes of Local Units), (unpublished doctoral dissertation, University of Zagreb, 1965).

regions took place, particularly in Romania and Yugoslavia. These changes are summarized on Table 6.3.[43]

The changes during the mid-fifties and subsequent periods did not fit into as common a pattern as was the case in the 1949–50 reforms. This probably reflects the growing independence from the U.S.S.R., notably in the case of Yugoslavia where a complete break had been made in 1948 on the eve of implementation of its *oblast* reform. Subsequent Yugoslav events have particularly been associated with a decentralization of decision-making to the commune (*općina*) level in an attempt to implement the old Marxist concept of the withering away of the state and its replacement by the self-governing local community.[44] Romania and Czechoslovakia reduced their initial spans of provinces by nearly half, but in contrast Bulgaria almost doubled its provinces, which put its initial span of control on a par with the other countries. Romania later reversed its policy and again markedly increased its numbers of provincial units. At the district level there was a notable reduction in the average size of units in Bulgaria, Hungary, Poland and Yugoslavia, but increases in size took place in Czechoslovakia and Romania. From Table 6.4 it can be seen that none of the countries had district units comparable in size to the then recently consolidated regions of the U.S.S.R., in contrast to the marked similarities observable in 1949–50. The average spans of districts within provinces did not change appreciably, except for an increase from 6 to 32 in Bulgaria. Spans of local units did increase generally, especially in Czechoslovakia where districts now controlled 111 instead of 50 communes on the average.

One common thread observable in several programmes of change was the institutional shift from centralized management of industry and agriculture to more decentralized decision-making. In 1957 Khrushchev had initiated a radical reform in the Soviet Union, and similar changes appeared in the other states, although Yugoslavia's decentralization obviously was not related to the U.S.S.R. programme. Rusinova indicates that reappraisal of the role of people's councils in administrative regions was begun by local parties in Poland in October, 1956, in Romania in December, 1956, and in Bulgaria in July, 1957.[45] The modifications of provinces and districts in Bulgaria and Czechoslovakia appear to have been particularly related to the reforms. The Bulgarian provincal (*okrugi*) were even termed 'economic-administrative' units on the model of areas of economic council (*sovnarkhoz*) jurisdiction in the U.S.S.R. Until management was recentralized in ministerial hierarchies in May, 1963, the Bulgarian province councils were given control over all industrial, agricultural, construction and commercial enterprises within their boundaries, and were also responsible for education, cultural activities, public health and the direction of commune governments.[46] The new provinces in Czechoslovakia were similarly given a high degree of economic autonomy:

> ... the people's committee system of Czechoslovakia was made responsible for the development of agricultural production, local industry, warehouses and shops, construction

TABLE 6.4 – *East-Central and Southeast Europe: Characteristics of Provinces and Districts: 1959–1960*

Country	Provinces	Average km²	Average Number of Districts per Province	Districts	Average km²	Average Number of Communes per District
Bulgaria	27 *Okrugi*	4,074	32	867 *Obshchinas*	127	—
Czechoslovakia	10 *Kraje*	12,800	10	108 *Okresey*	1,185	111
Hungary	19 *Megyes*	4,900	7	128 *Jurases*	726	25
Poland	17 *Voivodships*	18,900	21	396 *Powiats*	785	15
Romania	16 *Regiunes*	14,800	9	146 *Raionuls*	1,623	29
Yugoslavia	75 *Kotari/Srezovi*	3,400	11	839 *Općine*	305	—
U.S.S.R.	144 *Oblasts* etc.	153,400	12	1,711 *Raions*	13,090	23
Ukrainian S.S.R.	25 *Oblasts*	24,000	10	251 *Raions*	2,400	24

For sources see Table 6.3.

enterprises, kindergartens, schools of general education, professional schools, hospitals, etc. In exercising these and other responsibilities the people's committee system will control 30·6% of the total state expenditures.[47]

Although Poland and Romania did not have striking internal territorial changes in the 1959–60 period, their subdivisions did undergo similar major changes in function. In 1958 local councils in Poland were given authority over agriculture, small-scale state and co-operative industry, retail trade, roads, automotive transport, housing, communal services, primary and most vocational schools and public health.[48] In 1961 a major new planning law was introduced emphasizing regional co-ordination. The December, 1960, territorial shifts in Romania were followed by a law of March 20, 1961, establishing in each province (*regiune*) special economic councils to co-ordinate local industry with enterprises managed by central authorities and to study problems of general economic development and resource use.[49] The recent introduction of the 39 new provinces (*judeteţ*) has been accompanied by more specific delegation to people's councils of 'all local activity of the state in the economic, sociological, and administrative fields, as well as in construction and the defence of socialist property'.[50]

The resurfacing of nationalism in East-Central and Southeast Europe also had its impact upon the territorial-administrative patterns. This has been particularly true in Yugoslavia. The Soviet-style provinces (*oblasts*) of 1949 were eliminated in 1951 in favour of direct control of districts (*srezovi* [Serbian] *kotari* [Croatian]) by the strengthened national republic governments. More recently, the raising of the Shiptar region of Kosovo to a state of autonomous control comparable to a republic in response to ethnic political agitation has likewise forced a restructuring of bureaucratic frameworks of control.[51] The creation of a Magyar province in Romania and the granting of a true federal status to Slovakia reflect similar pressures. There were a number of other motivating factors for the changes that took place in the 1950's and 60's in the countries under discussion, and it is likely that such modification will continue into the future. In some instances, changes were for purely political reasons. The author was told on good authority, for instance, that in Yugoslavia, some Yugoslav districts were actually created in order to rusticate officials who had outlived their usefulness in central positions. Also, one should not minimize the personal rivalries for leadership and positions of eminence within the élite ruling groups. Some projects appear to have been the result of a quest for personal prestige derived from seeing an innovation carried through to its finality. Empire-building at the local level may also have been behind the merging of individual communes, particularly in Yugoslavia where most changes have been of a piecemeal nature. The establishment of a new group in power with new policies, particularly in the economic sphere, also may have led to some of the changes that have taken place.

One motivation for change that should not be discounted completely is the desirability of changing numbers of administrative units and job categories simply in order to release energies and ideas. Anthony Downs has noted the tendency in American corporations to shuffle posts within their higher and intermediate levels of bureaucracy for just this reason, although operating units at the lowest level generally are not changed as frequently because of the reluctance to disrupt the day-to-day operations with the public.[52] Results are generally refreshing for a time. Rigid channels of authority are broken up and orders from above seem to be followed much more closely, since lower offices are uncertain of how much autonomy they may enjoy. Downs points out that such reorganizations are particularly significant in bureaucracies that have a weak feedback from their clientele, a problem continuously plaguing socialist regimes. Some sociologists have gone so far as to suggest that there is actually a cycle of reorganization. A recent study, for instance, has pointed out that the bureaucracy of the United States Air Force in charge of research and development has been reorganized regularly about every seven years.[53]

Almost all change appears to have been initiated at the centre, whatever the motivations may have been. One prominant Yugoslav official suggested to me that there is normally about a three-year time lag in Yugoslavia between the original idea for change and its actual accomplishment. Instigators of a proposal for change send the idea to officials at the local level who in turn respond with specific suggestions that they think those at the top want to hear. When change finally does occur, such as the consolidation of local units, there are a number of adjustments that, of necessity, must take place. However, few, if any, of the redundant cadres actually lose their jobs, although there may be considerable padding of payrolls until the work load of the new administrative unit catches up with its capacity in Parkinsonian fashion.

THE FUTURE

The process of decentralization of decision-making and reduction of party influences in economic affairs is bound to have an effect on patterns of spatial interaction and development. The present situation of rigid fitting of economic and social functions to political areas has been one of many handicaps to development, because of its minimizing of production economies of scale and agglomeration.

As enterprises gather increasing powers to make their own investment and operations decisions, the present trend towards regional uniformity of industrial development with its associated patterns of circulation will be changed. An increasing premium will be placed on competitiveness, and locational decisions will be more and more based on considerations of comparative regional advantages. For example, it is likely that economic benefits to be derived from expansion of existing facilities to meet demand will take precedence over the social benefits sought in the past by establishment of duplicate plants elsewhere in underdeveloped regions. If COMECON is able to become a true common market, such tendencies will be intensified. At the same time, it is likely that many uneconomically sited 'hothouse' developments will falter, much as the inefficient enterprises in France and Belgium have run into problems when subjected to free foreign competition.

The present experience of Yugoslavia is instructive. The emphasis upon profitability in operations has already created significant adjustments, including regional unemployment and a shift in economic geographic trends. *Općina* councils faced with the increasing need to rely upon their own funds rather than federal investment allocations and subsidies are finding it necessary to co-operate with each other in specializing and sharing facilities. In the process, traditional patterns of spatial organization are reappearing again. A recent journal has commented on the re-establishment of the old coherent regions of Slavonia, Pelagonija and Pomoravlje as a result of such spontaneous regional co-operation.[54]

The quest for profitability, coupled with the ease of merger or partition of administrative units in Yugoslavia, has led also to an interesting reappearance of pre-war organizational patterns of local government. After studying the network of districts and communes in the Krapina area of Croatian Zagorje, Pusić reached the conclusion that as a result of decentralization in 1955, the area's traditional administrative patterns that had a 'natural' form based on circulation around major settlements had in essence reappeared again, despite the radical territorial perturbations of 1946–9.[55] He notes that the old imperial units had thus been able to survive two world wars, a fundamental change of government, and a socialist revolution. Perko-Separović has observed that a similar readjustment has occurred in other places in Yugoslavia, as the communes which numbered more than 3,900 in 1953 have been enlarged steadily until in 1965, at a total number of 548, they were in a majority of cases nearly identical with the 407 districts (*srezovi* and *kotari*) of 1947.[56] They were not only focused on the same urban centres, but included 80% to 90% or more of the villages found in the pre-war administrative units.

One can expect similar re-establishment of traditional patterns of spatial organization in the other countries, particularly if decision-making is decentralized to the enterprise level, rather than to the commune or district. Such re-establishment would result in part from the fact that fundamental change has not occurred in basic aspects of regional advantage, particularly in transportation facilities. Post-war developments, of course, will not all wither away. Once established, most industrial plants, auxiliary facilities, pools of skilled labour, etc. have become new regional assets of their own. Thus, the post-war partitioning of territory for effectiveness of administration will have lasting effects on the geography of East-Central and Southeast Europe. Processes of areal functional organization of the economy and society have been introduced which have modified and supplemented patterns of industry, agriculture and circulation established by the competitive development and somewhat similar bureaucratic influences in the days of the great empires.

NOTES

[1] Representative recent studies of bureaucracy include the following: Anthony Downs, *Inside Bureaucracy*. A Rand Corporation Research Study (Boston: Little, Brown, 1967); Michel Crozier, *The Bureaucratic Phenomenon* (Chicago: The University of Chicago Press, 1964); K. K. White, *Understanding the Company Organization Chart*. American Management Association Research Study 56 (New York: American Management Association, 1963); Joseph LaPalombara (ed.), *Bureaucracy and Political Development* (Princeton: Princeton University Press, 1963).

[2] Among recent works of political scientists concerned with East-Central and Southeast Europe are: Ghita Ionescu, *The Politics of the European Communist States* (New York: Praeger, 1967); H. Gordon Skilling, *The Governments of Communist East Europe* (New York: Crowell, 1966); and R. V. Burks, *The Dynamics of Communism in Eastern Europe* (Princeton: Princeton University Press, 1961).

[3] Theodore Shabad, 'The Administrative-Territorial Patterns of the Soviet Union', *The Changing World: Studies in Political Geography*, ed. by W. Gordon East and A. E. Moodie (Yonkers-on-Hudson, New York: World Book, 1956), pp. 365–85; Jack C. Fisher, 'The Yugoslav Commune', *World Politics*, Vol. 16 (1964), pp. 418–41; Ronald A. Helin, 'The Volatile Administrative Map of Rumania', *Annals, Association of American Geographers*, Vol. 57 (1967), pp. 481–502.

[4] Andrzej Wróbel, *Województwo Warszawskie: Studium Ekonomicznej Struktury Regionalnej* (Warsaw Voivodship: A Study of the Regional Economic Structure), ('Prace Geograficzne No. 24', Warsaw, 1960), p. 139.

[5] W. Ivor Jennings and Harold Laski, *A Century of Municipal Progress* (London, 1936), p. 32.

[6] The generic terms 'province', 'district' and 'commune' are used throughout this paper to facilitate comparisons among the different countries. Usually the official designation in the local language is given also. Classed as 'provinces' are all administrative regions of general government whose officials are subordinate directly to central authorities. Similarly, 'districts' are regions with officials answering to provincial administrators, and 'communes', in turn, have officials responsible to district authorities. In some

instances districts have been made into provinces, or communes into districts, without changes in the form of designations of the administrative areas involved, due to the elimination of an entire intermediate level of units.

[7] Eduard Táborský, 'Local Government in Czechoslovakia, 1918–1948', *The American Slavic and East European Review*, Vol. 10 (1951), pp. 207–9.

[8] F. E. Ian Hamilton, *Yugoslavia: Patterns of Economic Activity* (New York: Praeger, 1968), p. 11.

[9] Helin, *op. cit.*, pp. 487–93.

[10] Rude Petrović, *Prostorna Determinacija Teritorijalnih Jedinica u Komunalnom Sistemu Jugoslavije* (Spatial Delimitation of Territorial Units in the Communal System of Yugoslavia), ('Studia i Monografije, Knjiga 1'; Sarajevo: Ekonomiski Institut Univerziteta u Sarajevu, 1962), p. 25.

[11] Merle Fainsod, *How Russia is Ruled* (Cambridge, Mass.: Harvard University Press, 1956).

[12] *Ibid.*, especially pp. 327–52.

[13] Eugen Pusić, *Lokalna Zajednica* (The Local Community), (Zagreb: Narodne Novine, 1963), p. 101.

[14] The data on provinces and districts omits cities and towns whose urban governments had administrative rankings equivalent to territorial units. Following the 1948–51 reforms urban settlements in the various countries had independent status as follows:

Country	Cities and Towns ranked as Provinces	Cities and Towns ranked as Districts
Bulgaria	1	6
Czechoslovakia	1	9
Hungary	5	62
Poland	2	46
Romania	8	27
Yugoslavia	0	89

[15] In Poland the number of communes per district was increased to an average of 26 in 1954 following the elimination of the country's 3,000 *gminas* and their replacement by 8,787 *gromadas*, consolidated from their previous total of 40,000. The very large number of communes per district in Czechoslovakia reflects the distinctive Czech settlement pattern of numerous small villages.

[16] Such an upper limit, or maximum 'span of control', was apparently not given explicit recognition in the literature of the

period discussing the reforms. The only reference to it as a factor in regionalization that the author was able to find was in the stenographic minutes of the Sabor of Croatia which noted the problems of central organs in trying to supervise too great a number of provinces (*kotari*) (*Stenografski Zapicnici Sabora NRH*, Series III, Book 2; Zagreb, 1950, p. 159).

[17] In a recent survey by the American Management Association of 66 companies, a total of 39 had unitary leadership at the top. The number of principal subordinates directly under the chief executive of the firm varied as follows:

Number of Firms Reporting	Number of Principal Subordinates to Executive
0	1
0	2
1	3
1	4
3	5
2	6
4	7
8	8
2	9
4	10
3	11
6	12
4	more than 12

Source: K. K. White, *Understanding the Company Organization Chart*, p. 61.

[18] A term used by Harold Stieglitz, 'Optimizing Span of Control', *Management Record* (September, 1962), as noted by White, *op. cit.* Fesler notes that most federal agencies, 'mindful of the span of control', have kept their initial regionalization networks of field service areas to less than 20 units. James W. Fesler, *Area and Administration* (Tuscaloosa: University of Alabama Press, 1949), p. 7.

[19] James G. Bowland, 'Geographical Decentralization in the Canadian Federal Public Service', *Canadian Public Administration*, Vol. 10 (1967), p. 329.

[20] Downs, *op. cit.*, p. 58.

[21] Kazimierz Dziewoński, 'Problems of Regional Structure of Poland', *Przegląd Geograficzny*, Vol. 32, Supplement ('Special Issue for the XIX International Geographical Congress, Stockholm, 1960'; Warsaw, 1960), p. 121.

[22] Leszek Kosiński, 'Population and Urban Geography in Poland', *Geographia Polonica*, No. 1 (1964), p. 84.

[23] Leszek Kosiński, 'Studies on the Spheres of Influence of Small Towns in Poland', *Problems of Economic Regions: Papers of the Conference on Economic Regionalization in Kazimierz (Poland), May 29–June 1, 1959* ('Prace Geograficzne No. 27', Warsaw, 1961), p. 208.

[24] See the maps of commuting patterns in the Kraków region in Stefania Mańkowska, 'Der Pendelverkehr in der Wojewodschaft Krakow als Element der Ökonomisch–Geographischen Rayonierung', *ibid.*, p. 313.

[25] Fesler, *op. cit.*, p. 7.

[26] G. Bora, 'Nekotoryye Problemy Issledovaniya Ekonomicheskikh Raionov v Vengrii' (Several Problems of Investigating Economic Regions in Hungary), *Problems of Economic Region* . . ., p. 70.

[27] *Ibid.*, p. 73.

[28] 'The fact of establishment of . . . a center has substantial significance, because the attraction of further capital investments strengthens its role in the future.' Leszek Kosiński, 'Studies on the Spheres of Influence . . .', p. 208.

[29] The chairman of the executive committee of the Yugoslav commune (*općina*) of Pakrac in Slavonia pointed out to the author that there had been very little intercourse between his town and the rival city of Bjelovar before the latter was named the centre of a province (*kotar*) in which both were situated as a result of post-war redistricting. After establishment of the province, however, the dirt road between the two towns was paved, as were other routes in the area leading to the capital city of Bjelovar. Moreover, a daily bus service was established for the first time between the two towns. It carried an average of forty persons per trip, almost all of whom had official business in the new province centre. The bus service was discontinued after the province lost its functions.

[30] Kazimierz Dziewoński, 'Theoretical Problems in the Development of Economic Regions (Within One Country)', *Papers, Regional Science Association*, Vol. 10 (Zurich Congress, 1962), p. 58.

[31] Dziewoński, 'Problems of Regional Structure . . .', p. 121.

[32] Ivan Zakhariev and Marin Devedzhiyev, *Teritorialno Razpredeleniye i Ikonomicheska Efektivnost na Mlekoprerabotvatelnata Promishlenost v NRB* (The Territorial Distribution and Economic Effectiveness of the Milk Processing Industry in the People's Republic of Bulgaria), (Sofia: Bŭlgarska Akademiya na Naukite, 1965), p. 96.

[33] *Ibid.*, p. 26.

[34] Kazimierz Dziewoński, 'Economic Regionalization', *Geographia Polonica*, No. 1 (1964), p. 172.

[35] Jack C. Fisher, *Planning in Poland* (Ithaca: Cornell University Press, 1966), p. 240.

[36] Ionescu, *op. cit.*, p. 130.

[37] Jack C. Fisher, *Yugoslavia – A Multinational State: Regional Difference and Administrative Response* (San Francisco: Chandler, 1966), p. 128.

[38] It is difficult not to draw a parallel between the rushed decision-making on a desk-top map by individuals who had enjoyed no pre-war responsibilities for governing and who lacked a clear perception of the nature of the country outside of the capital cities where they had spent most of their lives, and the partitioning of Africa in Berlin in 1884 by persons who had never been there.

[39] Mihail Haşeganu, *Geografia Economica a Republicii Populare Romîne* (Economic Geography of the People's Republic of Romania), (Bucharest: Editura Ştiintifica, 1957), p. 316.

[40] Petrović, *op. cit.*, p. 33.

[41] Miroslav Blažek, *Analyse Géographique de la Regionalisation Administrative* (A Geographic Analysis of Administrative Regionalization), (A paper presented to the Commission on Methods of Economic Regionalization meeting in Brno from September 7th–September 12th, 1965; Brno, 1965), p. 17; Helin, *op. cit.*, p. 498. Deva has subsequently grown to a population of more than 28,000. *The Statesmen's Year-Book, 1968–1969* (New York: St. Martins Press, 1968), p. 1378.

[42] See the discussion in Dziewoński, 'Problems of Regional Structure ...', especially pp. 122–4.

[43] The changes in Romania are particularly well described in Helin, *op. cit.*

[44] Petrović, *op. cit.*, p. 28.

[45] S. I. Rusinova, *Gosudarstvennoye Ustroistvo Sotsialisticheskikh Stran Yevropa: Territorial'no-Natsional'naya Organizatsiya* (The State Structure of the Socialist Countries of Europe: Territorial-National Organization), (Leningrad: Izdatel'stvo Leningradskogo Universiteta, 1965), p. 107.

[46] *Ibid.*, p. 108.

[47] Carl Beck, 'Bureaucracy and Political Development in Eastern Europe', in LaPalombara (ed.), *op. cit.*, p. 280. The quotation is taken from *Politkia Ekonomie*, No. 6 (June, 1960).

[48] Rusinova, *op. cit.*, p. 109.

[49] *Ibid.*, p. 110.

[50] Victor Tufescu and Constantin Herbst, 'The New Adminis-

trative-Territorial Organisation of Romania 1968', *Revue Roumaine de Géologie Géophysique et Géographie, Série de Géographie*, 13 (1969), pp. 26–37.

[51] '"Nationalist" Manifestations in Kosovo-Metohia with Grave Political Connotations', *Radio Free Europe Research* (October 29, 1968).

[52] Downs, *op. cit.*, p. 57.

[53] *Ibid.*, p. 166.

[54] *Komunist* (January 19, 1967), p. 3.

[55] *Kotar Krapina: Regionalni Prostorni Plan*, (ed.) Branko Petrović and Stanko Zuljić (Zagreb: Urbanistički Institut Narodne Republike Hrvatske, 1958), p. 134.

[56] Irina Perko-Separović, *Veličina Lokalnih Jedinica* (The Sizes of Local Units), (unpublished doctoral dissertation, University of Zagreb, 1965), pp. 197–200.

Comments

JOSEPH VELIKONJA

In his quite complex paper, Professor Poulsen attempts 'to examine the impact of administrative organization and decision-making upon the economic, social, and political geography of the countries of East-Central and Southeast Europe' and analyses 'to what extent the geographic elements of these countries have had modifying effects upon structures of bureaucracy'. Within the space/time context of the paper, his attempt implies nothing less than an analysis of the relationships between the communist system of government and the geographic mosaic of the 'Shatter Belt'. The clear focus on the spatial dimensions of administrative operations, emerging in recent years, some 30 years after Professor Whittlesey's original call,[1] must be welcomed by political geographers, and Professor Poulsen deserves to be applauded for his contribution.

It is gratifying to see the author recognize the problems of the pertinent material and the limited number of interpretative studies covering this ground ... and presenting his thesis as indicative rather than exhaustive. I fear, however, that the problems facing a student of this area are even more severe than Professor Poulsen suggests. Truly we do not have enough factual material, we lack interpretative sources ... but, unfortunately, we lack also the interpretative ability to grasp the complex problems of his complex area *as a unit*, and the lack of theoretical infrastructure necessary for reasoning in that direction is appalling.

The highly meritorious goal at which Professor Poulsen aims would be difficult to accomplish even if only one country were to be thoroughly examined. The difficulty will increase when we admit that Eastern Europe is not a homogeneous region but rather an area, to use words of a friend of mine, 'where similarities are exceptional and differences statistically normal'.

The impression of homogeneity an observer sometimes gets reflects more the formal institutional parallels and similarities rather than true similarity in operational systems or behavioural codes. Even during the period of Stalinism, of 'cult of personality', and, later on, during the era of 'cult of non-personality',[2] the persistent attempts to apply previous Soviet experiences and methods homogeneously over the whole region were not too successful; the regional differences in the levels of social, economic and political development and the variations in national style and temperament required considerable regional adjustment in proposed techniques.

When dealing with the relation of bureaucracy to reality, there is an even more difficult problem within each of the communist countries. The primary source material in a society based on the principle of command economy and, in a broader sense, of a command society, is the pool of governmental decrees, orders, resolutions, operational directives, etc., which the late Professor Northrop[3] would call the *positive law*. However, in every human society the *living law*, i.e. the set of principles in reality directing the behaviour of the population, shows considerable differences and deviations from the *positive* norms, the differences varying in character and degree from one area of human activity to another. During the period of a drastic social change, the positive law shows, quite naturally, an almost absolute flexibility, while the living law retains its traditional tenacity. The gaping gorge between the two is well known to every native and forms one of the basic compass directions for an individual's behaviour, ... but it may and probably will be invisible and often unpenetrable to a foreign scholar with only a limited access to the hearts of the people.[4]

As you have noticed from my introductory remarks, I find it of little value to discuss individual details or my disagreements with some of his facts. Rather I intend to focus my remarks on selected methodological problems, some theoretical issues, and few interpretative conclusions. When referring to actual situations, I shall aim my comments primarily – though not exclusively – on Yugoslavia, fully aware that this material and this country is by no means representative in any sense of the other East European countries, and hence is not to be interpreted as a regional model or example.

What problems do we face when trying to determine the character and extent of the impact of Communist regimes on the geographic totality? To compare the contemporary 'reality' with one that existed some 30 years ago within completely different internal and external circumstances is a quite useless intellectual exercise. For a more precise attempt, one should try to compare the contemporary picture with a hypothetical situation that would exist today in the same region, had the region been administered by a different social and political regime. Comparison of developmental trends and their results is understandably speculative and its conclusion extremely tentative.

A second major difficulty in dealing with countries of Eastern Europe is that in the official literature announced plans and actual accomplishments are frequently merged. The assessment of the actual implementation and degree of success is difficult to make. Schemes which are still on the drawing board and those which have been implemented with greater or lesser success appear to be presented as identical. This is particularly relevant in operations where field checking is not only visual and where the discovery of less than complete success could be interpreted as a criticism of the organization rather than of the implementation.

Administrative and regional structure in Eastern Europe today reflects three basic forces:

1. the legacy of the past organizations;
2. the demands of contemporary system;
3. the projections for future socialist–communist society, or as Jack Fisher summarizes in his paper: past practice, social aspirations and currently fashionable solutions.[5]

All three sets of forces are ultimately interwoven so that they are often difficult to separate. Furthermore, they exist and operate from premises which are not comparable to the United States or British public administration or regional organization. They are conditioned by the economic as well as social and cultural-political legacy of the past and of the present, not unlike those in the West in general and in the United States in particular.

The Yugoslav case in its extreme complexity well illustrates the underlying principles and the basic difficulties. Here the administrative system emerged out of the legacy of Central European administrative structures and procedures which had evolved through two centuries of post-absolutist administration in the realm of Austrian and later Austrian and Hungarian Empires, in conflict with the legacy of the Turkish rule, slowly pushed out of the Southeast European area. These two distinct legacies, one essentially accepted as honest and operational, though paternalistic, the other corrupt and despised, produces two attitudes towards administration: the direct rule of the Austro-Hungarian area clearly established lines of responsibility and division of administrative power, on the other hand the indirect and less predictable administration of Turkish areas (extended at one time or another to the Una–Sava–Danube area of Southeast Europe), has not been completely eliminated from major areas of Yugoslavia, Albania and Bulgaria.

The territorial organization of today reflects rather strongly the war experience when the subsidence of previously legally established territorial administration was substituted by entities of local power (*ljudska oblast*) often long before the power had legal sanction for its operation. The transfer of responsibilities from partisan military organization to the territorial political administrative bodies, mostly local, is to my view the most important innovation in the Yugoslav modern history, since this 'power from the people' later was responsible for the establishment of workers' councils and territorial communes in spite of the delay of the 1945–53 period during which their role was minimized. The transfer of ownership from the state to the public, represented in the Yugoslav case by local communes, is being presented as a step towards the declining role of the state in the communist society. The post-war administrative organization shifted from the centralized federation of the 1946 constitution to a step-wise decentralization in 1953. The 1963 constitutions and 1969 amendments delegated of administrative responsibility to the lower level administrative entities, in essence communes and workers' councils, while the higher level bodies retained the control over the 'fundamental goals and values, and the broad policies governing most functional fields in which the local agencies are involved'.[6]

I am inclined to accept James W. Fesler's recently expressed warning with regard to decentralization, that (*a*) centralization–decentralization suggest dichotomy and polarization and is a poor substitute for the full continuum between the two poles – we do not have a term that specifies the middle range where centralizing and decentralizing tendencies are substantially in balance; (b) power is a complex phenomenon and its distribution difficult to measure; (c) it is difficult to differentiate degrees of decentralization even within a single country at a given time.

The consequence of this lack of precision is the tendency to 'link, then merge and confuse, decentralization and democracy'.[8] Although Professor Poulsen in his review of Eastern Europe does not accept these false premises, he does not reject them either. The assessment of relative value and restricted application of decentralization is only partially reviewed. The stress on bureaucracy and bureaucratic administration – without defining it within the context of Eastern Europe – leaves one to wonder if the term is used as a synonym for formal administration, or does it imply the negative overtones used in East European political literature, meaning with the term of bureaucrat and bureaucracy the essence of anti-social type of technocrat to whom the administration is his goal and the goal of administration is of no concern. Equating bureaucracy with decision-making obscures rather than clarifies the situation.

In assessing the operational impact, it is also relevant to discover the intensity of impact at different hierarchical levels. The hierarchy itself is of lesser significance. As an illustration, the disappearance of *srez* in Yugoslavia left very little consequences, as little as their creation in the 1920's.

The two poles of possible spatial organization by which the implementation of ultimate goals can be carried out, socialist or non-socialist, is therefore only an additional attribute, though socialist central government often tends to speed up the implementation of the centrally adopted goals through rigidly established and party-controlled machineries. Human organization, on the other hand, embraces a considerably wider range of activities than the bureaucracy or administration can cope with. Although the socialist leaders aspire to guide the totality of people's behaviour, they in fact embrace only a portion of the total man.[9]

The value assessment of decentralization as compared to a centralized system is often biased by our ingrained preference for decentralization, which at least gives the illusion of people's participation in the political decision-making.

The three components, indicated earlier, namely the legacy of the past, the contemporary demand, and the projections for the future, form a complex mix in different parts of Yugoslavia in spite of the homogeneous nature of the contemporary system. It is also interesting to notice, that the struggle and tension in one area and apparent conformity in another do not reflect the *actual* degree of tension. The old Montenegrin saying '*Bog je visoko a car je daleko*' (God is in heaven and the Tsar is far away) indicates in part the lack of concern for temporary discrepancies in some areas, such as Montenegro, while open and often bitter discussions go on in Slovenia, even though a reluctance to accept innovations is stronger in Montenegro than it is in Slovenia.

Dr Fisher's paper given at this conference provides an answer and documentation for a number of assumptions expressed by Professor Poulsen and also provides empirical data which are not available elsewhere. The administrative operation in some East European countries is not a *taboo* – decision-making is not secretive and hidden – I am referring to the 1969 situation – as Fisher amply proves with his numerous studies and as it is proven with the great volume of studies in each of the Eastern European countries, with the major exception of Albania.

It has to be further understood that interwar administration in East-Central and Southeast Europe did not emerge through local struggles and alliances, but has been imposed from above with some, often limited, discussions at the lower level. In the case of Yugoslavia, not only the territorial administrative units, but also their heads, have been appointed rather than elected (mayors of communes, heads of districts – *srezovi, veliki zupan or ban*).

In Czechoslovakia the elected administration on the local level was actually appointed by a 'district *hejtman*' appointed by the central government.

When trying to assess the effects of the territorial administrative organization upon the geography of a country under consideration, one is tempted to accept the old saying 'plus ça change, plus c'est la même chose' as the basic rule dominating the local, predominantly rural, areas. However much more pronounced the effects at the higher hierarchical peaks may be, abrupt changes adopted at the higher levels tend to turn into gradual transformations when reaching the village level, as Halpern has demonstrated in his study of Orašac.[10]

In conclusion, any analysis of the territorial administrative organization and its impact faces, first of all, the difficult task of assessing the operation content of the region. Secondly, it must include what extent the operational set (economic as well as social) is supported or hindered by the bureaucratic structure at its different operational levels. The myths and realities, hopes and accomplishments,[11] all intimately and intricately tied to the operational structures, make a thorough and especially a precise assessment of the actual conditions very difficult.

NOTES*

* The comments have been prepared by the author with the assistance of Mr Joseph Tvaruzka, Visiting Lecturer at the University of Washington. The responsibility for the statements, however, remains my own.

[1] Derwent Whittlesey, 'The Impress of Effective Central Authority upon the Landscape', *Annals of the Association of American Geographers*, Vol. 25 (1953), pp. 85–97.

[2] The era of the 'cult of non-personality' refers to the post-Stalin period in the C.S.S.R. when the country was dominated by the mediocre leadership led by Novotny and their confused, unprofessional politics.

[3] F.S.C., Northrop, *Philosophical Anthropology and Practical Politics*, (New York: Macmillan, 1960).

[4] The discrepancy between the positive and the living law develops not only through the spontaneous behavioural pattern of the people; sometimes the official machinery practices a living law contradictory to the positive norms but tolerated under certain circumstances. The existence of '*tolkach*' (pusher) in the U.S.S.R. and of the formally illegal practices of the communist parties and secret police groups illustrate the point.

[5] Jack C. Fisher. See Chapter 8.

[6] Eugen Pusić and Annemarie Hauck Walsh, *Urban Government for Zagreb, Yugoslavia*. Praeger Special Studies in International Politics and Public Affairs (New York: Frederick A. Praeger, Inc., 1968).

[7] James W. Fesler, 'Approaches to the Understanding of Decentralization', *Journal of Politics*, Vol. 27 (1965), pp. 536–66; specific reference on p. 537.

[8] Fesler, *op. cit.*, p. 545.

[9] For general discussion, see Mihailo Popović; *Problemi drustvene strukture*. (Belgrade: Kultura, 1967).

[10] Joel M. Halpern, *A Serbian Village* (New York: Harper and Row, revised edition, 1967). See particularly Chapter 13, 'Orašac and Twenty Years of Change, 1955–1966', pp. 301–39; Joel M. Halpern, *The Changing Village Community* (Englewood Cliffs, N.J.: Prentice Hall, 1967); P. Marković and Darinka Kostić, 'Structural Changes in the Yugoslav Countryside in the Postwar Period, 1945–1961', John Higgs (ed.) *People in the Countryside* (London: The National Council of Social Service, 1966), pp. 237–59.

[11] Joseph Velikonja, 'Yugoslavia Revisited', *Liverpool Geographical Society Bulletin*, 1969.

Comments
LESZEK A. KOSIŃSKI

I would like to divide my comments in this paper into three parts and to discuss (1) the basic trends, (2) the problem of an administrative area as an economic region, (3) the background of changes in administrative divisions.

(1) In the paper an interesting distinction was made between the trends towards centralization and decentralization on one hand, and unification versus preservation of the existing patterns on the other. It seems to me that the centripetal and centrifugal forces so vividly discussed in the present literature of urban growth can also be found in a study on administrative divisions in Eastern Europe. The existing patterns and their changes definitely reflect the intricate interplay of those two forces. The resulting situation changes in time as has been indicated in the paper. One can argue perhaps that the tendencies towards centralization or decentralization are of regional or national origin and they result in highly diversified administrative structure throughout the area. On the other hand, the tendency towards greater unification can be said of foreign origin, and this tendency acts against the inertia of the existing patterns. It seems that the factor of inertia is powerful enough to warrant a highly diversified system of administrative divisions in Eastern Europe.

(2) The author rightly points out that in a system of centralized economy, in a system where the public ownership of means of production prevails, there is a tendency to use the administrative areas as economic regions. In all countries of East-Central and Southeast Europe this tendency is very markedly pronounced. It is also reflected in the scholarly interests of economists and geographers who are much more concerned with problems of economic regionalization than scholars in other countries. It seems, however, that there is a constant contradiction between the rigidity of administrative divisions, even if they are frequently changed, and the dynamic and constant changes of economic situations. Hence, there is a continuous problem of two networks differing from one another. In

addition, the factor of inertia should also be mentioned with some old centres continuously attracting new investments and serving as central places for larger areas than would have been expected from their position in the administrative network. In any discussion on the reform of the administrative division those old centres enjoy privileged positions.

I can quote here two examples: after 1919 and the unification of Poland the former Russian division into *gubernias* was abolished, and in several cases, the seats of local governments were moved to other cities. Consequently, old centres of *gubernias* were degraded. However, even if they lost their administrative functions they retained enough central functions to serve as central places of some importance. When in the mid-1950's the discussion concerning the changes in administrative divisions was resumed, the names of those former seats of *gubernias* were again put forward, e.g. Łomża, Siedlce. This indicates the inertia that lasted for nearly forty years. After that long period of time, those towns were still considered to be important nodes affecting spatial organization.

Another example concerns the large villages in Upper Silesia selected by the Nazi authorities as centres for investment and development. In the late 1930's and early 1940's those centres were developed more than any others in that area. According to the regional plans based on Christaller's concept, those centres were to be selected central places, but at the same time centres of Germanization. After the war when the area was incorporated into Poland the places selected in the 1930's were again selected as nodes of development in the late 1950's and early 1960's. The factor of inertia was definitely present here.

(3) Finally, it seems to me that Professor Poulsen underestimates the role of studies made before the administrative changes were introduced. Central authorities frequently called upon geographers to prepare some rationale for the changes. I know that Polish and Czechoslovak geographers were very actively involved in the projects of administrative reforms. In Czechoslovakia several studies were published by M. Střida and others. Those studies should be added to the bibliography.

Remarks

Following Professor Poulsen's paper Dr Karger opened the discussion stating that there were basically two reasons for new patterns of administration in Eastern Europe. The first, which was related to internal political reasons, had as its objective the breaking down of older regional traditions. Changes for this reason were not unique to socialist countries but had also taken place in revolutionary France and in Russia under Catherine II. The second reason, he said, was economic and originated in the experience of economic development in the Soviet Union. The same principles that had been successful there were introduced into Eastern Europe in the 1950's, i.e. the experience of western Siberia was used in the reorganization of Upper Saxony.

Dr Fisher then said that it was clear that the structure and operation of local administration is a function of the local level of development. In a book by Dr Eugene Pusić, it was shown that the lower the level of *per capita* income the higher the number of people employed in local administration.

Dr Kasperson then asked whether, by using the example of Dr Whitney's work on China on the relationship of subcentres to the capital, it was possible to analyse a number of stages through which a country passed in patterns of administrative structure.

Dr Ivanička explained that major changes had been made in the administrative structure of Czechoslovakia in 1960 when nearly two-thirds of the former local town centres had lost their functions. He continued that in 1960 the reorganization of territorial divisions was made on the following basis: (1) the improvement of the existing network of districts; (2) a reduction in the size of the former large regional units and a doubling of their number; and (3) the introduction of a two-tier administrative division in which the national government dealt directly with the districts which in turn had their powers of self government increased.

Dr Karcz pointed out that getting rid of administrative

functions and units was always difficult because of built-in local resistance, as in the 'Hundred Counties of North Carolina'.

Dr Hamilton then raised several points. He posed the question of whether the increasing awareness of the need for economic efficiency in Eastern Europe had resulted in any changes in the pattern of administration or the kinds of services offered and whether there was often a definite conflict between economic, political and social ends involved there.[2]

Dr Hamilton indicated that two issues seemed to be relevant under 'regional structure': (1) the significance of the administrative boundaries for planning – how far are plans considered in terms of local units from the point of view of co-ordination and allocation of projects, and (2) the effect of the regional resources and potentials upon actual decisions. Do the administrative regions here have any real significance, and could a problem of this magnitude be incorporated in the papers?

Finally, Dr Hamilton expressed disagreement with Professor Velikonja's charge that it is not a rewarding task to try to compare pre-war capitalist with post-war socialist East-Central and Southeast Europe, but rather to compare pre-war capitalist with a projected non-Communist, non-Soviet dominated postwar East-Central and Southeast Europe. He said that the latter can only be hypothetical and raises, in any case, serious questions as to whether, even without U.S.S.R. backing, East-Central and Southeast Europe would not have become significantly socialist (as West Europe and Austria have) with a very large degree of state intervention. He quoted Phyllis Auty who once wrote that 'wars are great levellers' indicating her feeling that socialization in this area (all things considered, including a reactionary peasantry) was inevitable.

Professor Poulsen first answered Dr Hamilton saying that on questions of decisions of regionalization there were very few functional links between the speculations of geographers and economists and what actually took place. He said that there was a welcome trend in research (for example the work of Pusić and Fisher in Yugoslavia) which investigated field service regions. Here they describe what existed rather than what should exist. Despite assertions to the contrary, it did seem that ideas behind designing regions were rarely very rational and were more often influenced by political rather than economic ends.

In answering Dr Karcz, Professor Poulsen agreed that there was always an immense problem in changing boundaries of counties or cities, not the least aspect of which was the fact that in most areas there was usually effective opposition by local officials themselves to any change in the *status quo*.

Professor Poulsen said that he was interested in the points raised by Dr Ivanička on changing administrative units once again in Czechoslovakia and pointed out that one of the problems in that country had always been the lack of comparable regional units. He felt, generally, that the role of geographers in changing administrative regions was very limited; they had not been listened to in the Soviet Union. In his own study of regions in that country, he had concluded that despite small changes there was an enormous degree of similarity with the past system of organization. Despite great fluctuations, the present *oblast* network in the Soviet Union is little different from the *gubernia* system under Catherine II. He pointed out the great centrality of provincial centres in the Soviet Union, where the chief provincial city is nearly always twice as large as any other centre in the province.

In referring to Dr Krager's points, Professor Poulsen said that although there had been an attempt to break down the identity of the Länder in the DDR, this was not the case in Poland and Hungary where the traditional provinces retained their identity. The speculative ideas of Mendeleyev and Semionevtian-Shanski had been adopted by the radicals of Pre-Revolutionary Russia, and implemented, to some extent, in the Gosplan macro-regions of the 1920's, but these unwieldy units lasted only a few years and it is doubtful today if administration has been ever taken out of the hands of the old *gubernia* centres, nearly all of which became the capitals of *oblasts* and republics in the 1930's.

Professor Poulsen admitted that Yugoslavia was a maverick in administration and that centralized decision-making did not have the same impact there as elsewhere in Eastern Europe. Rather, the Yugoslavs had pioneered development of the small self-governing commune as the basis for most decision-making.

7

Nature versus Ideology in Hungarian Agriculture: Problems of Intensification

FRED E. DOHRS

Despite significant innovations in the management and operation of a collectivized farm system during the past decade, Hungarian agricultural productivity remains at a low level; yields are poor and *per capita* increases have been small. The flexibility of planning, operating and payment methods offered Hungarian collectives has to some degree improved labour efficiency, but actual production increases have been modest. Concern continues to be expressed by Hungarian leaders about the serious nature of the agricultural problem, yet the current Five Year Plan (1966–70) continues the pattern of limited capital investment and a very deliberate pace of agricultural growth.[1] Such policies take only limited advantage of the natural resource base.

Detailed field evaluation and published analyses, however, recognize the relative richness and varied character of the natural resource base for a productive agriculture.[2] This assessment remains valid despite the 1968 drought, the most serious of the century, and the floods of 1965 which delayed work on many thousands of hectares of quality cropland, so that the best return could not be achieved. Climatic, soil and surface conditions, nevertheless, continue to offer a framework, which if coupled with better organization and management – intensification – should substantially increase output and raise the incomes of the 38% of the population directly dependent on agriculture for an income.

With an industrial raw materials and power base much lower than neighbouring countries, and despite far greater industrial development than now exists, Hungary must look to a more productive agriculture as an important element in an improved standard of living. This study suggests that a more intensive agriculture, effectively utilizing the best of the natural endowment is possible, even within the obstructive framework of collectivization, and would contribute materially to the Hungarian economy.

Although rural population densities in Hungary imply the high labour inputs characteristic of some types of intensive agriculture, it appears that in a collectivized and nationalized agricultural system, despite large numbers of farm workers involved, actual labour inputs remain low. Members of collectives, generally, continued to work only as much as is required on collective land, preferring to put in many hours of personal time on their own private plots. Collective farms (called co-operatives in Hungary) are the major users of land in Hungarian agriculture and also the major producer of agricultural products. They occupy 80% of the cropland and produce about 72% of the gross agricultural productivity; in addition they enjoy an independence and autonomy of decision-making virtually unknown in other East European and Soviet collectives.[3]

At present there are about 2,840 collective farms. These are divided into three major categories. Some 500 of these are of first quality and are called 'good' or the 'best' farms. The middle 1,400 or so farms are classified as average or mediocre – the best of the average approach the quality of the good, and the worst are just barely able to break even. The balance of 900 farms, amounting to about 32% of the total, are classified as 'poor' collectives: some located on physically marginal land; others very badly managed, and, as a result, the production yields and income are very far below average, producing only 20% of the gross output of collectives.[4] Every year these operate with large government subsidies, not only for their production, but more importantly to provide some sort of income for the members of the collectives. State aid is derived chiefly from the profits of the 500 good farms to distribute among the 900 poor farms. Thus the good farms are literally being taxed to keep the poor farms solvent, and their members alive. These poor farms constitute a very serious drag on the whole economy, especially because the profits from the good farms cannot be ploughed back into productive agriculture and make the good farms even better, but must be siphoned off to pay for the necessary subsidies for the poor farms. It is not simply a matter of saying 'the poor farms must start shaping up and start producing'. These farms are serving the social purpose of providing a living – poor though it may be – for a large number of human beings, for whom there is no alternative occupation available at the present time. The farm, however inefficient, cannot simply be closed down, or the land allocated to other more efficient farms, because this fails to answer the question of what happens to the people who are living there? It is clear that something must be done about these poor farms, but whatever can be done, and because there are so many of them, it is going to be a long-term solution if indeed there is a solution for all of them, and for some, perhaps many, there may not be any solution except continuing deficits.

The average statistical collective has an area of 2,500 acres of cropland, nearly 4 square miles. These are by no means small farms – especially if we think in terms of intensive agriculture. In addition they have some areas of non-cropland, either permanent pasture, swamp or woods, averaging 1,100 acres per farm. There are on the average farm 315 members of the collective; there are often one or two and possibly three, but usually not more than two members per family, so there may be 200 to 250 families living on this farm. This means about 9 arable acres per collective member, implying a labour intensity which does not exist at present.[5] This is an extremely low figure for contemporary agriculture in what should be a fairly well advanced agricultural economy, and is one measure of the inefficiency of the operation.

To these data must be added another figure which is equally crucial and in many ways shocking: the average age of the collective members, 56 years. This is the *average* age; many are older, and probably more than are younger because of retirement age limits of 65 for men and 60 for women, whereas the younger could be in their 20's or 30's and thus alter the average significantly. A median figure for the age problem is not available but it probably is a little higher than the average. These two figures, land per member and average age of members, focus the agricultural problem: large numbers of relatively old people working very little land.

In addition to the collectives, there are State Farms operating 16% of the cropped area, as well as a small private farming sector on a little more than 3% of the arable area. Because these two farm systems work less than 20% of the arable land, this study will be concerned only with intensification in the collective sector. Neither the private sector nor the state farms can be considered as particularly efficient farm operations, and much of this study can be applied to many of the state farms.

From a brief introduction, it is not only clear that greater intensification of agriculture is essential for increased economic well being, but also there is little question that substantially greater intensification is possible. The primary concern here is the character and pattern of intensification of Hungarian agriculture which can give the most substantial returns especially in the short run, but sometimes in the long run as well.

IRRIGATION AND WATER USE

The abundant water resources of Hungary are at present significantly under-utilized, and greater use of suitable surface and ground water for irrigation makes possible control of one of the important variables in agricultural production, and almost always results in greater productivity. Hungary not only has large amounts of surface water in the Danube and Tisza Basins, but there are also great quantities of ground water from renewable sources at minimum subsurface depths.

The principal climatic factor in water use and the need for irrigation is, of course, precipitation. On the basis of long-term averages of annual precipitation, the situation for Hungary appears to be ample, with nearly all the country receiving 20 inches or more (Fig. 7.1). In these latitudes (46°–48°N.), even on continental climate margins, this would appear to be sufficient for an effective, intensive and productive corn/hog economy. The seasonal distribution of precipitation alters this situation very little, with even smaller areas on the average, receiving less than 12 inches during the April–October period. Furthermore, only very small areas receiving these summer minima are also areas having less than a 20-inch annual average.

The evapotranspiration rate, however, greatly alters the otherwise apparently sufficient rainfall for intensive cropping. Throughout the Great Alföld plains area of eastern Hungary and well into the trans-Danube (Dunantul) region in the west and the Little Alföld of the northwest as well, average evapotranspiration rates are equal to or greater than average annual precipitation. The correlation with local relief is very close so that those areas having the lowest evaporation rates are also

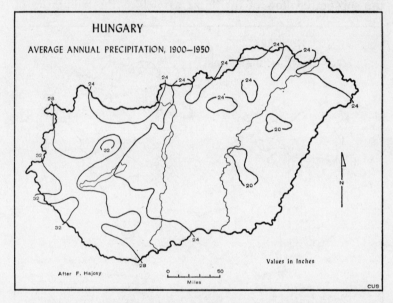

FIGURE 7.1 – HUNGARY: AVERAGE ANNUAL PRECIPITATION, 1900–50

those areas of uplands and mountains which severely limit agriculture.

As a result, even if annual average precipitation could be depended on as the rainfall regime for every year – which it most emphatically cannot – the serious early summer drought of 1968 underscores this. Even so, it must be recognized that large parts of Hungary must be classified as moisture-deficient for agriculture. A detailed agro-climatic pattern based on precipitation and evapotranspiration data, has been prepared, and those areas classified as moisture-deficient for agriculture have been delimited (Fig. 7.2). The extent of these regions demonstrate that to maintain the moisture requirements of a productive intensive agriculture consistently from year to year, a carefully controlled programme of irrigation over wide areas is mandatory.

The most obvious source of irrigation water in Hungary is the massive flow of the Danube which, in most of its course through Hungary, flows north to south. To this can be added the substantial flow of the Tisza as well as the tributaries of

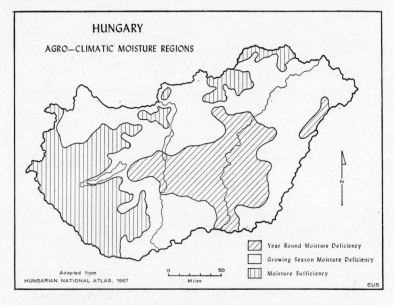

FIGURE 7.2 – HUNGARY AGRO-CLIMATIC MOISTURE REGIONS

both of these major streams. Data for the Danube show a mean annual flow at Bratislava of 2,080 cubic meters per second and at Budapest, 2,330; the Tisza at Szolnok has a flow of 600 cubic meters per second.[6]

Average annual flow data imply an abundance of water available for irrigation, and the annual regimes of these streams is such that highest levels correlate closely with the growing season, May through September, and especially well with the moisture requirements of corn. Furthermore, total drainage of all parts of Hungary is through the Danube and its tributaries (including the Tisza), so that the bulk of any water diverted for irrigation in Hungary will return to the Danubian system, although with high evapotranspiration rates some of the diverted water will be lost.

Up to the present, however, only a small amount of Danube and/or Tisza water has been diverted for irrigation (Fig 7.3). The total area (January 1, 1969) is 1,056,000 acres, and plans call for an increase to 1,102,000 acres during 1969.[7] This is out of a total of some 15,000,000 arable acres, or approximately

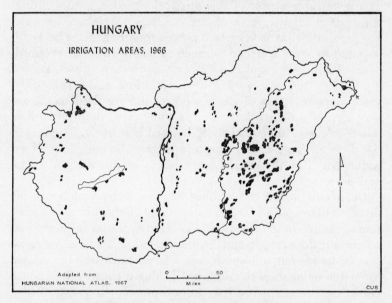

FIGURE 7.3 – HUNGARY: IRRIGATION AREAS, 1966

$7\frac{1}{2}\%$, an extremely low figure for a country which has nearly two-thirds of its cropland in areas which must be classified as moisture deficient. Various estimates of the irrigation potential range from 15% (Sarfalvi, 1964) to 35% (Sulyok-Schulek and Bacsi, 1965);[8] the latter figure appears to be more realistic, and even that may be slightly understated.

There are, however, several serious factors, natural and institutional, which continue to limit or inhibit a substantial increase of irrigation in Hungary. One of the most important natural factors at present is the serious flooding along the

Danube floodplain and adjacent lowland areas. These are of great extent and lie no more than 5 to 10 metres above the mean water level of the river; much the same is true for the Tisza floodplain and the Great Alföld. With river maxima occurring in June and July, these areas are frequently subjected to periods of several weeks of standing water, long enough to destroy existing crops or to prevent planting of other crops for autumn maturing. Flood control measures constructed to date have not prevented very extensive flooding.

A favourable aspect of this physical situation is that lifting river water to the level of the fields is a relatively simple problem, and has a very low power requirement. The head required to bring water to as much as one-third of the cropped lands of Hungary requires a lift of no more than 15 feet. Power inputs required are very low; calculations show that only three kilowatt hours of electric power are required to lift an acre-inch of water 10 feet,[9] and in some areas the inexpensive power-free hydraulic ram could be used effectively. The major cost item in large-scale utilization of Danube and Tisza river water would be the capital construction of distribution works and pipe.

Determination of the amount of river water which can be diverted to irrigation use is an important element in an evaluation of the irrigation potential. We have seen that the period of maximum flow corresponds to the period of maximum irrigation demand, but minimum water levels must be maintained to enable navigation to continue. Taking a conservative 8,500 cubic metres per second as a midsummer flow rate,[10] the Danube carries enough water to supply an inch of irrigation water to 100 acres of cropland every second. Irrigation of a major part of the Danubian floodplains by even a major water diversion would have no important effect on navigational levels. To a lesser degree this is also true for the Tisza floodplains, although much more of the smaller flow of the Tisza is currently being used for irrigation, and navigation requirements are modest.

Another abundant supply of irrigation water is available in ground water which nearly saturates a large part of the Hungarian lowlands, a part of the geologic Pannonian Basin. Throughout these lowlands, the depth of the water table level varies from about 1-4 metres (Fig. 7.4). Ground water at these shallow levels is often not potable because of local contamination (although in many places the water is used for domestic needs), but almost all is suitable for irrigation. The supply of ground water in the basin appears to be virtually inexhaustible, because the Quaternary sand and gravel layers function as gigantic aquifers which are regularly recharged by the abundant precipitation falling on the higher elevations

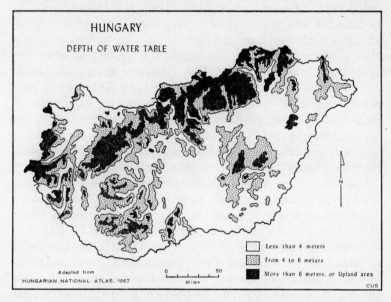

FIGURE 7.4 – HUNGARY: DEPTH OF WATER TABLE

forming the margins of the basin. The ground water is readily accessible, and at very low costs; the calculations shown earlier for raising river water 3 metres above level can be applied directly to pumping ground water, as a water table level of 3 metres includes large areas of the Hungarian lowlands.

To date, most of the developed irrigation in Hungary is utilization of river rather than ground water. There is, as there has been for many years, a small amount of very small-scale irrigation – by pail or similar hand vessel – of house-gardens and orchards, now chiefly on the private plots. Small, gas-driven portable pumps, which could pay their costs very quickly in increased yields on private plots and orchards are almost entirely absent from the Hungarian rural scene, although nearly every farm and private plot lies only a few feet away from an abundant supply of irrigation water.

The potential for irrigation whether from surface or ground water is very high in Hungary today. In addition to the flooding problem discussed earlier, there are further natural and institutional factors working against realization of this potential. A very important natural factor is the problem of drainage. In an area which is so nearly flat, with the extensive flood plains of the large rivers and their tributaries, as well as a very high water table, it might be thought that getting water on the fields is much easier – and possibly cheaper – than getting it off the fields, quickly, and at the proper time. Drainage is particularly difficult from irrigated fields during the high water of June and July although these are often periods when crops need substantial moisture. Research on this aspect of irrigation, however, suggests that drainage is a much less serious problem than high water table and recurring flood conditions suggest.[11]

Because of the very high evapotranspiration rate in the drier areas of Hungary, fairly large areas of alkaline soils have developed and are a factor in irrigation agriculture. Some $7\frac{1}{2}\%$ of the arable land has alkaline soils, varying from chemically neutral to strongly alkaline types. Distribution correlates to some extent with low precipitation and poor drainage. Draining techniques, liming and other amelioration methods have made possible effective cropping with irrigation practical on all but the most alkaline *solonchaks*.[12] The small areas of the latter have very little effect on agricultural productivity.

Both drainage and alkalinity have been successfully overcome in the present irrigated areas, which are sufficiently widespread to include some areas having the problems presented. It can be presumed, therefore, that extension of irrigation to much wider areas will present no serious problems which cannot be overcome.

Institutional factors have been of far greater importance in limiting development of irrigation. Most important of these is the familiar communist dogma that the bulk of capital investments must go into the industrial sector, a condition so well-known as not to require elaboration here. Hungary is no exception to this generalization, and the last as well as the current Five Year Plan maintain this imbalance.[13] This is perhaps a more serious factor in agrarian Hungary than in other more industrialized or potentially more industrialized countries of Eastern Europe.

When considering the costs of irrigation, however, several elements must be considered. As we have seen, most of Hungary's agricultural areas are climatically *marginal for agriculture*; they are not arid or even semi-arid in climate. Irrigation in Hungary means providing only the moisture margin between that which is present in natural precipitation and soil moisture, and that which is required for a successful crop – and at the proper time and in the amount required. In some crop years, then, only a few inches may be needed, and in others substantial amounts – as much as 15 to 20 inches – may be required during the growing season. In order to be able to supply the maximum water necessary in *any* year, the irrigation system must be built to a high capacity. For the average precipitation year, demands on the system would be well below its capacity. In effect, then, there is an investment requirement for an irrigation system which would not be used to capacity in most crop seasons, but such stand-by capacity is not unusual for irrigation systems.

A case can be made for the greater flexibility of a system based on ground water with smaller and more mobile or flexible pumping and distribution units which can be used more readily to meet local requirements. A combination of this plus a modest surface water system might well meet maximum water demands with minimum capital inputs. Details of such a balanced system, however, lie beyond the purview of this paper, although it is clear that application must be on carefully selected land; further detailed field work and research are required

Even with limited capital and long-term plans, the regime has tended to invest each year in that section of water control which is most desperate at the time – flood control following floods, and irrigation following droughts. As a result, a project may frequently have limited value for several years after completion – and return on the investment may be delayed.

A further limiting factor on rapid expansion of irrigation is a shortage of technically trained and competent personnel. Irrigation agriculture is far more exacting than the usual farm labour common to present farming practices on collectives as well as private plots. Water must reach the crop in correct amounts and at the proper time, then must be drained off suitably to prevent damage. Because of the high water table and low stream gradients, dangers of continued soil saturation are chronic problems and require much more attention, care and precise timing in water application and drainage than is usual in irrigation systems elsewhere. Specialist opinion in Hungary suggests that few peasants are equipped to make the critical timing decisions essential in anything but the smallest and simplest irrigation works. Prospects for expansion of suitable training appear to be limited. One of the important reasons is the somewhat advanced age of the farming population, and although efforts are being made to make farm life more attractive to young people, success is limited. Even if substantial capital investments are made for development of irrigation, without accompanying large-scale technical training, crop losses through mishandling rather than gains may be incurred.

Although only indirectly related to irrigation and water for irrigation, there are other water sources of some importance. These are the widely distributed subsurface thermal waters – many of an artesian nature – which have enjoyed world fame for their medicinal and curative properties. While generally unsuited, because of mineral content, for irrigation, they nevertheless have another property – temperatures from 105 to 165 degrees Fahrenheit, which might be used for local heating and extension of the growing season. An intriguing problem, but one on which little substantive research has been conducted as yet. A geographer with the Hydrological Institute, Budapest, has done the initial study to date and has proposed a modest field project to test the idea.[14] If at all successful, thermal waters may make a substantial contribution to a further intensification of agriculture.

When one attempts to evaluate realistically the possibilities of the development of irrigation agriculture to something approaching the scale indicated here, it is easy to be optimistic.

The image, if not vision, of Hungary functioning as an Indiana-sized truck farm and feed lot, producing in great quantities for the growing urban markets in both East and West, is difficult to dispel, especially when tramping the fields of one of the more efficient collectives in the company of an enthusiastic agronomist. But a realistic appraisal must recognize that there is little likelihood of the regime allocating sufficient capital and resources to bring about the programme outlined here. More likely, then, is repetition in the future of conditions which took place in the early stages of the devastating drought of 1968, when in County Györ 70 out of 230 small irrigation plants were inoperative because of spare parts shortages and other technical difficulties.[15] In this aspect of the conflict between nature and ideology, it is clear that ideology is losing, in the sense that continued inefficiency is resulting in continued low production.

THE PRIVATE PLOT MODEL

In one sense highly intensive agriculture already exists in Hungary on the private plots, and they offer a model for greater intensification which might be applied beyond the currently very restricted areas of the private plots. The private plots are highly labour-intensive, and probably excessively so, although little empirical data exists to indicate actual values for labour inputs. But the character and the conditions of private plot farming clearly require massive amounts of hand labour to achieve known production results.

Recognizing that high labour inputs can be an effective method of intensification under some agricultural conditions, a projection from this model should indicate some potentials. It is a well-known, if imprecisely documented, fact that the private plots produce well out of proportion to the amount of land involved, and especially in relation to land farmed on collective or state farms. Even knowing that significant quantities of livestock fodder, and small amounts of other needs, including fertilizer, seed, etc. are 'liberated' or otherwise acquired from the collective and used on the private plots, the productive performance on this land is unquestionably disproportionate and in some cases, even phenomenal (Table 7.1).

The question must therefore be asked: what are the possibilities of applying the private plot model over wider areas and achieving not only intensification on more land but also greatly increased production? The role and importance of

TABLE 7.1 – *Role of the Hungarian Private Plots*

Item	% of National Total
Arable Land (1968)[a]	9·6
Value of Products (1968)[a]	21·5
Cattle (1968)[b]	42·0
Of which, Cows (1968)[a]	48·3
Hogs (1968)[b]	36·0
Poultry (1968)[b]	75·0
Milk (1967)[c]	37·0
Of Exports to Western Europe (1966):[d]	
Beef	35·0
Poultry	24·6
Eggs	50·5
Fruit	27·3
Wine	26·0

Sources:
[a] *Nepszabadsag.* January 22, 1969.
[b] *Nepszabadsag.* November 6, 1968.
[c] *Figyelo.* May 29, 1968.
[d] *Magyar Mezogazdasag.* January 24, 1967.

private plots in Hungarian agriculture has increased significantly in recent years, and they are receiving increased attention and benefits from the regime in some ways, but not in greater size.

Working from these figures (see Table 7.1), if the area of Hungarian private plots were doubled, and farmed with the efficacy of present operations, then an output of approximately 42% would be the result, and proportionally higher until, with private plots occupying half of the arable land, present gross production would be equalled. This sort of calculation should not be carried too far, however, as it can lead to wildly unrealistic conclusions. For example, the current estimate for private plots in the Soviet Union suggests that they produce one-third the value of the total Soviet output on 3% of the arable land, thus implying that by tripling the private plot area, the present gross value of all agricultural production could be equalled – and on less than 10% of the arable land! Rather than pursue such comparisons, however, perhaps it should be asked: what is the optimum size for a private plot worked by a collective member and family in the present manner which could sustain current yields and livestock growth? That is, recognizing the labour-intensive character of present private plot farming, at what size plot would diminishing returns set in?

Present laws allow at least a minimum of 0·285 hectares for each member working a minimum of 150 ten-hour days on the collective land, and up to a maximum of 0·57 hectares for 300 ten-hour days on the collective.[16] In theory, then, a family with three collective members (though the present national average is 0·44 hectares – 1·1 acres – per peasant family) could have a private plot of 1·71 hectares. This is quite unlikely as there are few jobs on most collectives – even dairy farms – offering 900 days work per year, and the practical likelihood that one family would have three of these jobs is remote indeed. Although a workday is very flexibly defined,[17] the intent of current regulations is to increase the commitment of the collective member to work more on collective land, yet provide opportunities for slightly larger private plots as a means of ensuring gross agricultural output.

The vast majority of collective members will not and cannot be working 300 days, and 200 is a more likely average. Peak labour demands on all farms come during the crop season with harvesting calling for the maximum, and commonly require additional labour from nearby cities and towns. Winter and early spring labour requirements are minimal, except on dairy and some livestock farms. Other than caring for privately owned livestock, the same conditions obtain on private plots.

Doubling the area of private plots would take another 10% of the arable land which would be of qualities comparable to the local collective. As mentioned earlier, there are some 9 acres of collective land worked by each member, with a substantial amount of mechanization, although most work on row crops continues to be done by hand. Doubling the size of the private plots would reduce this average by about 1 acre per member, and increase the private plot about the same amount. Working an additional acre of land privately, even entirely by hand labour, should not present any serious difficulties to the farm family, although there might be problems in some cases. Making the increases voluntary could overcome this latter problem.

An important step which shows to some extent the serious concern of the regime for the importance and increased efficiency of private plot productivity, was the introduction in 1967 of small garden tractors for use on private plots, especially those having orchards and vineyards.[18] Although only 500 of these machines were imported from Czechoslovakia in 1967 on an experimental basis, the willingness to provide any power machinery for use on private plots is an important forward step for a communist regime committed to the collectivized concept. Even so, 500 tractors per year can hardly be viewed as a substantial measure for meeting the needs of a large part of the private plots for mechanization. Even after several years of imports, less than 1% of the private plots are supplied with these tractors.

The small garden tractor or the 'one-axle' tractor could be the instrument for revolutionizing private plot agriculture and would make farming on enlarged plots very efficient. One-axle tractors are now used in large numbers on small farms in Austria, Switzerland and West Germany, and increasingly in private agriculture in Yugoslavia, where yields are frequently no larger and often smaller than from Hungarian private plots. They are remarkably flexible in performing all types of field operations and for local hauling and transport as well. Their initial costs are relatively high – three to five hundred dollars – but the return on the investment is quickly recovered. Even in personal transportation alone – walking, bicycling or horse and wagon – the use of one-axle tractors could save enormous numbers of man hours which might better be used for productive labour on the fields.[19]

Widespread use of one-axle tractors presents an ideological dilemma which could not be resolved by any 'bending' of the collectivized concept, but would have to be adopted on primarily pragmatic terms. One possibility discussed by the author with a senior agricultural official would be to keep ownership of these tractors in the name of the collective, and lease them on short or long terms to the membership, thus maintaining the principle of collective ownership of the means of production. This would be similar to the assignment of collective fields for individual or family working under the *Nadudvar* incentive systems, now a common practice throughout Hungary. With a small tractor readily available, doubling or tripling private plot area and increasing production at least proportionally could be accomplished with ease.

Although at present there appears to be only limited concern by the regime for this and other problems of intensification, in a few years substantial changes may be required, simply to maintain existing levels of productivity. The advancing age of collective membership presents a very critical problem, but at the same time offers an opportunity to make essential changes in the agricultural structure. Earlier it was stated that the present *average* age of collective members is 56 which means that many are approaching retirement age (65 for men and 60 for women), and many others (more than 200,000) have already retired and maintain their collective memberships and private plots to which they are entitled. Furthermore the reduced birth rate during and immediately after World War II has meant that the 20- to 30-year age bracket shows a significant gap at present, a gap which is of importance to industrial growth as well as agriculture.

Before the end of the current Five Year Plan (Dec. 31, 1970), estimates indicate that there will be a shortage of more than 100,000 trained agricultural workers (collective members) and at least as many less skilled workers as well.[20] At present there does not appear to be any prospect of being able to obtain trained personnel, either by keeping young people on the farms, or getting others to return from the cities, although serious efforts are being made to promote the attractive features of farm life as well as its increasing social 'respectability'.[21]

Although this farm labour crisis lies several years away, there is no question but that significant alterations of the farm structure will be required. In replying to questions about this developing problem, many specialists suggest that it will be solved simply by replacing the people with machines.[22] The plans and programmes for either building or buying the required machinery and other equipment is, however, not underway at present, nor does it appear to be projected, as yet at least, for the future. Lip service, but not production or capitalization service, is being paid to the need.

While it is convincingly clear that agriculture can and must be intensified, and that furthermore the private plots offer a usable (though far from ideal) model for high labour input intensification over larger areas of land, it is also equally clear that whatever intensification methods are to be used, they will probably not be capital intensive in character.[23]

Within the restrictions of low capital inputs, however, expansion of private plot areas offers reasonable prospects of substantially increased productivity. The actual enlargement of private plots themselves would require little additional capital, but could be accomplished within the administrative framework of the collective itself. There is no likelihood that such a move would be resisted by the collective farm membership, who continue to hold strong views on the collective principle – present collectivization levels were reached only as recently as 1961.

Operational efficiency and increased productivity would be greatly enhanced, however, if the one-axle tractor programme, initiated in 1967, were to be substantially expanded – to the limit of capital available. Using the 500 dollar per unit cost figure suggested earlier, the annual import of 500 units would be 250,000 dollars. Increasing this 100 fold, to 50,000 machines, would have a cost of 25 million dollars, a substantial sum in the Hungarian capital budget, but nevertheless, only 10% of the 1969 allocation to agriculture.[24] Fifty thousand tractors would mean only one tractor for approximately 15 really active collective members, which is too high a ratio for the most effective use, but which would nevertheless give tremendous impetus to the whole spectrum of Hungarian agriculture. Furthermore, as the tractors would be owned and 'leased' by the collective, their use could be made more effective through careful allocation. In addition, the export sector (perhaps the most important reason for intensification) could be greatly strengthened in the same manner.

Finally, in considering combining the two means of intensification – increased labour intensity through larger private plots, and small scale, modest mechanization – two fundamental points must be restated: (1) In adopting a sliding scale for allocation of private plot area (even though based on labour inputs to the collective), the regime has approved a mechanism which legitimizes further increases of private plots in the interests of increased overall agricultural production. (2) The ideological objections to providing mechanization and power for use on private plots were overcome with the initial 500 tractors imported in 1967, despite the stated implication of an experiment.[25] Furthermore, numerous precedents exist for using larger equipment for some work on private plots.

Although some ideological reservations must remain concerning both of these changes (as they certainly do, even today, in some quarters for the widespread application of *Nadudvar* and related incentive systems), it would appear that a regime, pragmatic and willing enough to make these earlier decisions, in principle and on a limited basis, should be able to apply them far more broadly.

Enlarging private plots and making tractors available, however, do not directly overcome the continuing and increasing problem of ageing on the collectives. Although specific supporting data are not available, it can be suggested that should these changes be adopted and implemented, farming as a way of life in Hungary would become much more attractive to young people. The use of efficient machinery on an area of land large enough to provide a substantial, if not handsome income – an income which, however, would be well in excess of that of unskilled factory labour – would be a very important way of holding younger people on the farms.[26] Conversely, the reduction in drudgery and heavy hand labour made possible by introduction of small tractors should result in continued higher productivity by the older collective members. Although the present migration patterns to the cities, especially Budapest, does not suggest it, there are few farms more than 60 miles away from a city of more than 100,000 population, so that many attractive features of the urban scene are increasingly accessible from rural areas, especially as private transportation by motorbicycle and car increases.

The combination of elements – ageing of rural population, regime decisions and mechanization, private plot size, the flexibility of leasing and payment methods under the *Nadudvar* system, limited amounts of capital, and the need for increased agricultural output – all of these are coming together at a time calling for serious decisions. Transforming the private plots offers a way of assuring increased output with a minimum of ideological compromise and allocation of scarce capital resources.

OTHER INTENSIFICATION

The two types of intensification examined in this study – irrigation and private plots – are two of several which could be applied effectively to the Hungarian agricultural problem. They appear to offer the most favourable prospects for rapid and substantial increases in output. Two others are worth mentioning.

Chemicalization is only just beginning in Hungarian agriculture. Although use of fertilizer is increasing rapidly, present consumption averages approximately 10 pounds of active materials per acre of arable land, an extremely low figure. Widespread applications of insecticides – except in the vineyards and orchards – are quite limited, and herbicides are being introduced on a limited scale. Increased inputs of all of these, but especially fertilizer, would increase yields substantially. Present plans call for minor increases in production, import and application of fertilizers during the next several years, but no large increases are envisaged. Any further chemicalization in Hungarian agriculture could be accomplished effectively within the structure of either or both intensification programmes considered earlier.

It has long been something of an article of faith among communist regimes that they should grow sufficient wheat and other grains to meet the bread requirements of the country. Usually this has been accomplished by requiring every farm to plant a certain percentage of the arable land with wheat; in Hungary, until 1968, each farm was required to plant about 30% of the land area with winter wheat. Although this requirement was rescinded in 1968, agricultural specialists expect that this level will be maintained because in Hungary, self-sufficiency has a chauvinist as well as a communist character. The difficulty is that much of the wheat is grown on some of the best land which might be better used for higher income-yielding crops. Although no longer a mandatory crop, plans call for a gradual reduction in wheat acreage as yields increase, so as to maintain a relatively constant output. Failure to produce to this level would be costly either in hard currency imports from the West or politically through imports from the Soviet Union. Both 1968 and 1969 wheat crops were well above domestic requirements, and some exports were possible.[27]

CONCLUSION

In this evaluation of major intensification through irrigation or the private plots, feasibility seems clearly evident. But two questions must be asked: first, does the regime really have the resources to accomplish either or both of these projects? Without going into an analysis of detailed budget allocations, the answer must be a qualified yes, qualified because the key manufactured elements in both programmes – pumps and one-axle tractors – are not produced in any quantity in Hungary today, although the capability certainly exists. Initially, substantial imports would be required to implement these programmes, but they appear well within the practical economic framework existing in Hungary today.

Second, will the regime make the required decisions and take the necessary steps to bring either or both of these programmes into being? Here, the answer must be a qualified no, qualified because the regime has already taken many of the necessary introductory steps, freeing the agricultural structure farm more than has happened in other communist countries. But it may not have reached a point, as yet, where the ideological wrench or subordination of vested interests, even for the sake of agricultural production can, in reality, be undertaken.

Finally when one looks at the Hungarian situation either from the perspective of more than 4,000 miles distance, or when wandering through rural Hungary and feeling the potential of a rich and productive garden, it seems difficult to believe that these decisions can be long delayed. Certainly, if positive decisions are not forthcoming, the 1970's will be difficult not only for Hungarian agriculture, but for the whole Hungarian economy as well. It may be then that they will be taken in the near future simply because there is no really effective alternative.

SUMMARY

Any programme suggesting courses of action which require substantial allocation of scarce resources is bound to be in conflict with all other possible courses of action, not only alternatives in agriculture, but also in other parts of the economy as well. Communist regimes commonly make many decisions of an economic nature primarily on political or ideological criteria – Stalin's original collectivization decision was of this sort. Furthermore, if collectivized agriculture is viewed as it is by many chiefly as an instrument of political control, the alternative of maintaining the *status quo*, however poor it may be, is always available.

The important aspect of the present situation in Hungary, however, is that in recent years collectivized agriculture has been released from many of the shackles of the rigid Stalinist model. Several decisions have been made by the regime which can be used as models for further steps along the same lines. These decisions must be construed as important progress away from the collective as primarily an instrument of political control and a recognition, though limited, of the collective as a means of achieving increased agricultural output.

It is somewhat fallacious to view the proposals made here as being of short-term importance only. Almost invariably, irrigation systems are viewed and constructed as having long- rather than short-term production advantages, and in this, Hungary is no exception. Enlargement of private plots would indeed result in substantial immediate production gains, but when coupled with the mechanization provided by one-axle tractors (as increasingly appears to be happening in Hungary), long-term benefits are assured.

Reservations have been expressed about projecting production results from increased private plot areas – and with some justification. Yet, the fact remains that within the whole spectrum of communist agriculture over the years, the private plots stand out as the only consistent high yield producer. We cannot, however, base any projections on any actual record of increased area because of regime resistance to enlarging the plots. But surely, the private plots are a better measure of the potential of intensive agriculture than the record to the present of unwieldy collective and state farms.

The present condition of Hungarian agriculture needs to change. Intensification, the primary concern of this study, offers a practical method consistent with ideological acceptability.

NOTES

[1] Statute No. 2/1966, Plan Law. Imre Demeny, Minister of Agriculture and Food, statement that only 50% of investment demands for farm modernization can be satisfied (*Magyar Nemzet*, January 21, 1969). The writer is greatly indebted to the research staff of Radio Free Europe for many translations of Hungarian publications.

[2] The best single source in English is M. Pecsi and B. Sarfalvi, *The Geography of Hungary* (London: Collets, 1964). In addition, the series, *Studies in Geography in Hungary*, Academy of Science, Budapest, has many excellent articles on Hungarian resources and agriculture. *Magyaroszas Nemzeti Atlasza* (Hungarian National Atlas), (Budapest, 1967), has a wealth of basic resource and production information in maps and tables. In September and October, 1966, October, 1967 and September, 1968, the writer visited a number of collective, a few state and several private, farms. Almost invariably, with senior farm administrators, agronomists, technicians and peasants, the comments to the visitor were, 'Hungary is a poor country; we have few resources; please don't judge our poor agriculture by your American standards.' On one collective, for example, after suggesting a trade of 70 hectares of stony glacial till of my family's Michigan farm for the same area of the rich Hungarian silt loam we were standing on, so that my farm would be more profitable, there were looks and comments of disbelief. The notion that Hungary's agricultural resources could be the basis of a highly productive and remunerative agriculture seemed not to be an acceptable hypothesis except among a few specialists in the Institute of Agricultural Economics of the Ministry of Agriculture.

[3] A detailed study of the Hungarian *Nadudvar* and other incentive forms and systems appears in the writer's, 'Incentives in Communist Agriculture: The Hungarian Models', *Slavic Review*, Vol. 27, No. 1 (March, 1968), pp. 23–38. Briefly, the *Nadudvar*

system assigns specific fields of collective land to famileis or small field teams of members with full responsibility for cultivation and harvest. In addition, a specified percentage of the harvested crop is *guaranteed in advance* to those working the field; as a result, the collective member is no longer a residual claimant on collective output, but receives a substantial income whatever the production. Other incentive forms include sharecropping, also with guaranteed returns, as well as several flexible payment methods of substantial bonuses for increased production.

[4] *Nepszabadsag.* August 2, 1969.

[5] *Statistical Yearbook* (Budapest: Hungarian Central Statistical Office, 1967).

[6] *Magyaroszag Nemzeti Atlasza, op. cit.*, p. 28.

[7] *Hajdu-Bihari Naplo.* January 16, 1969.

[8] Pecsi and Sarfalvi, *op. cit.*, p. 236. Sulyok-Schulek and Bacsi, 'Kombinalt Ontozes A Folyok Teli Vizhozamanak Felhasznalasaval' (Combined Irrigation by Utilizing Winter Discharge of Streams), *Vizugyi Kozlemenyek*, No. 3 (1965), p. 349.

[9] Based on two horse-power motor on a two-inch suction pump.

[10] Pecsi and Sarfalvi, *op. cit.*, p. 46.

[11] 'Belvizvedelmi Ertekezlet Szegeden' (Szeged Conference on Surface Drainage), *Vizugyi Kozlemenyek*, No. 1 (1966), pp. 16–49.

[12] M. A. Nagy and E. Korpas, 'A Hazai Szikesek Talajfoldrajzi Vazlata' (The Sik or Alkaline Soils), *Kozlemenyek A Szegedi Tudomanyegetem Foldrajzi Intezeteol* (Szeged, 1956).

[13] The 1961–65 Plan had investment funds of 43% for industry and 19% for agriculture; the 1966–70 Plan calls for industry, 40% and agriculture, 21%.

[14] The writer had a long discussion with this colleague which led to the idea of using light-weight, low-silhouette plastic 'tents' as a means of controlling the warmth from thermal waters. These tents, in effect, would be low cost greenhouses heated by thermal water and planted to crops – vegetables, fruit, flowers – which could be matured to reach West European and other markets at the peak price. The Environmental Research Laboratory, University of Arizona, is conducting research with such tents in Baja, California. Two hundred square metre plastic structures cost less than fifty cents per square metre (*Christian Science Monitor*, June 3, 1969). Other American agronomists see the Hungarian project as quite feasible so long as costs of heat can be kept low; thermal water should provide extremely low cost heat.

[15] *Kissalföld.* May 7, 1968.

[16] This method was under consideration by the regime in 1966

and operated experimentally on one farm; as substantially increased labour inputs resulted, the method became policy beginning January 1, 1969 (*Vas Nepe*, October 6, 1968).

[17] This is rarely measured in hourly terms at present, but rather in job accomplishment, livestock fattening on private plots and a variety of other methods. The present collective law states that each collective establishes its own measure of work units in any 'suitable way', and suitable can be, and is, interpreted very broadly (*Magyar Hirlap*, July 20, 1969).

[18] *Szabad Fold*, December 11, 1966. Not only are exports of fruit and wine important, but nearly 65% of the vineyards and 55% of the orchards are on the private plots.

[19] A study of the massive losses of time and in production through the continued use of horse-drawn transport on Hungarian collective farms would be revealing, and would help to explain continuing low labour productivity. There are some 300,000 horses on Hungarian farms at present.

[20] *Nepszabadsag*, September 20, 1966. This figure represents an average of some 35 workers per collective farm.

[21] *Nograd*, December 14, 1968.

[22] Comments made to the writer during the International Symposium on the 'Effects of Industrialization on Agricultural Population in the Socialist Countries', Hungarian Academy of Science, Budapest, October, 1967.

[23] *Petofi Nepe*, November 29, 1968.

[24] *Szabad Fold*, November 3, 1968. On May 12, 1969, an agreement was signed with a West German firm to manufacture small tractors for Hungarian private plots, at an initial rate of 500 per year (*Magyar Hirlap*, May 13, 1969). Credits will be available for purchase of these tractors at a reasonable price ($400–$500 at official exchange rates) by collective members. The latter represents a significant pragmatic step away from ideological limitations on private plot productivity.

[25] *Szabad Fold*, December 11, 1966.

[26] The average age of collective members correlates closely with the efficiency and income of the farm. For example, on two farms with very high member incomes, the *Nadudvar* farm has an average age of 40, and on the farm where the initial experiment on sliding scale private plot areas was conducted, the average age is 31.

[27] *Magyar Hirlap*, August 1, 1969.

Comments

JERZY F. KARCZ

In discussing Professor Dohrs' paper the following points are raised:

(1) Despite the number of labourers involved, the inputs and outputs of the agricultural system are low; therefore, the productivity of the system is low. It probably has to be read into this context that the quality of agricultural labour is low.

(2) Investment in irrigation. In some years, irrigation would need all the capacity which one might create. The question of investment is dependent on the time limit. The capacity one would build is also a question of the time limit. It is a matter of whether it would be economically valid to construct an irrigation system which might be used only 2 out of 10 years, an economist's approach, or build to the maximum need no matter how often you used the maximum output, an engineer's approach to the problem.

(3) One must consider the private plots within the context of the entire agricultural economy today. They could not produce as they do unless they received seed, etc. from the socialized sector of the economy (collective farms). All the technical crops are placed on collective farms and not on the private plots.

(4) When considering the subsidies which the large 'good' farms ultimately give to the 'poor' farms, one must realize that the poor farms cannot be shut down because even though they produce so inefficiently, that which they do produce is needed. They are not exempt from the procurement quotas. The increase in industrialization creates increased wages, creating a high elasticity of demand on food items. Thus, rather than jeopardize the balance of payments which would result from importing food to fulfil the demand, the poor farms must be kept in operation to supply what the good farms cannot. Thus, the good farms are not allowed to plough back their profits to increase their production, but rather must subsidize the poor farms.

(5) Dr Dohrs' point that the increased mechanization and size of the private plots is a good idea in order to increase the

productivity on the national scale is a good plan because it does not sacrifice the socialist's principles. If these plots increased their productivity, the good farms could return their profits to create a larger total product. Possibly, after production on the good farms increased, there might be a voluntary disbandonment of the private plots.

There are, however, alternatives other than those expounded by Dr Dohrs. His plan is good only if you are interested in improving production today and not necessarily for the future. He fails to consider such things as the development of agricultural services. Or can you let the agricultural system improve on its own initiative? The ultimate question is – Who's going to pay for the future benefits? Who will pay for the present costs of these future benefits? Governments alone can give the final answer. All economists and geographers can do is make the politicians aware of the consequences of their actions.

Remarks

Following Dr Dohrs' paper Professor Harris offered two comments: The first considered the age of the people on the farms, stating that it is evident that the age of farm labour is important. Restating that Dohrs' paper suggested that over 50% of the farm labour force will retire within a decade, he pointed out that the universal process of the transfer of the rural population to the urban environment is much more highly advanced in Hungary than elsewhere as seen by the raw figures. It shows that the working age group (young) 20 to 25 years old, is largely transferred to the urban sector already, at least in Hungary. This reveals a very significant figure which will have profound effects on the potential for intensification of agriculture.

Besides looking at farms as productive units, Professor Harris indicated that we must also consider them as 'retirement havens' in terms of social welfare. In terms of retirement havens it might be considered cheaper to return the old to the farms than to maintain them in the urban environment.

Professor Karcz indicated that this certainly is so, but that it is true in the United States also.

Professor Harris's second comment referred to Dr Dohrs' evaluation as to the proper procedure of intensification, saying that one cannot generalize about so large and diverse an area. Some regions do not need to be irrigated since they flood regularly. He concluded that one must deal with specific areas, and the needs of that particular area alone.

Professor Zakrzewska then posed the question: 'Are there any incentives for the young to return to the farms?'

Professor Karcz replied that he had seen playgrounds for children in kindergarten in Czechoslovakia which were full of farm machinery so that the children might become familiar with this type of life. He said that scholarships were also being given to people to go to high school and university in the hope that they would return to the farms. However, he continued, once the young have been introduced to the dynamic cultural pull of the city it is difficult to get them back on the farm, and therefore the cultural pull of the urban environment is an important factor.

Professor Harris then pointed out that around Budapest there is a significant division of labour; the men go into the cities to work, leaving their wives on the farm, and themselves return to the farm on weekends. If there is a shortage of labour because of the men leaving for urban industrial complexes, surely this shortage could be replaced in part by machinery.

Professor Romanowski replied that this is done, to some extent, but that in Czechoslovakia they still take factory workers to the fields during the harvests as well as taking the Czechoslovakian army. This, he said, is not a very efficient method.

To which Professor Karcz added that this system is not good for either industry or agriculture.

Professor Kosiński disagreed with Professor Karcz's statement, finding it a profitable scheme for providing the urban labourers with an opportunity to work on the farms during harvest. He granted that it could halt industry if it were not well planned, but only the blue collar workers would be involved and the urban workers would gladly accept this change in their otherwise routine jobs. He felt that it was all a matter of organization on the part of the government.

Professor Karcz pointed out that the increase of mechanization has reduced the necessity for large amounts of human labour and that co-ordination and planning is now needed. With regard to mechanization he felt that there is sufficient machinery to eliminate the need for human labour, but the lack of repair parts caused problems in using machinery; without parts the need for human labour rises. If parts were available, there would be little need for the increased human labour, except in emergencies.

Professor Karcz further commented that farmers should concentrate on factors of production (seeds, fertilizers, etc.) in certain areas and that through increased use of technology and more modern methods of farming, better results could be achieved. One cannot expect the poor, old farmers to produce. In other words, farming must be improved a small area at a time on the more productive areas, using increased technology.

Professor Hoffman pointed out that Bulgarian cereal farms are rather poorly developed while their vegetable farms are of high quality and that historically they have been, in effect, co-operative farming. He wondered if this could be the reason for the better vegetable farms in Bulgaria. Professor Harris indicated that this is true, but also that prices are higher for vegetables.

Professor Velikonja added that he felt that credit should be given to the higher level of quality in the Bulgarian vegetable farming rather than explaining it merely by price considerations.

On a broader scale, Professor Fisher pointed towards the better forms of transportation, i.e. better road and railroad systems which have helped in Eastern European agriculture. As an example, he described the road system in some sections of Yugoslavia as so good that the children of the farmers can go back and forth to the University of Belgrade easily. Therefore they do not become isolated either on the farm or in the city as happens in other parts of Eastern Europe.

Professor Kristof mentioned that the vegetable gardens in Romania were worked by Bulgarians who had been forced to leave because of collectivization. He said that the old-age farms do not produce as much, but neither do the elder farmers work many days. Traditionally they were sharecroppers, but this is no longer the case.

Professor Zakrzewska mentioned the transient sharecropper system which has developed in Hungary whereby the city folk work for six months in the summer and autumn on the farms, and during the six months of winter they work in the city. In return they are paid 50% of the crop, and since the real farmers only receive 40% of their produce, the young people are drawn to the farms.

Professor Harris added that men who live and work on farms during the summer often work as construction workers in cities during the rest of the year.

Professor Hoffman then asked if the older people have any problem as far as adaptation to new techniques are concerned, to which Professor Fisher-Galati replied they do not; they have problems of adjustment which they have always had, and therefore this is not new.

8

The Emergence of Regional Spatial Planning in Yugoslavia: The Slovenian Experience   JACK C. FISHER

Urban and regional planning, and those agencies assigned to this responsibility, have significantly increased their functional authority in Yugoslavia as a result of economic and administrative decentralization. This has occurred with a corresponding loss in prestige and authority for central economic organs.[1]

The strands linking economic and physical planning have been analytically weak in both planned and unplanned economies. The integration of local physical development plans with the national financial plans is as difficult for the East European countries as in the presumably unplanned Western states. The general financial plans of the Central Planning Agency had no explicit spatial or regional dimensions even though they had definite regional consequences. The plans, calling for allocation by branches, were 'regional' only to the extent that the existing spatial distribution of the branches caused the allocation to be skewed in some regional direction. Further, in Yugoslavia, the interrelation between Federal plans and Republic plans and between Republic plans and commune decisions is broadly indicative. There are few direct clearances and little controlled plan implementation. In addition to the common information of shared forecasts and time perspectives, which in Yugoslavia is highly variable from agency to agency, indicative planning needs regional co-ordination of decision units and uniform administration, both of which were difficult to achieve previously on the basis of federal standards in the multinational Yugoslav state.

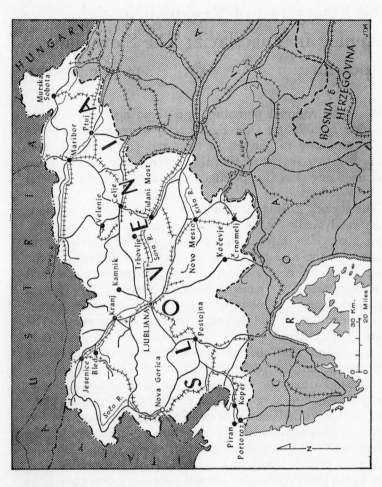

FIGURE 8.1 – THE SOCIALIST REPUBLIC OF SLOVENIA, YUGOSLAVIA

THE PHYSICAL PLANNING PROCESS

Let us examine the physical spatial planning process. The so-called requirements of physical planning methods are an amalgam of average past practice, social aspiration and currently fashionable technical solutions. To set the stage for later examination, it may be useful for the reader to review some of the characteristics of traditional physical planning as it exists in the United States and as it may now be evolving in East-Central and Southeast Europe.

The two major decision areas for urban and regional physical planners are land use and capital investment in infrastructure. Land use decisions involve amount, location and other elements of demand by type of use. The prescriptions of the land use planners are based upon projections of demand (to meet the test of reasonableness) and upon norms which are independently validated (the standards of the profession). Capital budget decisions of the physical planners are concerned chiefly with priorities, since the levels of the budgets are matters of economic determination. Priorities are set in accord with judgments of future need (based on obsolescence of existing facilities or deficiencies in meeting standard conditions). Physical planners, however, have contributed little to the development of criteria of choice among competing projects, leaving this issue to welfare economists or public administration experts.

Physical planners' first needs usually are for economic and demographic projections extended over the expected life of the physical plan. These constitute constraints on physical planning as well as estimates (when multiplied by standards or by trends of usage) of probable demand. Along with land-use data, which is for the most part descriptive, economic and demographic studies comprise the physical planner's traditional information trilogy.

The physical planner has no general theory of the best strategy of infrastructure investment, the best locational pattern, or the best environment for living. In practice, these follow from the types of knowledge: the social planning determination of needs and the economic determination of available means. The minimum consumption levels of social services acceptable to the community are set by social policy. These minimum standards contribute to the requirements of physical planning, though because of the social aspiration levels embedded in them they should not be confused with engineering requirements. Nevertheless, the socially determined requirements serve as elements in physical planning production functions along with engineering possibilities.

In practice, physical planners customarily equate requirements with demand. In modelling the economy, one can treat the physical planning model as a submodel generating demands for construction activity, floor space, etc., almost independent of the main economic dynamics. On close examination, the demands turn out to be relatively arbitrary determinations which provide little feedback information to the economic model. At worst, they have little force, and are soon modified by circumstances, while at best they are unexplained constraints on the operation of the economic system.

The economic counterpart of these requirements can be conceived as the set of average coefficients of, for example, the ordinary input-output table, which provide a summary statement of practice in the recent past. However, there are no evaluative elements. If the input-output table is sufficiently disaggregated by sector and is further disaggregated in space, it may provide use inputs to the physical planner. But location decisions are often enterprise decisions: disaggregation of even finer grain than that typical of economics to date may be necessary in order to translate into manageable generalizations the seemingly capricious behaviour of economic units in isolation. Leontief has recently urged such disaggregation,[2] and Orcutt has for several years been testing techniques for handling highly disaggregated systems. Disaggregation should be further supplemented by actual flow data – another long-time dream of empirical economists. Such data are especially important to physical planners who need measures of physical as well as economic flows. Even a highly disaggregated system of physical flow data would fall short of integrating the conceptions of economic and physical planning.[3] Nor do industrial complex and cost-benefit analyses provide fully effective planning linkages. Further, locational analysis badly needs a time dimension, almost all existing studies are cross-sectional, and the theory lacks time dynamics. Is a presently optimal location best in the long run? In Yugoslavia, as elsewhere, 'Physical plans are, by their very nature, long-term periods. Renovation and continuous prolongation of these plans is carried out in connection with the preparation of medium-term plans.'[4] But we have done little in any country to analyse changing land-use patterns in terms of *underlying locational strategies and to prepare operational time series of land and locational change*.

Finally, the integration of economic and physical planning in a social planning system requires a broader conception than the economist's rationality.[5]

POST-WAR EVOLUTION

Marked regional variation and a strong political desire for 'socialist' institutions stimulated the evolution of a unique administrative mosaic which though institutionally uniform throughout Yugoslavia is substantively diverse from place to place in response to varying historical traditions, social values and levels of development.[6] The country was, in 1945, and is now, in 1969, run by one party – the Communist Party. There has been in the intervening years, however, a distinct shift towards the increased mass participation in the decision-making structure of both local socio-political units and economic enterprises.[7] This attempt to create a compromise between Western democratic institutional traditions and proletarian dictatorship lies at the root of Yugoslavia's post-war institutional development. Visible party intervention in local affairs has receded more and more into the background, party members have assumed or are expected to assume a new role of guidance, stimulation and suggestion, rather than direction of administrative intervention. The concept of 'self-management' of one's enterprises evolved, in part, to stimulate increased motivation towards one's work through a collective structure with which each could identify in contrast to reliance on individual motivation induced under conditions characteristic of capitalism.

It must be emphasized that a fundamental change in approach evolved in the period of 1963–5, the whole philosophy of which was manifested in the April, 1963, Constitution, and secondly, the period which has elapsed since the Constitution's adoption. The first period stressed an approach towards integration based upon institutional standardization and local administrative autonomy. Thus, territorialization of the country developed with major powers and responsibilities, political, administrative and economic, placed in the hands of local bodies directly responsible to local political authority under indirect federal control and supervision. This approach led, however, to a number of negative tendencies. As it was found during the initial post-war period that cultural self-determination without increased local administrative authority was meaningless, so it is now realized that economic integration – at least at the level of each republic – must precede further development of local autonomy, and well before social integration can be expected to take place.

Between 1956 and 1963 each commune had the right, implied in the existing political theory, to expect a living standard comparable with the most advanced areas of the country. This implied that the earnings of the advanced North should be properly channelled to the South. Furthermore, it was stipulated that economic development must occur without significant interrepublic migration (in large part due to ethnic differences) which suggested by implication, that an equal distribution of agricultural and industrial activities, proportional to population distribution, was necessary for each of the basic regional divisions of the country. This obviously was, if not impossible, certain to produce inefficiency.

The period since April, 1963, has witnessed the gradual development of a new policy of decentralization which stipulated that each republic or commune (*općina*) should develop its economy in accordance with existing natural, financial and local resources with only moderate and reasonable support from external (federal and republic) institutions. Federal investments (which are viewed as political) were supposed to decrease.[8] The reform implied that local enterprises were to be given greater discretion in disposal of surplus and greater access to capital, independent of local political and administrative bodies. The new policy of restricting inter-regional capital transfers appears to be an improvement or stimulus to reinvestment where both surplus and returns can be expected to be higher, i.e. in the North. The fund for financing the economic development of underdeveloped republics and regions will insure that capital continues to be made available, on a competitive basis, in the South.[9]

Thus, the economic reform of 1965 further stimulated the republics to step into many co-ordinative roles that were previously the responsibility of the federal government. One may say, today, that only defence and foreign affairs are truly federal functions and even these federal activities are increasingly responsive to republic pressure. The post-1965 political tendencies of Croatia and especially Slovenia called for increased independence in the regulation of economic and social service activities at the expense of federal or all-federation regulations and codes. As a result of the drastically increased autonomy of the republics, the interesting hypothesis could be advanced that only the large Yugoslav corporations (with numerous plants often located in more than one republic) are the major force working for integration in Yugoslavia today. It is the large industrial entities that have profited from the Reform and it is the operation of market factors that stimulates and allows them to ignore the otherwise all-prevailing pull of local interests and ethnic bias.

Yugoslavia swung from a Soviet-style industrial fundamentalism, in which heavy industry was deemed essential for the economic and political development of each region, to a market version of efficiency in which investment funds are to be freed, so far as possible, to seek their own opportunities for profitability. Many physical planning objectives, such as conservation of amenities, ecological balance, and minimizing of undersirable neighbourhood effects of development are endangered by this emphasis, and the declared objectives of the social planning system, which includes certain redistributive goals, may be slighted. A kind of suboptimization may result in the longer run in the course of achieving high-priority short-run economic targets.

In a relatively affluent economy, the investment in social equipment can be defended on purely redistributive grounds, without regard to considerations of economic efficiency. In Yugoslavia, which is struggling to overcome a poor material base, it is necessary to advance social planning considerations on grounds of long-run economic efficiency. The measurement of the contribution of infrastructure projects to national economic growth is conceptually manageable, though operationally very difficult. Over any long-run future, it is not possible to forecast without it. Yugoslav social planners will need to resolve long-term *v.* short-term conflicts.[10]

REGIONAL SPATIAL PLANNING IN YUGOSLAVIA

Regional spatial planning in Yugoslavia does not have a long tradition. The activity originated as an extension of urban planning around the end of 1950 in various urban centres of the country. At that time, the main problems were: (1) the general lack of experience limited professional planning practice to only a few institutions, and (2) that regional planning evolved from urban physical planning producing 'regional planners' with an inadequate professional preparation (mainly architect-designers with little or no exposure to the social sciences, or statistics).

In view of the emergence of regional planning from the highly architecturally oriented urban planning field, a brief history of urban planning in Yugoslavia is appropriate to understand the current problems of regional planning in the country.

The larger Yugoslav cities have a long tradition. Many of them (especially Ljubljana and Zagreb) were built, or were reconstructed and expanded, on the basis of elaborate designs or urban plans. The Austrian tradition of strict building control positively influenced the development of cities in those areas of Yugoslavia under Austro-Hungarian control. Between the two world wars various cities grew rapidly (especially Belgrade and Zagreb). Limited professional planning experience and the inadequate organization of planning services resulted in the chaotic urbanization of the zones of larger cities (for example, Zagreb). There were attempts at international co-operation (an international competition for the new centre of Zagreb was held in 1931) but without visible or viable results.

World War II produced new problems. A wider social

involvement was demanded from the small number of existing urban planners. There was need to reconstruct destroyed settlements and build new settlements for the large population caught in the migration which occurred at the end of the war. It should be noted that the temporary government of Croatia, during the occupation in 1944, organized an urban planning service on territory freed from German control. The first plans for the reconstruction of destroyed settlements were thus formulated and this Croatian effort, together with work started on the reconstruction of Belgrade, is generally recognized as the beginning of urban planning practice in the new Yugoslavia.

Immediately at the end of World War II, in almost all cities of the Yugoslav Federation, institutes for urban planning were created. Founded in the period prior to 1947, they carried out the spatial planning function as administrative agencies and components of ministries of the republican governments. During the 1945-52 period Yugoslav planning was a carbon copy of the Soviet system of planning. It was thought that all social and economic action could be planned and controlled from one centre. Under the system of centralized administrative decision-making there was neither a structured role, nor the time, for the comprehensive analysis which is central to the spatial planning approach. The urban planning agencies functioned more like operational administrative services and less like professional research and planning institutions. This situation produced a stagnation in the development of urban planning. In all republics, with the exception of Croatia, republican institutes for planning were abolished. The larger institutes were broken up into smaller drafting bureaus, which, due to demand, were more engaged in designing specific objects and less with urban planning.

An important event, influencing the further development of urban planning in Yugoslavia, was the change in the system of financing of urban planning activities. In the period from 1953 to 1955 all institutes of architecture, drafting and urban planning in Yugoslavia were required, through their own efforts, to be self-financing. On the one side, this restricted their activity as every drafting office had to produce sufficient income to cover expenses. From the other side, this situation had the tendency to produce serious professional work for institutes of urban planning as these institutes finally ceased to be administrative agencies dependent upon and responsive to their place in an administrative hierarchy.

With the exception of Croatia, by 1955, no large institute for urban planning existed in Yugoslavia. The professional specialists for urban planning were still mainly architects even though a few other disciplines were now involved, especially geographers and economists. The interest in and the need for urban planning grew. This need occurred as centralized, so-called, 'social' (economic) planning both at the federal and the republic level, was gradually restricted in function and lost its former competency. Thus, the individual communes and cities found it necessary gradually to rely upon their own plans and projections of future economic and spatial development.[11]

During the preparation of the first urban plans for the largest Yugoslav cities (Belgrade, Zagreb), the need was felt to examine various elements of the plan within a regional perspective or scale. The transportation network, water reserves, recreational areas were all viewed from a regional scale. As a result of this it became standard practice for urban physical plans to have a so-called regional scheme (see for example the 1950 plan of Belgrade and the 1953 master plan of Zagreb). This was the introduction of regional planning considerations into operational practice in Yugoslavia.

At this time, in the mid-1950's, when the question of institutionalizing regional planning in Yugoslavia was actually debated, the largest concentration of urban specialists existed in Croatia. A group of urban specialists was brought to the Urban Planning Institute of Croatia in Zagreb, and working closely with planning institutes in the main centres of the Republic (Zagreb, Rijeka and Split), undertook studies designed to advance the theory and practice of urban planning in Yugoslavia. In 1955 there was founded the Association of Urban Planners of Croatia with widespread professional participation. Soon there developed a network of republican associations and somewhat later a Union of Urban Planners of Yugoslavia. Efforts were made to secure responsible legislation that would insure and define the specific competency of both urban and regional planning. In view of the complex system of planning in Yugoslavia and the fact that 'regional planning' was treated as a part of socio-economic planning, but not physical planning, a group of Zagreb planners proposed the term 'regional spatial planning' as a special form of physical planning in Yugoslavia. This term is generally used in Yugoslavia and has entered into planning legislation.

Urban and regional spatial planning laws were adopted in various republics and all republics established administrative departments. Characteristically, urban and regional planning is viewed as an activity which does not fall under the competency of the federal government.[12] Because of this the organization of urban planning offices among the various republics varies sharply. For example, in Slovenia there was founded an Urban Planning Institute, which by name is a similar institute to the one which had had a continued existence in Croatia, even though by function and work load they are distinct. With time, the organization of urban and regional planning activities in the various republics changed; it is not possible now to speak of a general Yugoslav organizational scheme or a general methodology and practice.

At the time efforts were under way to secure legislation for urban and regional planning, initiative developed to prepare the first regional spatial plan in Yugoslavia. In the Urban Planning Institute of Croatia the first such plan was prepared for one district in northwestern Croatia. Even though the area was relatively small, the presentation of the plan was based on certain methodological principles and stimulated further work in the field of regional spatial planning.[13] This first regional plan was discussed at the original Congress of the Federal Association of Planners of Yugoslavia and its methodological approach accepted as the direction for future work.[14] At almost the same time, but with a more narrow focus, regional plans for Rijeka and the adjoining coast area were prepared.

Regional spatial planning in Yugoslavia, because of numerous problems, is restricted in practice. Certainly there are insufficient specialists, especially those with other than an architectural background. This, however, is related to a more major problem, the financing of planning projects. Regional planning requires a strong research and analytical base.

Unfortunately the required research input for regional planning is not recognized as scientific research work. Because of this it was not possible to obtain or utilize the funds available for scientific research work for operational planning projects. Funds had to be secured from the commune or the republic. The commune as a territorial frame for planning is too small, and financially too poor to cover plan preparation expenses. It is often difficult to organize the joint action of several communes for each commune may well have needs considered more important than the funding of regional planning studies. Because of this, regional planning efforts are not undertaken in those areas where such an approach may be the most useful but rather where funds may be more easily raised.

Significant new developments occurred after 1960. In Slovenia the first work was published on regional planning theory.[15] In Serbia the question of regional planning for northeastern Serbia became actual with the development of the Djerdap hydroelectric and navigation system. A regional plan was developed for the southern Adriatic coast of Yugoslavia, an area which includes parts of Croatia, Herzegovina (part of the Republic of Bosnia and Herzegovina) and Montenegro. This was a co-operative project with the United Nations. Further, the regional plan of Istria was completed and work was started on regional plans for various areas in Croatia. A regional plan is under preparation for the metropolitan area of Zagreb, which, by the number of institutes involved and the concentration of economic potential and population included in the planning area, is probably the largest regional metropolitan planning venture yet undertaken in Yugoslavia.

Significant steps in the regional spatial planning field occurred in Slovenia. Efforts were taken to expand the theoretical basis of both urban and regional planning through the introduction of advanced social science concepts and techniques. The American-Yugoslav Project in Regional and Urban Planning Studies was organized in the summer of 1966 and administered on the Yugoslavia side of the Urban Planning Institute of Slovenia. A number of co-operative efforts were undertaken by Slovene and American specialists including a study of the 15 commune metropolitan areas of Ljubljana discussed later in this paper. The activity of the Project included at various times and in varying forms individuals and institutes from the rest of Yugoslavia and from other European countries.

The thesis that regional spatial planning requires a wider territory, specifically one entire republic, suggests the need for a corresponding organizational-territorial form. Around 1955 there was discussion about the need for 'land plans', which would cover the area of an entire republic. A Zagreb group of planners proposed that a single system of planning require various levels of plans: (1) the all-state plan (Yugoslavia), (2) 'land plans' (republics), (3) regional plans, (4) urban plans (cities and settlements), and (5) detailed urban plans (neighbourhoods). Such a system implied that the all-state plan for the country as a whole would be concerned with the network of highways, water routes, railroads, etc., and urban plans would be concerned with all the details of urban life and social standards.[16] The initiative to develop republic regional plans was not all successful. At the moment, republic plans are again under preparation in Serbia, Croatia and Slovenia. In view of the fact that the territory of Slovenia may be viewed for certain planning purposes as a single regional entity, the Slovenia activity in regional spatial planning is perhaps the most advanced.

Slovenia led the other republics by first recognizing the importance of regional physical planning co-ordination and programming at the republic level. The reader unfamiliar with Yugoslavia should keep in mind that Slovenia is the most developed of the republics, with a *per capita* income roughly double the national average, and with a highly developed physical infrastructure.

There developed a recognition at the republic level that infrastructure planning had to be better co-ordinated and the physical planning function significantly improved. The Republic of Slovenia has taken the lead in this direction. The goal was to create a Republic mechanism for the rationalization of infrastructure investment, rationalization of locational decisions, and broad equity through provision of a minimum consumption level of social goods. Despite the inherent problems of physical planning, the consequences of the reform and accompanying administrative reorganization has made inevitable the situation in which local physical planning organs must bear the principal burden of this rationalization, in principle, if not in fact.

This short introduction into the Yugoslav practice of regional spatial planning suggests that there are various centres which have developed more or less independently. Further, this implies that the methodological character of planning in the various centres differ greatly. Because of this, it is not possible to speak of a single Yugoslav theory and practice of regional planning.

THE EXAMPLE OF TOURISM

Since tourism development is one of the key growth sectors in Yugoslavia and is so obviously related to regional planning, a brief discussion will be made of the investment patterns in tourism. This section is based on work completed by a staff member of the American-Yugoslav Project and suggests the type of analysis needed to supplement the traditional physical planning orientation of Yugoslav urban and regional planners.

Table 8.1 presents the information available in planning documents, empirical observation, and the best estimates by officials in Slovenia on the current status of tourism development. It should be noted that this example is used as the impact of tourism, has not been well thought out in Yugoslavia or Slovenia, nor has a planned, co-ordinated programme been established for the development of tourism in the country and Republic. Thus, though tourism is developing outside of 'planning organs responsibility', it is suggested that tourism development should be one of the chief concerns of regional planning agencies. Tourism and its related services has become one of the most dynamic and attractive sectors of the Yugoslav economy for post-Reform investment, both because of the steadily expanding volume of tourist expenditures in Western Europe and because of a growing response to Yugoslavia's increasingly liberal policies to induce tourist travel along its Adriatic Coast. Internally, tourism is attractive as an investment focus because of the relatively high proportion of earnings in this sector which are received in foreign (negotiable) currencies

The Emergence of Regional Spatial Planning in Yugoslavia

and because the foreign exchange retention quota[17] allowed on these earnings is the highest in Yugoslavia.

The year 1969 produced an income from tourism of some 300 million dollars in Western currency for Yugoslavia. The 1969 summer season produced 25% more tourists than in the

TABLE 8.1 – *Yugoslavia: Tourism: The example of Slovenia*

Existing Conditions	– exceptional potential (natural) surroundings, international crossroads; – underdeveloped touristic facilities and services;
Current Trends	– building of hotels, motels and some of the equipment, e.g. cable cars; – relatively rapid increase of tourist traffic; – emphasis on transit tourism (about 90% of all tourist traffic entering Yugoslavia flows through Slovenia);
Current Planning Policies	– initiation and development of regional plans for tourism (networks of motels, winter-sports-centres, etc.); – better equipment and facilities on highways to 'catch' the transit flow;
Main Planning Problems	– underdeveloped touristic potential; – unco-ordinated development; – conflicts with other land uses; – inadequate conservation programmes;
Specified Planning Objectives	– to increase (rapidly) income from tourism in hard currency; – to build hotels and equipment; – to catch the transit flows of tourists; – to increase winter tourism;
Supposed or Anticipated Extreme Manifestations	– a major dependence on tourism for the national economy of Slovenia; – whole villages converted into tourist attractions; – national parks with recreational activities.

first 7 months of 1968 and 29% more foreigners on the Adriatic Coast. However, 'at a time when the entire Mediterranean is expecting further tourist expansion, even greater than this year, at a time when on tourist maps, thanks to increasingly numerous chartered planes, new names and places are being inscribed, probably only Yugoslavia today does not know how tourism will develop after 1970 for it has no plans, nor any instruments of economic policy.'[18] Nor, might it be added, does it have the administrative means to link physical development of tourist facilities to any regional co-ordinative scheme or overall financial investment policy; this despite a major United Nations Special Fund Project in the country.

Tourism development is obviously most affected by the availability of investment funds (Table 8.2). The regional distribution of tourist credits was heavily concentrated in Croatia, which possesses most of the Yugoslav Adriatic Coast, the prime tourist attraction in Yugoslavia, with Serbia, Slovenia and Montenegro having roughly equal secondary shares and Bosnia and Herzegovina and Macedonia having negligible participation.[19]

TABLE 8.2 – *Yugoslavia: Regional Distribution of Bank Credits for Tourism*

	Investment in 1967		
	Mils. ND	% share	Mils. $U.S.
Croatia	541·9	54·5	43·4
Serbia	178·0	17·9	14·2
Slovenia	131·3	13·2	10·5
Montenegro	123·3	12·4	9·9
Bosnia & Herzegovina (B & H)	18·9	1·9	1·5
Macedonia	1·0	0·1	0·1
	994·4	100·0	79·6

The situation can be more precisely examined by establishing a matrix of intra- and inter-Republic credit flows for investment in tourism (Table 8.3). For this purpose, funds invested by the Yugoslav Bank for Foreign Trade, the Yugoslav Agricultural Bank and the Yugoslav Investment Bank are treated separately as Federal, rather than Serbian flows.

Since the diagonal of this matrix represents *Intra*-Republic credits, we may find *Inter*-Republic flows by subtracting the major diagonal from the matrix total: 994·4 − 250·6 = 743·8. Hence, 75% of all credits for tourism since 1967 were granted across Republic boundaries and only 25% of such investments were internally or self-financed. From this we can observe that those regions with the greatest tourist attraction, Croatia, Slovenia and Montenegro, do the least amount of self-financing – that is, they are heavily dependent upon outside

The Emergence of Regional Spatial Planning in Yugoslavia

TABLE 8.3 – *Yugoslavia: Matrix of Credit Flows of Investment in Tourism, 1967*

Bank credits issued	Croatia	Serbia	Bank credits received Slovenia	Mont.	B & H	Mace.	FED	Total
Croatia	104·6	22·8	64·2	69·5	8·9	0	0	270·0
Serbia	145·9	135·1	25·8	25·9	3·7	0·2	0	336·6
Slovenia	41·7	0	3·8	0	0	0	0	45·5
Montenegro	0	0	0	0	0	0	0	0
B & H	0	0	0	0	6·3	0	0	6·3
Macedonia	0·1	1·0	0	0	0	0·8	0	1·9
FED	249·6	19·1	37·5	27·9	0	0	0	334·1
Total	541·9	178·0	131·3	123·3	18·9	1·0	0	994·4

Notes to matrix
(a) In the absence of specific information about the total of 195·0 invested by 'General export' and 'Interexport' companies, this amount has been distributed on the assumption of the overall regional shares pattern.
(b) It has been assumed that the residual 1·0 of Macedonian investment was allocated entirely to the adjacent republic of Serbia. With these two assumptions, the flows from Croatian banks to the other republics become residuals necessary to reach the specified column totals.

capital and its mobility across Republic boundaries. But the actual situation is even more complex than this. As might be expected because of its possession of the choice coastal sites, Croatia received 1·6 dinars from outside (mainly from Serbia and the Federal banks) for every internal dinar's credit. Serbia, on the other hand, was largely self-financing, receiving only one in four dinars from outside, supplied about equally by Croatia and the Federal banks. Macedonia was almost completely self-financing in tourism while Montenegro, at the other extreme, received all its tourism investment from outside. Slovenia invested in Croatia more than 10 dinars for every one she invested internally, but surprisingly received more than

Credits received as % of those issued		% self-financing
Croatia	201	19
Serbia	53	76
Slovenia	287	3
Montenegro	—	0
B & H	302	33
Macedonia	53	80

100% of this amount *back* from Croatia and another 60% besides from Serbia. This is a bit startling in view of the fact that these two regions possess what might be viewed as competing tourist areas and, further, raises the question as to the relative evaluations of investment opportunities in the two regions. Evidently, Slovenian banks consider tourism investments more attractive in Croatia than at home – and opinion

TABLE 8.4 – *Yugoslavia: Bank Credits Received* (millions of new dinars)

Bank credits issued	Bank credits received							
	B & H	Mont.	Croatia	Mace.	Slov.	Serb.	FED.	Total
B & H	×							
Montenegro		×						
Croatia			×					
Macedonia				×				
Slovenia			21·77		×	11·48		33·25
Serbia					1·84	×		
FEDERAL					42·39		×	
Total					44·23			+10·98

which is evidently *not* shared by Croatian, Serbian and Federal banks. To investigate this further, we examine the 'Slovenia' row and column from the national matrix in some greater detail by drawing on data on Inter-Republic flows of tourism investment supplied by the Social Accounting Service in Ljubljana. A matrix of investment flows between Slovenia and the other five Yugoslav republics has been formed for all industry taken together and for ten industrial and economic sub-sectors for the year 1967.[20] This consolidated matrix of ten sectors indicates that Slovenia received a total of 134·5 million new dinars of investment funds from other republics and the Federal banks, and itself granted 43·4 million new dinars in investment resources. However, the largest proportion of these funds were received from the Federal banks (117·4) so that if the Federal flows are excluded from consideration, Slovenia would show, instead, a *net outflow* of funds of about 26·4 million new dinars. From this we might tentatively conclude that the Federal banks still play an important role in redistributing investment resources among the Republics and, in the case of Slovenia, serve to counter a trend towards the flow of substantial amounts of investment funds away from Slovenia. From the general matrix described above, the tourism sector is extracted as shown in Table 8.4.

We see that Slovenia received a total of 44·2 million new dinars from other regions (4% from Serbia and 96% from Federal banks) for investment in tourism and related facilities and issued 33·3 million new dinars *to* other republics (65% to Croatia and 35% to Serbia) for the same purposes thereby enjoying a *net inflow* of tourism investment of 11.0 million. The results can be expressed in relation to Slovenia's overall structure of investment flows as follows. (Note that these statements relate only to investment flows and do not include the self-financed investment in Slovenia.)

1. Funds for investment in tourism in Slovenia coming from outside Slovenia (44·2) accounted for 33% of *all* receipts of outside capital investment in Slovenia in 1967.
2. The *net* inflow of tourism investment in Slovenia (11·0) accounted for 12% of *net* investment inflows into the Republic for *all* sectors of industry.
3. Slovenia's credits for tourism (33·3) accounted for 76% of *all* credits granted by Slovenia for *all* sectors of investment in other republics.
4. Finally, we can show that, among the credits granted and received for tourism investment in Slovenia, there is a sharp difference in the technical structure between the *inflows*, which are heavily concentrated in buildings (96%), and *outflows*, only 41% of which are granted for buildings and the remaining 59% for other uses. This raises some interesting questions about the differences in the nature of tourist facilities and needs in the different Republics which should be examined by specialists in tourism planning.

The analysis revealed a changing pattern of inter-regional distribution of these funds. In 1964, all of the tourism investment coming from outside Slovenia (2·3 million new dinars) originated in Serbia; in 1967, as we have noted, the bulk of incoming funds are invested by the Federal banks. As for credits issued by Slovenian banks for tourism investment in other regions, in 1964 these went entirely to Macedonia in a single investment of 6·9 million new dinars; in 1967 nearly five times this amount was distributed almost evenly between investment in Croatia and Serbia with nothing going to Macedonia. The *technical structure* of the investments going *to* Slovenia has been heavily concentrated in construction in both years. From this very sketchy evidence we may surmise that the 1965 economic reform *has* produced, in the tourism sector, both a surge in total investment in this sector to meet the burgeoning demand for tourist facilities (aided also by the increasing contributions of private and individual as well as social undertakings) and a significant 'deterritorialization' of these funds in search of the *best* profit opportunities.

The analysis of tourist financial trends, contrasted with the orientation of the physical planning agencies, suggested to Slovene authorities the necessity of much greater attention to physical planning co-ordination and development of tourist facilities in order to properly channel the resources pouring in to the Republic. It may seem strange but the implications of Yugoslav tourist investment into Slovenia was not apparent before this analysis was prepared. There was a strong criticism of Croatian banks by Croatian economists after the report was made public.

SLOVENE PLANNING REALITY

Since the drive for decentralization in the context of the new socio-economic system, and most particularly since the 1965 economic reform, planning functions in Yugoslavia have been separated and allocated to different decision-making bodies. Spatial planning responsibilities have fallen almost completely to the communes with, in the most recent period, increased republic responsibility or co-ordination while economic planning now rests with the enterprises (*podjetje*).

Because the recent enabling legislation for regional planning in Slovenia considers comprehensive planning (planning for social, economic and spatial elements) to be too extensive an undertaking within the context of present conditions, the law is limited to an investigation of only the spatial aspects of regional planning. Under the law, however, regional spatial plans are expected to contain elements which necessitate consideration of, and conformance with, the social and economic requirements of each region. These requirements (that is the use of socio-economic starting points and directives for the space economy, space management and landscape preservation) are the results of a gradual realization that spatial planning efforts may be justified only as they satisfy social and economic aspirations and objectives of the regional population.

This gradual awakening to the problem of economic and spatial planning separation and the institution of legal means for their integration has given rise to a new problem. This problem is one of determining an adequate process of integration. The enabling legislation has set forth the substantive elements for spatial planning, but the development of procedural elements has been left to the community.

With the passage of the Regional Spatial Planning Act and the Urban Planning Act of 1967, the Assembly of Slovenia authorized a systematic programme in the field of spatial planning.[21]

Following the passage of both acts, the Bureau of Regional Planning was established with the authority and substantial funding to prepare a regional spatial plan for the territory of the Republic. Thus, for the first time, in Yugoslavia, the institutional and financial framework for regional (republic) physical planning was established with strong political backing and large financial co-ordination authority. The framework for government policy-making and administrative responsibilities in the preparation of the regional spatial plan are the following:

The *Assembly of Slovenia* has a permanent Committee for Urban Planning, Communal Affairs and Housing.

The *Executive Council* of the Republic has established a 'Commission for Regional Spatial Planning'. This Commission is headed by a member of the Executive Council, and its members consist of republic deputies, representatives of organizations which are responsible for the preparation of the plan, and representatives of those governmental branches and professional institutions which will co-operate in the preparation of the plan. The Commission is the body of the Executive Council which formulates policies for submission to the Executive Council and the Assembly in the form of questions to be discussed and decided upon by them.

Within the organizational framework of the *Republican Secretariat of Urban Planning* there exists the 'Bureau for Regional Planning' which co-ordinates all the work, including various research studies of other professional institutions assisting in the preparation of the regional spatial plan. The Bureau has full responsibility for evolving the regional plan. The Bureau has a small staff to develop the plan's data requirements and to co-ordinate the work contracted among several institutes. Co-operating agencies providing funds and services to the Bureau for Regional Planning include the Road Fund of Slovenia, the Water Economy Board of Slovenia, the Power Association, the Chamber of Commerce of Slovenia and the two Slovene Foundations: The Boris Kidrič Fund and the Boris Kraigher Fund.

By the end of 1970, it is anticipated that the Bureau will submit the proposed spatial plan for the territory of Slovenia to the Parliamentary Commission for review. The main elements of the regional spatial plan according to the provisions of the act will comprise:

– socio-economic goals, objectives and guide-lines for the economy and the suggested spatial organization of the Republic and landscape preservation;
– general concepts for the basic transportation, water economy and power network;
– land use distribution for the basic categories of agriculture, forestry, tourism, housing, transportation and infrastructure; and the designation of various preserved and reserved zones.

DEMOGRAPHIC CONDITIONS

This section provides insight into some of the spatial changes which have occurred in Slovenia and the Ljubljana urban region. It also articulates a number of functional activities

The Emergence of Regional Spatial Planning in Yugoslavia 323

which are dominant in the Ljubljana region and are currently undergoing significant change in the spatial pattern as a result of post-reform policies.

All planning work must be based on population assumption. In comparison with other European countries, Yugoslavia has a high fertility rate and a relatively high infant mortality rate because of the demographic structure in the underdeveloped south of Yugoslavia. Croatia and Slovenia have demographic conditions approaching those of Western Europe. Until 1961, Slovenia's birth rate had been constantly decreasing.[22] From 1961 onwards the decrease had halted. It is not clear whether this halt is a transitional phenomenon caused by more marriages and earlier marriages, or represents a change in desired family size. The specific rate of marriage (that is, the number of marriages per marriageable, or single adult inhabitant) is much lower in Slovenia than in Yugoslavia as a whole. There are two reasons for Slovenia's low rate: (1) its population is typically 'old' and has a high percentage of marriageable people; (2) it has a disproportionate ratio of females to males and, consequently, a large number of unmarried females. Slovenia's general marriage rate (the number of marriages divided by the population) is higher than the Yugoslav average.[23]

Slovenia's mortality rate began decreasing about 1880, nearly one hundred years later than in the northern and western parts of Europe, and by 1960 was close to the European average. Infant mortality has been relatively low in Slovenia, compared with the rest of Yugoslavia, but the mortality of Slovenian males in the 15–24 age group is above the Yugoslav average. The percentage of deaths in this age group due to accident and violence is disproportionately high.[24]

Traditionally, there has been a large out migration from Slovenia, but during the period from which the data has been taken (1961–4), there has been a net immigration.[25] Slovenia's population has grown at a much slower rate than that of the rest of Yugoslavia, mainly because of low fertility and excessive external migration.[26]

Between 1920 and 1963, Slovenia's population increased by 25% and Yugoslavia's population increased by 57%. The project developed by Rogers indicates that by 1998 Slovenia's population will have increased another 32% and Yugoslavia's 39%.

If present trends continue, Slovenia's total population over the next 30 years will increase by about half a million people. Thus, by the end of this century, over two million people will reside in Slovenia. Of this total, about a third will be located in the Ljubljana Metropolitan Region.[27]

Migration within Slovenia has and will continue to alter the regional distribution of its population. Almost 3,000 people from the rest of Slovenia migrated into the Ljubljana Metropolitan Region during 1963. Only about 8,500 migrated into Slovenia from the rest of Yugoslavia during the same time period.

THE LJUBLJANA URBAN REGION

The detailed analysis upon which the above discussed was based, formed a major input into subsequent interpretations about the future of the Ljubljana urban region. As a step in identifying the existing spatial configuration of the Ljubljana urban-region, a number of specific functional sectors were described. Each sector was examined with respect to existing conditions, current trends, current planning policies, main planning problems, specified planning objectives and anticipated or possible extreme manifestations. The preparation of the tables was the result of extensive field work on the part of our Slovene partners, review of appropriate statistical and cartographical information and access to special 20-year projections of activities and anticipated needs prepared by each ministry of the Slovene government. The tables allow the reader, at a glance, to differentiate between the existing situation and trends, the perception of the future as suggested in the official planning documents and future possibilities for each sector as suggested by Slovene officials in a series of conferences conducted for this purpose in Ljubljana. This work is preliminary and is deliberately displayed here in its non-quantitative form.

The tables placed in the appendix provide insight into the spatial characteristics and socio-economic trends of the other more traditional functions existing in the Ljubljana metropolitan area. Similar quantitative studies exist for each functional area such as the one briefly presented above for tourism.

Though it is not possible to provide a detailed review of the tables here, a number of characteristics will be self-evident. First, the tables clearly confirm the fact that physical planning controls are weak and that existing 'plans' are more idealistic than realistic. In almost every case a strong dichotomy exists

TABLE 8.5—*Yugoslavia: Summary of Developments in Four Functional Areas*

HOUSING	LAND ECONOMICS AND DEVELOPMENT CONTROL	TRANSPORTATION	WORK
Current trends Small, scattered developed sites for predominantly apartment construction but more one-family houses being built in last few years	Scarcity of nationalized land and rapidly increasing land prices	Increase in motor bikes and 30% increase in car ownership per year	Unskilled low-paid jobs filled by individuals from the South while highly skilled Slovene labour migrates to Western Europe
Main planning problems Building enterprises governed exclusively by market considerations cannot supply larger and lower income units. One-family housing units are in demand and very few are available or in current plans. Administrative control is difficult, speculation on the increase and unplanned development apparent	Weak or almost non-existent land use controls	Insufficient space planned for transportation and related uses; increased congestion and spiralling traffic accident rate	Industrial pollution raising rapidly as industrially 'suitable' land decreases

between planning goals and current problems (a situation common to the United States). The implications suggested by the tables can be backed up by some of my previous publications, project reports and numerous other studies prepared by Yugoslav scholars. Let us look briefly at four functional areas extracted from the tables in the appendix.

Careful examination of the tables will provide the reader insight into the spatial characteristics and socio-economic trends of the Ljubljana metropolitan area of Slovenia. A variety of quantitative studies of the area exist for Project purposes. This set of 'qualitative' tables and indicators would provide perhaps a deeper insight into the 1969 reality of administrative trends and social conditions than more refined quantitative indices. The tables were prepared and revised on the basis of official documents and prolonged discussion.

They also demonstrated that a rapid spatial transformation of Slovenia is under way and that this is occurring outside of the public planning process. Illegal construction, pollution, congestion are increasing at a time when public authorities have the least funds and weakest administrative control mechanism. Construction, both private and 'public', is proceeding in an unrestricted fashion and legally possible fines are too low to restrain even an ordinary Slovene worker. Unco-ordinated growth is occurring in a number of different functional areas. The political fragmentation of the Ljubljana urban region into five communes which are administratively autonomous and financially competitive further limit co-ordinated, guided action.

The demographic analysis and the tables were part of a general input base for the study of alternative patterns of spatial organization for the Ljubljana urban region. The major components of the study include:

(1) The development of a system of goals, objectives and policies (in part discussed here and presented in the appendix) to be used as a guide in the evaluation of alternative patterns. The major sources for these are their 'recognition' and analysis covering policies for the major sectors of the economy presented in the tables in the appendix. A survey of the 'environmental preferences' of residents in about 15 different types of neighbourhoods in the region was undertaken. A detailed sector analysis of housing was prepared.

(2) The elaboration and refinement of four basic alternative spatial patterns for the region, and their projected development at 5-year intervals until the year 2000 through the use of the 'Lowry' land activity projection model.

(3) The evaluation of the four alternative spatial patterns as projected by the model, in terms of the general community goals, objectives and policies developed above and presented in the objectives and more specifically as to how each would meet the objectives determined with respect to housing,

industrial location, transportation, agricultural preservation, infrastructure requirements and the pattern of commercial centres.

The Bureau of Regional Planning was created as a result of administrative decentralization, a corresponding erosion of the economic planning mechanism and the growth and stimulation of individual private action coupled with increased incomes. Regional planning, regional control and a regional financial investment strategy are needed. The Bureau of Regional Planning has evolved with increasing control over infrastructure investment and as a prime mover of intercommunal co-ordination. The main areas of concern which can only be effectively dealt with at this larger Republic regional planning scale are:

1. Land use control, including better co-ordination of building sites within and among communes, preservation of the Republic's scenic beauty under tight and enforced district zoning regulations and more stringent regulatory controls to prevent increased air and water pollution.
2. A fundamental requirement of the Bureau must be and is to develop a multifaceted, contemporary, Republic-wide, metropolitan-focused transportation system.[28]

Regional spatial planning, as regional planning is called elsewhere, must relate physical planning of regional infrastructure, to the general financial plans of the Federal and Republic governments, to the investment intentions of the decentralized enterprises, and realistically to fiscal constraints and stringencies of regulatory measures, i.e. to the political climate.

The tables have summarized the existing trends and perceived planning problems for selected key factors such as housing, land control, transportation, work, etc. The planning process of the past did not conceptually anticipate the change in life-style, evolving in the mid-1960's. Decentralization limited the ability of both the communes and higher governmental units to deal effectively with these problems, and the enterprises for a time were made independent of any administrative control. Beginning in 1966 and 1967, it was clear that the Republic, after the elimination of the District (*srez-kotor*), was the only possible unit capable of evolving territory-wide infrastructure planning, insure uniform and enforced land use development and control policies, and take the necessary legislative actions to insure that the enterprises would be guided by some acceptable minimum (republic) control. With the completion of the regional spatial plan of Slovenia by 1970, and its enactment as law, the Republic will have produced the framework for development guide-lines and insurance of intercommunal activity co-ordination, if not actual co-operation.[29]

Appendix

The following tables were adopted from *Toward a Methodology for Regional Planning*, American-Yugoslav Project in Regional and Urban Planning Studies, Ljubljana, 1968, Volume 1.

APPENDIX 1. HOUSING

Existing Conditions
- serious housing shortage;
- about 50% of the housing production in Slovenia built by building enterprises, 50% by private persons; from those built by enterprises about 75% built for the market and the rest on contract; most flats are of modest size (average: 60 m²/flat);
- financing coming from housing funds of working enterprises, and bank loans;
- very modest diversity of choice, non-availability of flats for rent;
- comparatively good building standards;
- all new housing for purchase;
- very low turnover of rental housing.

Current Trends
- small and scattered development sites for predominantly apartment construction;
- definitely more one-family houses being built recently;
- housing and flats tending to higher densities, better services and utilization of sophisticated building methods;

- new types of residential buildings emerging slowly: e.g. atrium houses;
- one-family houses being built as permanent structures (brick, reinforced concrete – buildings for two or three generations; only a very small proportion in pre-fabricated wooden panels and those similar);
- second dwellings – weekend house construction beginning to be significant;
- building for the market with consequences and implications such as larger purchases of developmental real estate and corresponding banking policies.

Current Planning Policies

- concentration of development (definition of large concentration sites, for the first time a 'Ljubljana communes' action programme is emerging);
- preparation of more comprehensive site plans (before individual locations are approved);
- the role of 'Housing Enterprises' and utility working organization in site development being stressed;
- 'neighbourhood unit' – still accepted and going strong, as a conceptual device for structuring services;
- emphasis on the rational sequence of developmental areas – phasing;
- attempts to increase attractiveness of organized forms of housing construction (apartment ownership, etc.);
- attempts at integrating activities in urban planning, design and ground preparation;
- no social housing programme planned at present.

Main Planning Problems

- building enterprises governed exclusively by market consideration and cannot supply a full spectrum of housing needs (for large and low-income families), and desired environmental amenities;
- availability of a sufficient number of locations prepared for one-family houses;
- non-synchronized construction of residential areas (housing first, facilities later);
- lack of concentrated (and sufficient) funds for large-scale projects;
- need of more 'programming' work in city planning offices for matching 'projects and design orientation'.

- cumbersome procedure of the acceptance of plans;
- land speculation;
- non-existence of market intervention meaning such as public-social-housing or apartments available for rent;
- illegal housing construction;
- re-design of urban plans to meet demands for single-family homes.

Specified Planning Objectives

- to increase spatial standards;
- to relieve the housing shortage;
- to rationalize housing construction with modern construction methods – prefabrication.

Supposed or Anticipated Extreme Manifestations

- 'suburbia' – U.S. type with accompanying patterns (high accessibility by individual car, traditional dense network of roads);
- industrially mass-produced individual homes;
- hillside building developments (to save arable lands);
- housing pressure towards the central area – return to high-density, high-rise housing constructions;
- many people intending to live permanently in secondary dwellings many of which are already built for such purpose;
- re-introduction of housing co-operatives.

APPENDIX 2. LAND ECONOMICS AND DEVELOPMENT CONTROL

Existing Conditions

- different land marketing system;
    - (*a*) nationalized land (with urban areas at time of Act – 1958) – purchase price based on 'just compensation';
    - (*b*) other non-nationalized land purchased at market price;
    - (*c*) land for public purposes – procedure similar for nationalized land.

Current Trends

- growing scarcity of nationalized land;
- increasing price of land.

Current Planning Policies

- growing awareness of importance of planning for future needs;
- a regional approach;
- land for communal and regional interests being safeguarded;
- designation of areas for special uses;
- policy of low land prices.

Main Planning Problems

- increased shortage of nationalized land for development;
- land development controls not tight enough in communes;
- local communal *v.* regional interests;
- illegal development.

Specified Planning Objectives

- the rational use of restricted space;
- to keep land costs down to levels which will not inhibit development.

Supposed or Anticipated Extreme Manifestations

- overall nationalization – tight controls;
- free land market – loose controls.

APPENDIX 3. TRANSPORTATION

Existing Conditions

- five classes of roads – (1) international, (2) interregional, (3) inter-city, (4) local, (5) rural; (1, 2 financed by republics – 3, 4 and 5, by commune);
- strategic position of the region on cross-roads between Central Europe, the Balkans and West and East Europe;
- dense network of roads and railways;
- few modern roads;
- rapid increase of motor traffic in recent years;
- small role of air transportation;
- buses are predominant urban and regional public transportation.

Current Trends

- transfer from bicycle use to the motorbicycle (motorbicycle has enlarged accessibility, daily commuting);
- increase of individual motor car ownership by 30% per year (possession of a car is a status symbol);
- one of the world's highest accident rates;
- movement of goods from rail to road transport;
- abolishing of local railroads (within the region in the last two years, two were discontinued: Ljubljana–Kočevje is under discussion);
- negligence in dealing with mass transport problems (attractiveness, efficiency);
- common use of individual garages because of high prices of vehicles and cold weather;
- car ownership approaching one commuter car per family, rest of family preferably within walking distance of services.

Current Planning Policies

- emphasis on improvement of roads, intersections;
- plans for motor roads (for entire Slovenia);
- segregation of different modes of movement (pedestrian, bicycle, motor cars);
- slow modernization of railroads;
- attempts at a more comprehensive and sophisticated approach (road construction *v.* transport problems).

Main Planning Problems
- reservation of excessive areas for potential future transportation lines for long periods;
- parking space;
- urban traffic;
- high casualties;
- lack of comparative studies in different modes of transportation;
- subjective and partial approach;
- slow road construction because of shortage of funds;
- priority determination effected by shortage of funds;
- indecision as to the type of road needed to serve both inter-city and inter-regional traffic (Freeways or Feeders).

Specified Planning Objectives
- to link and integrate present superhighways into regional road systems;
- to provide improved public transportation to all areas;
- to improve all major roads with construction of hard surfaces;
- to provide adequate parking areas in central parts of Ljubljana;
- to route through-traffic around the edge of the city centre.

Supposed or Anticipated Extreme Manifestations
- individual motor car ratio 1 : 3 or one car per family;
- new modes of public transportation (urban railway?);
- waterway: Danube–Adriatic Sea;
- new modes of mechanized movement.

APPENDIX 4. EMPLOYMENT

Existing Conditions

- 'tertiary' employment is predominant in the city of Ljubljana; in other towns of the region, manufacturing is the main source of employment;
- a high percentage of women in the labour force, with some sub-regional variations;
- the 'growth' industries are electronics and metal-machinery-equipment;
- a wide range of skilled labour availability within the region;
- manufacturing employment scattered throughout the urbanized part of the region;
- many of the service industries still deficient in meeting present needs;
- small farmers and others commuting to industrial jobs.

Current Trends

- increasing acceptance of low-paid jobs by emigrants from the southern republics (especially in the building trades);
- increasing emigration rate of skilled labour force to Western Europe;
- more industry for Ljubljana (an expressed policy orientation);
- growth of tertiary activities and a slow-down in expansion of quaternary activities (schools, health services) due to lack of funds;
- integration of working organizations into larger and stronger enterprises;
- co-operation with international firms;
- increasing activities of Yugoslav firms abroad;
- shortening of working hours and a switch to a five-day work week;
- concentration of large industries and warehouses in large outlying sites.

Current Planning Policies

- formation of industrial zones – 'parks';
- belts of light industry situated between residential zones;
- location of work places near home;
- support of tertiary activities development (commerce, transportation, hostelry, tourism).

Main Planning Policies

- air pollution, water pollution;
- reservations for future industrial land use;
- decision of which type of industry has comparative advantages;
- question of capital (availability, concentration and mobility);
- low wage levels.

Specified Planning Objectives

- to have full employment;
- to achieve a balanced structure of employment (men/women) in the region and in most towns;
- to safeguard enough space for expansion of working-areas (industries, services, etc.).

Supposed or Anticipated Extreme Manifestations

- automation;
- very low percent of labour force employed in industry (5–10%);
- regional industry parks for 'research' industries;
- foreign investment in labour-intensive industries concentrated in few regional centres;
- possible overproduction of trained professional and skilled workers causing increased migration to Western Europe, with replacement in the region by in-migrants from the South.

APPENDIX 5. CENTRAL ACTIVITIES TERTIARY: COMMERCE–BUSINESS–ADMINISTRATION

Existing Conditions
- concentration of central activities in Ljubljana;
- underdeveloped business potential in CBD.

Current Trends
- growth of demands for office space in Ljubljana CBD;
- idea for shopping centre in the CBD (and more elaborate projects);
- development of retail chains;
- prepacking techniques in central warehouses;
- development of department stores (increase in scale).

Current Planning Policies
- building up business and administrative centres;
- providing central services in residential suburbs.

Main Planning Problems
- lack of recognition of planning problems (and regional aspects) on the part of commercial enterprises;
- banks and insurance enterprises do not play their potential role in CBD development.

Specified Planning Objectives
- to strengthen the CBD in Ljubljana and in satellite cities.

Supposed or Anticipated Extreme Manifestations
- establishment of secondary centres, segregation of central activities;
- decentralization of mobile office functions (branch houses, banks, insurance offices, headquarters);
- formation of regional 'shopping centres', 'discount houses' (outside of central city).

APPENDIX 6. RECREATION, SECOND HOMES

Existing Conditions
- exceptionally attractive natural amenities surrounding Ljubljana (short distances to Alps and the sea coast);
- moderate use of existing recreation facilities within the city;
- weekend houses and summer cottages around Ljubljana – in 3 belts: (*a*) within 15 km, (*b*) as far as 80 km, (*c*) at sea coast; (*a*) partly for daily use, (*b*) Alps and on rivers, etc., (*c*) for weekend use;
- gardening very popular;
- extremely frequent weekend excursions due to the newness of having cars and the crowded housing conditions;
- tradition of Sunday trips (hiking, bicycling, driving).

Current Trends
- increasing leisure time;
- relatively high increase in number of secondary houses;
- sea/mountain vacation time: 50:50 occurrence;
- increasing gardening with single home ownership;
- demand for organized recreation, increase in games.

Current Planning Policies
- concentration and clustering of summer weekend houses;
- preservation of recreational areas within the urban region;
- integration of sports and recreational facilities into residential areas;
- development of recreation belt along Sava river, north of the city.

Main Planning Problems
- non-synchronized construction of facilities in residential areas;
- scattered development of summer cottages;
- provision of communal facilities (roads, water) to weekend houses;
- insufficient conservation measures.

Specified Planning Objectives
- to preserve landscape and natural amenities for recreational purposes;

- to increase the accessibility and availability of recreational facilities;
- to expand the spectrum of recreational facilities and activities.

Supposed or Anticipated Extreme Manifestations
- weekend-house family ratio: 1 : 10 or 1 : 5;
- large-scale national parks to prevent destruction of natural areas;
- 30-hour work week.

APPENDIX 7. AGRICULTURE (INCLUDING MEAT AND DAIRY PRODUCTS) AND FORESTRY

Existing Conditions
- majority of agricultural land in small holdings (subsistence farmers also holding industrial jobs);
- low productivity;
- large socialized farms on flat land, producing approximately 15% of total market product;
- lack of specialization in production;
- important products: cattle, potato, dairy, vegetables;
- significant role of forestry in the economy.

Current Trends
- creation of large-scale farms (mechanization);
- modernization of cattle breeding;
- orientation into 'monocultures' (including crops for industrial processing);
- depopulation of the countryside by urbanization process.

Current Planning Policies
- preparation of regional plans for agriculture and forestry;
- socialization of agricultural land in the flats (through acquisition);
- stabilization of prices for agricultural products;
- planning of quick-growing forest for cellulose lumber;
- conservation of large units of arable land (the best agricultural land should be protected from urbanization);
- conservation of forests in flat country;
- development of forestry in abandoned mountainous regions;
- vegetable and milkshed agriculture near cities;
- decentralized processing industries.

Main Planning Problems
- small pieces of arable land scattered in private hands;
- urbanization;
- abandoning of mountainous regions;
- redistribution and location of production centres in agriculture;

- conflicts between agriculture, urbanization and tourism;
- agricultural development policies being subject to exogenous forces (i.e. world market prices, tariff restrictions, etc.).

Specified Planning Objectives
- to preserve the best agricultural land;
- to round off the agricultural holdings for intensified production;
- to increase productivity of employment in agriculture.

Supposed or Anticipated Extreme Manifestations
- high mechanization (and industrialization) of agriculture;
- large changes in rural landscape as a result of depopulation;
- large drop in agricultural labour force;
- growing vegetables under glass;
- subsidized small farms.

APPENDIX 8. CONSERVATION

Existing Conditions
- gradual transformation of the countryside as a result of industrialization and urbanization processes;
- inadequate legislation;
- deterioration of landscape around cities;
- lack of means for proper conservation of cultural monuments;
- inadequate economic uses of buildings designated as monuments.

Current Trends
- improvement of legislation;
- growing public awareness of the need for conservational measures;
- regional approach, followed increasingly by professionals in related fields;
- emerging attention in the planning process.

Current Planning Policies
- protection of air, water, soil and mineral resources, plant and animal life;
- creation of national parks;
- implementation of the new law;
- interrelation of conservation and the planning administration;
- preservation of historical monuments and environments through activization or development of new commercial and institutional activities;
- preservation of open green spaces near and in urban areas.

Main Planning Problems
- lack of expressed social values;
- destruction of natural resources;
- scattered development (housing, industry, tourism, second homes);
- depopulation and 'ageing' of the population in the countryside;
- new functions for monumental buildings.

Specified Planning Objectives
- to revitalize monumental buildings;
- to increase accessibility and use of national parks and amenity areas;
- to include monuments and amenities as important elements of plans.

Supposed or Anticipated Extreme Manifestations
- emergence of broad popular support for conservation.

APPENDIX 9. WATER

Existing Conditions

- abundant resources of high quality water;
- high pollution of water in rivers (equivalent to the effect of a population of 9 million in Slovenia);
- no sewage treatment plants;
- delayed implementation of relatively strict legislation;
- water management agencies – local;
- specific problems arising from the soil conditions; moor, karst, gravel;
- a locally administered water supply system.

Current Trends

- growing awareness of the fact that water is not an unlimited commodity;
- continued increase of pollution of rivers;
- building of some treatment plants at industries;

Current Planning Policies

- conservation of water resources including potential resources on the regional scale;
- taxation of water used by industry.

Main Planning Problems

- lack of enforcement of legislation concerning water pollution;
- co-ordination of different uses of water, e.g. drinking-water, irrigation, flood control, navigation, hydroelectric power, recreation, etc.;
- lack of investment funds in specialized agencies;
- lack of funds for treatment plants.

Specified Planning Objectives

- to insure adequate quantities of high quality water for future uses;
- to modernize existing water supply systems;
- to rationalize industrial water demand (re-use);
- to study the feasibility of navigation;
- to create more possibilities of water recreation.

Supposed or Anticipated Extreme Manifestations
- maximum use of hydroelectric potential of all rivers;
- artificial lakes and basins for recreational use;
- navigation channel Danube–Adriatic Sea through Ljubljana.

APPENDIX 10. POWER

Existing Conditions

- shortage of electrical power;
- consumption *per capita* equal to European average;
- all potentially economic hydroelectric power already developed;
- hydro- *v.* thermo-electric power in Slovenia cca 50 : 50;
- inadequate distribution network.

Current Trends

- declining importance of coal mining;
- construction of high-tension network;
- predominant orientation on thermo-electric power;
- contemplation of oil, gas, nuclear and other new sources of energy.

Current Planning Policies

- increased production of power;
- concentration of production in large units.

Main Planning Problems

- insufficient funds for new development;
- disproportion of demand and possible present production (power demand presently exceeds supply).

Specified Planning Objectives

- to develop abundant and economically feasible sources of energy;
- to approach the equalization of production *v.* demand.

Supposed or Anticipated Extreme Manifestations

- development of nuclear sources of energy to the level of replacing most other sources of energy;
- substantially reduced costs for energy.

NOTES

[1] The writer has made an extensive review and analysis of Soviet planning principles in order to have a better understanding of their impact on Eastern Europe: see Jack C. Fisher and James Gibson, *Soviet Resource Development: Avowed Principles and Performance*, mimeographed version, Division of Regional Studies, Center for Housing and Environmental Studies, Cornell University, 1967. It is an underlying thesis of this paper as well as a subsequent one now under preparation on Czechoslovakia, that the 'principles' of the 1940's and 1950's will have increasingly less planning significance in the future as the influence of central economic planning decreases in importance and the authority of regional spatial planning bodies increases.

[2] W. W. Leontief, address to Economic Development Administration summer seminar, Williams College, Williamstown, Mass., August, 1966.

[3] Devising the ways and means of disaggregating economic systems is by no means an unheard of practice within planning circles. What is surprising, however, is that so little effort has gone into shaping the theory and construction of aggregate spatial models.

[4] Mimeographed Report by Marko Frković (Head of the Sector of Social Economic System in the Federal Institute of Economic Planning), edited by the FIEP, Belgrade, May 28, 1966.

[5] The intervention of planning is a political act which implies some ruling public purpose. As Tinbergen says, '... *Planning* will be defined as the preparation for political action, that is, action by public authorities. *Economic* planning then refers to the economic activities implied ...' (J. Tinbergen, 'International Economic Planning', *Daedalus*, Spring, 1966, p. 530). It is not necessary to give up the goal of rationality to include some non-economic objectives, but it may be necessary to expand traditional notions of economic rationality to include additional system ends. As the Yugoslavs move to a condition of greater economic affluence, these values may be expected to enter increasingly into the planning system. The French progression in emphasis from the First Plan to the Fourth Plan illustrates this extension of aims. 'The emphasis placed on providing social equipment, improving the living standard of less-favoured groups and developing a better balance between the different regions is one of the principal features of the Fourth Plan.' (*France and Economic Planning*, Ambassade de France, Service de Presse et d'Information, April, 1963, p. 25). Is it rational to attempt to

anticipate this drift and to make the ultimate realization of more complex objectives easier?

[6] It is not possible here to review the historical evolution which led to the institutional posture of planning as it will be described in the next section on Slovenia. The reader is referred to Jack C. Fisher, *Yugoslavia: A Multinational State* (San Francisco: Chandler Publishing Co., 1966) as well as Richard P. Burton, John W. Dyckman and Jack C. Fisher, 'Toward a System of Social Planning in Yugoslavia', *Papers, Regional Science Association*, XVIII, Vienna Congress, 1966, pp. 75–86. These and other articles by the author were used in the preparation of this monograph.

[7] After a prolonged evaluation of the communal system and the function of the commune, Dr Eugen Pusić poses the basic question of his research: 'Has the commune retained the characteristics of the territorial model of administration and does it differ significantly from the functional model of institutional self-management? On the whole, the exploratory test points, however tentatively, towards an affirmative answer to these questions. The idea of the commune as a component part of the self-management system may have to be reappraised. It is possible that the commune, even with the most radical measures of decentralization, remains a part of the traditional system of government based on the systematic wielding of political power over people, and, therefore, intrinsically in opposition to a society-wide system of self-management.' Eugen Pusić, *Patterns of Administration: A Theory and an Exploratory Test* (Ljubljana: American-Yugoslav Project in Regional and Urban Planning Studies, September 1967), p. 20.

[8] In fact, however, federal investment did not decrease during the economic reform period because of long-term obligations previously incurred by the federal government. Federally constructed projects under construction today, requiring vast funds, are: (1) Djerdap (co-operative Romanian-Yugoslav power and navigation system on the Danube); (2) the iron and steel works in Skopje; (3) the Belgrade–Bar railroad; (4) the port of Bar; (5) the bridge over the Sava in Belgrade (partly financed federally); etc.

[9] Contrast the philosophy of preceding paragraphs and pages with the past and present voices guiding Soviet, and, indirectly, East European development.

> One feature of the history of old Russia was the continual beatings she suffered for falling behind, for her backwardness. She was beaten by the Mongol Khans. She was beaten by the Turkish beys. She was beaten by the Swedish feudal lords. She was beaten by the Japanese barons. All beat her – for her backwardness: for

military backwardness, for cultural backwardness, for political backwardness, for industrial backwardness, for agricultural backwardness. She was beaten because to do so was profitable and could be done with impunity. Do you remember the words of the pre-revolutionary poet: 'You are poor and abundant, mighty and impotent, Mother Russia.' These words of the old poet were well learned by those gentlemen. They beat her, saying: 'You are abundant,' so one can enrich oneself at your expense. They beat her, saying: 'You are poor and impotent,' so you can be beaten and plundered with impunity. Such is the law of the exploiters – to beat the backward and the weak. It is the jungle law of capitalism. J. Stalin, *Problems of Leninism* (Moscow: Foreign Languages Publishing House, 1940), p. 365.

The belief in the necessity of defence through economic strength is still held.

Only a society which makes possible the harmonious co-operation of its productive forces on the basis on one single vast plan can allow industry to settle in whatever form of distribution over the whole country is best adapted to its own development and the maintenance of development of the other elements of production. Frederick Engels, *Herr Eugen Duhring's Revolution in Science (Anti-Duhring)*, trans. Emile Burns (New York: International Publishers, n.d.), p. 331.

It is impossible to conceive effective leadership without the existence of one single leading centre which is to be duly responsible for the development of the branch [of industry], for the unified policy of technical improvement, for an advanced organization and effective production, the right location and the rational employment of labour and technical cadres. *Pravda*, October 3, 1965, p. 2.

The reform of the Soviet planning system in October, 1965, stressed that the role of the State Planning Commission (GOSPLAN) was to achieve unquestioned authority through the reintroduction of centralized planning methods and through the fulfilment of one single state plan, despite the fact that increased initiative was proposed for enterprises by a system of economic stimulation instead of administrative manipulation. The role of the planning commissions of the union republics is to elaborate projects of the plans of the enterprises under the jurisdiction of the republics and to make propositions to GOSPLAN regarding projects of the plans of All-Union enterprises on their territory. Thus, the role of a republic's planning authority is only to make propositions and projects, not decisions. See *Pravda*, October 3, 1965, p. 2.

Full-scale communist construction calls for a more rational *geographic distribution* of the industries in order to save social labour and ensure the comprehensive development of areas and the specialization of their industries, do away with the overpopulation of big cities, facilitate the elimination of essential distinctions between town and countryside, and further even out the economic levels of different parts of the country. Communist Party, *Programme of the Communist Party of the Soviet Union adopted by the 22nd Congress of the C.P.S.U., October 31, 1961* (Moscow: Foreign Languages Publishing House, 1961), p. 68. The significance of industrialization here [backward regions of the country] is exceptionally great, for it not only facilitates the raising of the economies of all backward regions but also involves the formation of national cadres of the working class and the growth of the political consciousness and the culture of the population. The planned distribution of large-scale production and of industry throughout the country under socialism therefore serves as the material basis for the solution of the nationality question. R. S. Livshits, *Razmeshchenie chyornoy metallurgii SSSR* (Moscow: Izdatelstvo Akademii nauk SSSR, 1958), p. 12.

[10] Proponents of programmes for development of the southern regions, for example, feel that development of infrastructure of those regions was inadequate to allow them to contribute effectively to national and international economies, even during the period of the General Investment Fund. Thus, one writer asserts that 'During the period (1953-60) the construction of infrastructure projects in the underdeveloped territories was of insufficient proportions, which hindered their linking up with the rest of the country, just as it did the linking up with economic centres inside the territories themselves. Along with other adverse factors this meant that investment results were below those attainable in conditions of optimum infrastructure.' (Vladimir Pejovski, 'Yugoslav Investment Policy', *Medjunarodna politika* (Belgrade), No. 3 (1965), p. 34. This is a proposition, however, which could be tested only over a long period, if at all.

[11] See Jack C. Fisher, *Yugoslavia: A Multinational State* (San Francisco: Chandler Publishing Company, 1966), pp. 166-86.

[12] The Federal Institute for Urbanism, Housing and Communal Problems was created and attached to the Federal government but without any administrative responsibility over republican

institutions for planning. In the most recent period of intensive decentralization the institute ceased formally being a federal institute.

[13] Branko Petrović and Stanko Žuljic (editors), *Kotar Krapina, Regionalni prostorni plan* (Zagreb: Radovi Urbanističkog Instituta NRH, Number 2, 1958).

[14] Franjo Gašparović, Branko Petrović et al., 'Regionalno prostorno planiranje', *Prvi Kongres Saveza društava urbanista Jugoslavije*, Belgrade, 1957.

[15] Urbanistični inštitut Slovenije, *Osnove regionalnega in urbanističnega planiranja* (gradivo) (Ljubljana, 1962).

[16] See Branko Petrović and Stanko Žuljić, *Metodologija regionalnog prostornog planiranja* (Zagreb: Urbanistički institut NRH, 1956).

[17] The retention quota is the proportion of each dollar of foreign exchange earnings which an enterprise can keep *as foreign exchange* and spend at its own discretion. The remainder of each dollar must be 'sold' to the National Bank at the official rate of 12·50 new dinars per dollar. The retention quota for tourism and related industries is about 30% in contrast to an average rate for all industry of about 7%.

[18] *Vjesnik*, Zagreb, August 16, 1969.

[19] This analysis was made by data supplied by the Association of Banks of Yugoslavia and reported in *Ekonomska politika*, January 20–8, 1968, p. 82, and conducted by Katharine Lyall of the American-Yugoslav Project staff. For the full report see: Katharine C. Lyall, *Inter-republic Flows of Investment in Tourism – Yugoslavia*, 1967 (Ljubljana: American-Yugoslav Project, 1968).

[20] For a more detailed analysis of these data for Slovenia see: Katharine C. Lyall, *Structure and Policy of Yugoslav Banking: A Model for Evaluating Multiple Goals* (Ljubljana: American-Yugoslav Project, 1968).

[21] For a fairly complete list of material available on Slovenia see: *A Selective Bibliography on Regional Planning in the Ljubljana Region from 1966 to 1968* (Ljubljana: American-Yugoslav Project in Regional and Urban Planning Studies, 1968). Reference is also drawn to the two volume work on the American-Yugoslav Project demonstration study upon which some of this section is based: *Toward a Methodology for Regional Planning: The Proposal and Evaluation of Alternative Patterns of Spatial Organization for the Ljubljana Urban Region* (Ljubljana: American-Yugoslav Project in Regional and Urban Planning Studies, 1968), Volumes One and Two. See especially Braco Mušič and Danilo Goriup, 'Osnove za prostorsko planiranje na področju

bivšega Ljubljanskega okraja (The Basic Elements of Spatial Planning in the Area of the Former Ljubljana District)', *Urbanizem* (Ljubljana), No. 3/4 (1967), pp. 59-70.

The remaining section of the paper draws heavily upon the work of staff members of the American-Yugoslav Project in Regional and Urban Planning Studies located in Ljubljana. The original objectives of the Project were: (1) to contribute to the development of urban and regional planning as a professional field of international importance; (2) to create a method of work in regional planning that concentrates on and maximizes the interdisciplinary character of the field; (3) to achieve to the highest degree possible the formation of an integrated regional planning discipline (with concentration on such specific components as transportation, social planning, quantitative methodology, etc.); (4) to perform a training function for young professionals. In particular to assist Yugoslavia in the training of professionals in regional planning; (5) to lay the foundation for the possible development of a training and research centre in regional planning which might eventually assume an international character.

[22] This brief section is based upon the very extensive work accomplished by the American-Yugoslav Project on demographic and migration characteristics of Slovenia and the Ljubljana urban region: Silvo Kranjec, *Preliminary Report on the Study of Migration in Slovenia* (Ljubljana, 1967); Silvo Kranjec and Andrei Rogers, *Interregional Population Growth and Distribution in Slovenia* (1968); Silvo Kranjec, *An Analysis of the Aggregation Problem in Demography, Data for Yugoslavia* (1967); and especially, Andrei Rogers and Susan McDougall, *An Analysis of Population Growth and Change in Slovenia and the Rest of Yugoslavia* (Berkley, Cal.: Department of City and Regional Planning, University of California, Working Paper No. 81, June, 1968).

[23] Dolfe Vogelnik, 'Razvoj prebivalstva Slovenije zadnjih dvesto let z Jugoslovanske in Evropske perspektive' (The Development of the Population of Slovenia in the last Two Hundred Years within the Yugoslav and European Perspective), *Ekonomski zbornik* (Ljubljana, 1965), p. 113.

[24] Vogelnik, *op. cit.*, p. 114.

[25] *Ibid.*

[26] *Ibid.*

[27] Rogers and McDougall, *op. cit.*, 1968.

[28] Slovenia lost a major round in its efforts to develop a modern highway system by a decision of the Federal Executive Council during the summer of 1969 to exclude the Slovene highway

projects (Vrhnika–Postojna and Maribor–Celje) from consideration by the International Bank for Reconstruction and Development. This was a severe blow to the Republic's leadership but represents an increasing trend within the country's top leadership of concern or disapproval over recent Slovene policies. The highway projects to be submitted to the Bank by the Yugoslav government are: (1) Peć–Priština-Niš, (2) Bar–Ulcinj, (3) Sarajevo–Zenica, and (4) Belgrade–Novi Sad. It may be noted that with the exception of the Novi Sad–Belgrade route, all the remaining highway projects are located in the underdeveloped areas of Yugoslavia – the areas where the consequences of the Economic Reform were most severe.

[29] After prolonged debate a committee of the Yugoslav Federal Parliament produced a draft resolution on urbanization and spatial control; the first preliminary statement of its kind on the federal level; see: *Tezeo urbanizaciji i prostornom uredjenju* (Resolution on Urbanization and Spatial Organization), Savezna skupština, Komisija za urbanizam i uredjenje prostora, Maj, 1968. No uniform policies exist throughout the country and there is no federal law on urban and regional planning. A decision of the Constitutional Court a few years ago overruled the action of communes to expropriate land. The Court required Dalmatian communes to return at cost land previously nationalized. The impossibility of developing federal legislation has stimulated most republics to adopt their own planning laws and codes. Serbia was the last Republic to establish a Secretary (Ministry) for Urbanism, Housing and Communal Affairs, and the Republic adopted a regional planning resolution in 1967. For details of the proposed Serbian regional plan see: *Prostorni plan Srbije* (The Spatial Plan of Serbia), Republički sekretarijat za urbanizam, stambene i komunalne delatnosti, Belgrade, April, 1968.

Comments IVAN CRKVENČIĆ

Dr Jack Fisher has provided us with a serious contribution to the knowledge of regional spatial planning in Yugoslavia. He has spent a rather long time in Yugoslavia and during his stay gave special attention to a study of the function and structure of our cities. With his work he placed himself among those foreign scientists who are well acquainted with Yugoslavia and who are also correctly interpreting our realities.

The problems in which Dr Fisher is intensively engaged are not identical to my scientific research work, though as a geographer I took part in the efforts of the Yugoslav and especially of the Croatian planners to find a solution to the problems of spatial planning. The questions which were treated by Dr Fisher are indeed rather complicated and he rightly points out that the importance of urban and regional planning in Yugoslavia has grown as a result of the economic and administrative decentralization. I consider the information he has presented about the hierarchical relationship between the Federal authorities and the Republican and communal bodies for planning very useful, inasmuch as this relationship is still very confusing to the Westerner.

In his discussion of the Physical Planning Process the author raises one very significant point and I quote: 'We have done little in any country to analyse changing land use patterns in terms of underlying locational strategies and to prepare operational time series of land and locational change.' To me this is of considerable importance because he clearly indicates that one cannot plan exclusively with statistical and other figures received from authorities; that in planning, a basic social and economic analysis of the space is indispensable. I am of the opinion that every form of planning must include the social element, and Dr Fisher has taken this into account in his observation that 'the integration of economic planning and physical planning in a social planning system requires a broader conception than the economist's rationality.' This underlines the fact that planning requires close co-operation of a broad number of different scientific spatial disciplines than has been the usual case.

The author also presents a comprehensive picture of the political development, which influenced the emergence of the Yugoslav institutions for planning. As far as I know, this is the first comprehensive analysis of its kind among foreign scientists. But his statement that 'This attempt to create a compromise between Western democratic institutional traditions and proletarian dictatorship lies at the root of Yugoslavia's post-war institutional development' is somewhat puzzling, inasmuch as most planning institutions in Yugoslavia developed during the period after 1948 and are the result of the balance between unitarian and regional tendencies within Yugoslavia.

Regional spatial planning in Yugoslavia, based on urban studies, developed only after 1950 and therefore has a very short tradition. The work of the American-Yugoslav Project in Regional and Urban Planning Studies (organized in 1966) and administered on the Yugoslav side by the Urban Planning Institute of Slovenia must be mentioned, especially, for its impact on all of Yugoslavia. I believe that the special significance of this Project mentioned in Dr Fisher's paper lies in the new methods of scientific research which now has been introduced into our theory and practice and the wide consultations at home and in the international field as well as in the lectures by experts of different schools which were organized by this Project.

It was pointed out by Dr Fisher that Slovenia was the first of all republics to discover the importance of co-ordinated regional physical planning, primarily because Slovenia is economically the most advanced of all republics (taking into consideration its *per capita* national income). However, the regional spatial plan of Zagreb, which is also mentioned by the author, comprises a territory of the approximate size of Slovenia. To me it seems to be of greater importance that the same contemporary social and economic process of quantitative transformations comprises almost the whole of Slovenia, while in all other republics there are still considerable regional differences which cause difficulties within the framework of regional planning within each republic. I do not suggest, however, that there is no need for regional plans in those republics nor that they are without facilities for the realization of such plans.

In the discussion of tourism, which is of special interest to me, Dr Fisher indicates that tourism has become the key point of the economic growth of Yugoslavia and thus an important element of planning. Two tables developed by the American-Yugoslav Project cast new light on the inter-republic relationship of investment credits. The results of these investments have been a surprise, even here in Yugoslavia. The unequal relationship of granting and accepting loans between the various republics must, however, be observed from the point of view of inadequate distribution of the financial capital. It is evident that the granting of loans is not of permanent character and that this fluctuation may have its influence upon the regional spatial planning. In addition, it should be mentioned that the various republics are not only granting credits to the tourist industry of the other republics, but they are also building on the territory of the other republic hotels and their own tourist enterprises. These investments are sizeable and thus have certain influence on overall regional planning.

The discussion of Slovene Planning Reality points up the policy of regional spatial planning in Slovenia, of the demographic conditions and the research work in the urban district of Ljubljana. It contains valuable material and we can generally conclude that the work Dr Fisher presents makes a vital contribution to the understanding of policy and practice in regional planning in Yugoslavia.

Comments
JOSEPH VELIKONJA

The essay touches upon three related topics of the regional spatial planning in Yugoslavia, compared and analysed within the general scheme of planned and unplanned economics. The historical review of the institutional framework emphasizes the distinct character of the Yugoslav *'regionalno prostorno planiranje'* (regional spatial planning) which has shifted the focus of planning process from a spatial national economic planning of the early post-war period to the territorially oriented planning mechanism and processes of the late 1960's. The recent efforts extend beyond the administrative boundaries of major cities and their metropolitan areas. In spite of the fact that the framework is being implemented to the greatest degree in Slovenia, and partially in Croatia, neither the process nor the mechanism are restricted or unique to the Slovenian scene.

The second topic deals with the Intra- and Inter-Republic investment credit flows for tourism. The spatial overtones of these flows have been so far only marginally related to planning; they are more a result of spontaneous than channelled orientations, which have developed, as the author indicates, 'outside of planning organ's responsibility'. The significance of tourism as a growing sector of the economy is undeniable, but it has been left to unregulated initiatives, and repeatedly criticized in the Slovenian, Croatian and Federal Assemblies.

The third part presents the Slovene planning reality, primarily of the Ljubljana urban region. The sketchy form of the appendices which cover the present conditions, current policies, planned goals and objectives for ten sectors of the economy suggest a great wealth of significant information, but their only qualitative form does not provide a sufficiently specific summary for an evaluation at this point.

The Yugoslav planning went through repeated structural modifications of the regional planning process. The early post-war years, 1945–52, have been characterized by centralized economic planning (Kidrič era) in which the territorial physical planning results have been more implicit than explicit.

The administratively decentralized regional planning has gone through considerable transformation in the sixties, by shifting the planning decision-making from the republics to the communes, with slow withering away of the intermediate administrative bodies (districts). The communes have become the grass-root multi-purpose administrative institutions at the lowest level with responsibilities of economic and territorial planning, though often without adequate means. The fact that the general guide-lines for economic planning are provided at the federal level, while the territorial physical planning is done at the commune level, leaves considerable gaps at the intermediate levels. With the exception of the Zagreb urban region and more recently of the Republic of Slovenia, there are no regional administrative or planning bodies which by the nature of their territorial framework would automatically assume the decision-making for planning at the intermediate level.

The contemporary 1969 situation should be visualized just as a stage in the transformation trend of a decentralized system, in which efforts to co-ordinate communal services and planning are more a response to the needs felt by the independent communes than as a theoretically rigid imposition from above.

The financing of tourism and regional planning for the Ljubljana region illustrates quite lucidly what results can be expected when unregulated flows of investments take place without even being detected, primarily due to the decentralized and somehow clumsy accountancy and inadequacy of clear federal regulations. On the other hand, the co-ordination of the spatial planning for Ljubljana region, one of the highly laudable efforts, to which the American contribution in funds, organization and experience has been fundamental, demonstrates that the inadequacies of institutional infrastructures do not prevent the creation of a co-ordinated and sophisticated regional plan, with high probability of being implemented in spite of the less than unanimous support of the Slovenian scholars and administrators.

The administrative decentralization strengthened the communes and placed them in the focus of socialist transformation with many new functions which before the 1963 constitution and 1965 economic reform remained in the hands of higher institutional agencies.

The primary function of the commune is social planning, concerned with the social well-being of its legal residents, limited to the territorial perimeter of the commune. The spatial restriction is self-evident and often requires compromise and co-ordination with neighbouring territorial communes during the planning and the implementation stage. In theory and practice the compromise leads to less than optimal solution for each component territory while the benefit for the total area is being maximized. It is evident that the independence of communes has also led to rivalries and duplications in the regional spatial planning and lead to administrative consolidations, spearheaded by the administrative reorganization of the Zagreb metropolitan area.

The legislation regarding economic planning – under discussion at the time of writing – is unavoidably also going to affect territorial organization and is opening new rather than solving the existing problems.

The economic planning by enterprises is at the time of writing the most far-reaching endeavour with substantive territorial consequences which by-pass the framework of communes and the republics. Enterprises are planning expansions in new locations outside their present areas. The prevailing southward orientation of the planning efforts puts Slovenia and Croatia in the position of entrepreneurial decision-makers, which have a territorial impact far beyond the administrative territory of the decision-makers. The parallel between the American corporation planning which has territorial dimensions and the planning of enterprises, as indicated by Dr Fisher, run along similar lines, with the principal difference in the relationship between free-market type locational mechanism which is consistent with the American economic system and less consistent with the socialist planning of social benefit maximization.

The regional/spatial planning is evidently only in its initial stage and with the exception of the American-Yugoslav Project does not go far beyond an assessment of regional spatial properties and physical linkages. The intra- and inter-regional flow assessment for different sectors of the economy is only sketchy as evident in the presented case of tourism investment, where the magnitude of tourist flows have been assumed frequently, without an adequate analysis of other economic cost-benefit assessment or investment allocation. The credit flow from Croatia to Slovenia and simultaneous flow from Slovenia to Croatia, are in fact two complementary flows in opposite directions and for different purposes. It can be assumed, and empirical research confirms it, that the Slovene investment in Croatia is oriented primarily to the coastal seashore tourism – the few miles of Slovene coastline are unable to accommodate even the internal demand – stimulating tourist investment in Croatia, whose coast extends from the Gulf of Trieste to Dubrovnik, with the primary objective providing accommodation for Slovene sea-seeking vacationers themselves. At the same time, mountain recreation is only just beginning in Croatia while it is well established in Slovenia, and is sought by the Croatians, for whom the Slovene areas are easily accessible with the opening of modern roads. In the decentralized system of allocations, a rather uncontrolled flow of investment support originates from regional (normally republic) banks and has been rather spontaneously directed to diverse destinations, with – surprisingly – preference given to individual rather than communal investments; while the individual is in no way territorially restricted where he wants to use the allocated fund, communal agencies are generally limited to their own territories. The result is an anomaly for a socialist country, favouring individual rather than communal benefits. Tourism is just one relatively small though highly flexible segment of the economic structure, accounting for about 10% of the foreign currency procurement in the Republic of Slovenia.

The regional spatial planning, as illustrated by the case of Slovenia, deserves close attention as an experiment of broadly conceived and integrated effort, a blending of the Western experience and the socialist goal achievement. The essay gives us only a glimpse of the work under way and only suggests rather than documents the leading role played by Dr Fisher and his associates in the regional planning of Yugoslavia.

Remarks

Professor Alexander spoke of the validity of Professor Fisher's paper which had generated several questions such as: (1) To what extent do the traditional animosities and competition prevail in the new economic planning and how much intra-republic planning is there. Is this planning for Slovenia *per se* or as a part of Yugoslavia as a whole? (2) Is there in truth a real merging of the traditional physical planning approaches; economic location analysis and investment programming? Mention should also be made of some of the planning programmes in Zgornja Pivka with regard to future use of abandoned agricultural land. Both the forest sector and the cattle-breeding sector had presented plans for land use. (3) After the regional spatial plan is developed, how much will the Federal government have to say about its actual function and growth? (4) What are the future problems in integrating various regional plans? It may be fruitful to take individually each of the aspects Fisher studied (forestry, tourism, etc.) and make a comparison between what the Slovenes suggest and plans which might be put forward by an objective, rational, idealistic Yugoslav.

Professor Karcz took issue with respect to Fisher's mention of 'the economist's rationality' and reminded him that the discipline of economics was a method of approach and rational did not necessarily mean profitable. He added that to ignore long-term costs was a basic human problem and not one concerning only the Slovenes or the Yugoslavs.

Professor Ivanička then pointed out the problem of localization of industry from the point of view of space. He indicated polar growth areas such as Upper Silesia, Bratislava and Košice and also axial growth areas such as the Danube Basin – localization of industry on the basis of importing raw materials and re-exporting finished products. He also considered the problem of the whole territorial complex.

Several questions then were posed by Professor Velikonja. (1) Was there any investment between one republic and another? (2) If so, where did the money for this investment come from? (3) Is investment capital from tourism in Croatia and Slovenia going to individuals or to institutions and organizations? and, (4) How much are the communes forced to coordinate their plans with those of the republics?

Professor Fisher replied that it was only a matter of time until the republics would take over (and necessarily so) most of the powers of the communes. He went on to say how concerned he was about the probable social problems arising in Slovenia, i.e. he saw a possible polarization into a two-tiered Slovene society, one the well-paid management and administrative group and the other the unskilled labour supply. This was a result of the necessary continued inflow of workers from the south who were not being provided with good social conditions, e.g. adequate housing facilities. He was, therefore, pessimistic about the future, foreseeing an increasingly segregated society.

Professor Weigend then asked if the increase in the planning function within each republic and the decrease in the flow of investment from one to another would not accentuate existing conditions in each republic?

Professor Fisher agreed that this seemed inevitable. However, he continued, if a proper infrastructure (especially a transport network) could be developed, then the attraction of getting a better return on money in the south than in the north might lead to some investment by the north in the south.

Professor Hadžić, addressing his remarks primarily to Dr Weigend, reminded him that there was a credit fund set up to help only the underdeveloped republics, and that these republics could also get financial aid from the Federal government in order to reach at least a limited level of social conditions.

Professor Crkvenčić stated that since Yugoslavia would like to have larger enterprises which could compete with other countries, factories in different republics had amalgamated, and also that foreign investment was being spread among different areas. He also compared the post-war German policy of placing factories in agrarian areas in order to find workers, with the Yugoslav plan of introducing factories to provide jobs. He added that he knew how much money went to the republics and how little went to the communes, and hence the inevitable conflict between the two.

Mr Wilson cited the example of the integration of the shoe industry in order to compete in the world market, for example, with the U.S.A.

Finally, Dr Hamilton presented a few additional comments on the paper. He said that he did not think that the 'philosophy underlying the evolution of the country' was that each commune had the right, in theory, to expect a living standard comparable with the most advanced areas. He had no doubt that many naïve officials interpreted the idea of developing backward areas faster in this way, but it was never the philosophy. Before 1956, the same problem existed on the republic and, later, district levels.

With respect to 'Slovene planning reality' Dr Hamilton indicated that there is a need to stress (*a*) problems of co-ordination functionally and spatially; (*b*) relevance for the rest of Yugoslavia; and (*c*) the need for inter-republic co-operation.

9

The Belgrade–Bar Railroad: An Essay in Economic and Political Geography

ORME WILSON, JR.

Announcement on March 22, 1968, of a $50 million World Bank loan to help complete construction of a railroad from Belgrade to Bar on Yugoslavia's southern Adriatic coast focused renewed attention on a concept which had been alive some 90 years.[1] This paper aims to provide geographic perspective which may assist in assaying the significance of the Belgrade–Bar railroad, which is expected to be open for service along its entire route in 1973.

RAILROAD DEVELOPMENT IN YUGOSLAVIA

The first railroad line in the territory of present-day Yugoslavia was built on Austro-Hungarian territory in 1850 when Trieste was linked with Vienna via Ljubljana and Graz. Subsequently, in 1860, a line from Budapest was connected south of Maribor with the Vienna–Trieste line. Zagreb and Ljubljana were linked in 1862 and a line from Zagreb to Rijeka was completed in 1875.

By 1874, the Ottoman portion of present-day Yugoslavia saw completion of a line from Thessaloniki to Kosovska-Mitrovica which passed via Skopje.

With Austro-Hungarian support railroads began to be built in the Principality of Serbia in 1878. Initially a line was built from Belgrade to Niš where it forked: one branch going to the Bulgarian border at Caribrod (Dimitrovgrad) beyond Pirot and the other travelling southwards along the valley of the Morava to the frontier point with Turkey at Vranje. Austria-Hungary agreed to link this Serbian system with the Austro-Hungarian system via a bridge to be built across the Sava river from Belgrade to Zemun. The Belgrade–Budapest connection was completed in 1884. On August 12, 1888, an inaugural train travelled from Budapest to Istanbul via Belgrade, Caribrod and Sofia. Completion of the line across Bulgaria had been delayed by the Serbian–Bulgarian War of 1885 and the revolution in Eastern Rumelia in the same year.[2]

Austria-Hungary's move into Bosnia-Herzegovina in 1878 brought about construction of a narrow-gauge railroad system in that area. This system linked Slavonski Brod (served from the north by standard-gauge line) on the Sava with Sarajevo. It extended beyond Sarajevo to the Dalmatian coast at Metković, and thence to Gruz (port for Dubrovnik) and Kotor. A number of branch lines were extended off the main line of this system. Some of these penetrated extremely difficult terrain.[3]

During the period 1880–1900, favourable Austro-Hungarian laws encouraged construction of a considerable network of railroad lines in the Vojvodina and Slavonia. Additional impetus to railroad construction in these areas was provided by easy terrain and highly productive agricultural lands. Through development of this network, Belgrade and Zagreb became linked by railroad in the early 1890's. However, through traffic from Belgrade to Zagreb was not established.[4]

The Yugoslav state which emerged after World War I inherited about 9,000 km of railroads, some 1,600 km of which were in Serbia. Standard-gauge lines comprised about two-thirds of this total. Railroads in Serbia, however, were almost totally unusable due to war damage. Those in the former Austro-Hungarian provinces were in deplorably run-down condition because of wartime difficulties of maintenance.[5] The new Yugoslav state set itself the primary task of getting these lines back into operating condition. Additionally, it gave priority to transforming the through line Ljubljana, Zagreb, Belgrade, Niš, Djevdjelija and its Niš–Caribrod branch into first-class international routes. On April 15, 1919, the Simplon–Orient Express made its first Paris–Istanbul run over this line. The Thessaloniki–Athens section of the Simplon–Orient Express began operating in 1920. The Bucharest section of the Simplon–Orient Express began operation in 1923 travelling via

Vinkovci and Subotica in Yugoslavia to the Romanian–Hungarian frontiers.[6]

During the interwar period, new railroad construction in Yugoslavia included a standard-gauge line from Zagreb to Split designed to preclude over-dependence on the port-area of Rijeka, the main portion of which was then under Italian control. Additionally, a new line was built in Serbia linking Kragujevac with Kosovska-Mitrovica via the Ibar Valley. New lines were also built in Macedonia, one of which connected Veles (Titovo-Veles) with Bitolj.[7]

The end of World War II found Yugoslavia's railroads again damaged and dilapidated. Since World War II, much effort has been made to repair, expand and modernize Yugoslavia's railroad system. Sarajevo has been linked by standard-gauge line to the main Belgrade–Zagreb line. Furthermore, in 1966, Sarajevo became connected by standard-gauge line with the Adriatic port of Ploče. This line was financed in part by a 1963 World Bank loan of $35 million.[8] Another World Bank loan of $70 million, which was made in 1964, provided funds for the installation of electric traction on the main lines as well as installation of modern signalling and communication equipment.[9] Other important new standard-gauge lines have connected Doboj, Banja Luka and Dvor with the Zagreb–Belgrade line and linked them via Knin with the port of Split.

In summary, Yugoslavia's main railroad line runs northwest to southeast along valley routes. North of this main line the railroad system is well-developed over essentially easy terrain. To the south, much effort has been made since World War II to expand and improve the system. As a result in this southern zone, where the predominantly mountainous terrain has seriously complicated railroad development, the only sizeable area lacking a standard-gauge link with the rest of the country is the area south of Belgrade which the Belgrade–Bar railroad is to serve (Fig. 9.1).

Despite the post World War II progress, Yugoslavia has one of the least extensive railroad networks in Europe. With an area of 255,804 square km and approximately 9,000 km of standard-gauge lines, the average density is only 3·6 km of standard-gauge line per 100 square km. Furthermore, the

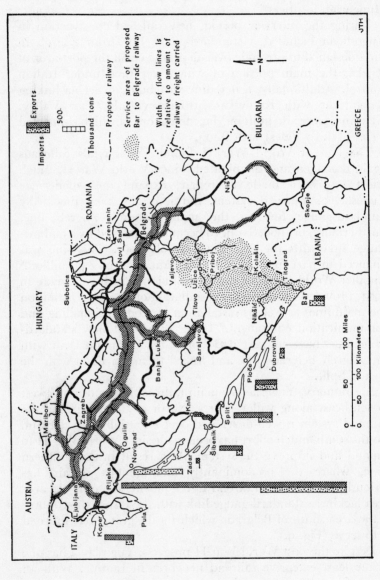

FIGURE 9.1 — YUGOSLAVIA: PATTERN OF FREIGHT FLOW: RAILWAYS AND PORTS, 1967, based on data in *Statistički Bilten, No.* 543, 1967

average density varies from some 6·2 km of standard-gauge line per 100 square km in Slovenia to approximately 2·8 km in southern Serbia and 0·4 km in Montenegro. These average densities compare with approximately 6·5 km of standard-gauge line per 100 square km in Austria, 6·9 in France, 9·3 in Hungary and 7·6 in Poland.

THE BELGRADE-BAR CONCEPT

The concept of a railroad connecting Belgrade southward to the sea is not a new one. The landlocked Principality of Serbia wanted reliable access to the sea. When Montenegro, under the 1878 Treaty of Berlin, gained access to the sea at Bar, interest in a railroad linking the two Orthodox and Serbian-speaking principalities was stimulated. However, since the Treaty of Berlin also provided for continued Ottoman control of the Sandjak of Novi Pazar which separated Serbia from Montenegro, this goal was impractical at the time.[10]

In the 1880's Austro-Hungarian influence on railroad building in Serbia was considerable and consideration was given for a time to the possibility of linking Belgrade with the Adriatic at Gruz via Visegrad on the Drina.[11]

In 1903, King Peter Karadjeordjevic ascended the Serbian throne. The return of the Karadjeordjevic dynasty led to growing manifestation of independence in the economic field. In 1905, Austria-Hungary sought to discourage this trend by embargoing further imports of Serbia's most significant export product, livestock. This Austro-Hungarian embargo, which lasted until 1910, came to be known as the 'Pig War'. Although Serbia managed to make arrangements with Turkey which enabled export of Serbian pigs and cattle to other markets via Thessaloniki,[12] Serbia's interest in access to the sea was intensified by the 'Pig War' as well as by Austria-Hungary's annexation of Bosnia-Herzegovina in 1908.

During the early 1900's, the question of a trans-Balkan railroad became a frequent subject of discussion at international meetings and in the press. Following the ascent of King Peter, who was a believer in railroad development, interest grew in the concept of linking the Danube by rail from eastern Serbia with the Adriatic, and then of connecting this line eastward by a bridge across the Danube with the Romanian railroads which in turn would join the Russian railroads.

Prince (later King) Nicholas of Montenegro also supported this trans-Balkan concept and advocated a line from Radujevac on the Danube to Knjaževac, Niš, Kuršumlija, Mitrovica, Podgorica (Titograd), Virpazar and Bar. Serbia, on the other hand, in the light of its worsening relations with Austria-Hungary and its improving relations with France and Russia grew to favour an alternate route from the Danube to the Adriatic. This would extend from Prahovac on the Danube to Knjaževac, Niš, Kuršumlija, Priština, Djakovica, and thence along the valley of the Drin across Albania to the port of Shëngjin (San Giovanni di Medua). To Serbian policy-makers this route had the advantage of obviating all need for negotiation with Austria-Hungary regarding the railroad since its course would neither cross the Sandjak nor terminate at Bar. In the Sandjak, Austria-Hungary enjoyed garrison rights, and at Bar, under Article 28 of the Treaty of Berlin, the Dual Monarchy had police and sanitary rights.[13]

Reflecting the intense political and economic interests of the era in the Balkans, a conference of French, Italian, Russian, Serbian and Turkish engineers discussed the Danube-Shëngjin route at Paris in July 1908. An international company with French (45%), Italian (35%), Russian (15%), and Serbian (5%) financial backing (100 million francs total) was formed to construct the trunk line from Shëngjin to the Danube.[14] The Porte, however, received with reserve the Serbian demarche of 1908 regarding this line which would cross Kosovo, Metohija and Albania, then parts of the Ottoman Empire. With regard to the Turkish reaction, however, it should be borne in mind that the Porte was then confronted with grave internal problems as well as with German- and Bulgarian-sponsored railroad proposals for the Balkans which competed in considerable degree with the Shëngjin-Danube proposal.[15]

Nevertheless, plans and work on the northeasternmost section of the Shëngjin-Danube line went ahead. The Balkan Wars and World War I were to interrupt this, but following the War, the Yugoslav portion of the line from the Danube as far as Peć in Metohija was completed.

After World War I, although Belgrade was no longer capital

of a landlocked state, steps were initiated to connect Belgrade via Montenegro with the Adriatic. The economic need for a direct standard-gauge line was stressed. In 1922, a contract was signed for an external 8% loan of $100 million for this purpose of which some $15 million had been floated in New York by the late 1920's.[16] The great depression, however, ended these efforts.

Nevertheless, surveys of a route from Belgrade to Bar were carried out and the course over which the Belgrade–Bar railroad was to run was essentially decided in 1936. Work on this route was begun in 1953, but lack of funds presumably stopped work in 1954.[17] All the same, the idea of completing the railroad did not die. President Tito, in a speech in Plevlje, on September 23, 1959, following resumption of work on the railroad, stressed the need to hasten its completion.[18]

THE SCOPE OF THE PROJECT

The Belgrade to Bar project involves completion of a standard-gauge, single-track, electrified line between Belgrade and Bar. The total distance from Belgrade to Bar is 476·1 km. The new line branches from a line at Resnik, 14·2 km south of Belgrade. The Resnik–Vreoci (7·3 km) and Titograd–Bar (53·0 km) sections have been in operation since 1959. When the World Bank loan was announced in March 1968, 371·4 km remained to be built, including some sections which had been under construction since 1959. The Belgrade to Bar project incorporates final signalling and electrification installations on the otherwise completed Resnik–Vreoci and Titograd–Bar sections. Additionally, telecommunications as standard on the Belgrade–Bar line are to be installed on the Belgrade–Resnik section. Fifty-eight stations and crossings are being built.

The Belgrade–Bar line will traverse mainly mountainous terrain, part of which is extremely rugged and difficult. There are to be 353 tunnels, totalling some 80 km in length. Some of these tunnels are 2 to 6 km long. Many bridges and viaducts will be required. The Mala Rijeka viaduct will be a major work for it is to span a wide gorge 200 m deep.[19]

The maximum gradient is 2·5% on the descent from Kolašin to Titograd over a distance of 59 km. On the Valjevo–Kolašin stretch, the maximum gradient is 1·8%, while from Belgrade to Valjevo it is 1·2%. Speeds of 75 to 80 km/h on curvy sections and of 120 km/h on straighter sections will be possible.[20] Journeys of seven hours or less from Belgrade to Bar are foreseen.

The Belgrade to Bar line will have electric traction with 25,000 volt A.C. 50-cycle current. This is the same electric traction system as is being used on the Sarajevo–Ploče, Dobova–Zagreb–Belgrade–Skopje, and Vrpolje–Sarajevo lines.

Electrification of the Belgrade–Bar line has been justified by the following considerations: (1) anticipated high traffic density in gross ton-kilometres per route-kilometre per day; (2) anticipated heavy power consumption in kilowatt per route-kilometre per year; and (3) the availability of low cost electric power from nearby hydroelectric plants. Additionally, the advantage of further standardization of major Yugoslav rail lines would seem a significant factor in the decision to electrify the Belgrade–Bar line.

Completion of construction is scheduled for 1972 at a total estimated cost equivalent to $211 million. The World Bank loan of $50 million will provide about one-third of the amount of $142 million included in contracts expected to be put to international bidding and consultants' fees. These costs will be financed by the Federal Government (85%) and the Republics of Serbia (8%) and Montenegro (7%). The World Bank loan is for 25 years at $6\frac{1}{4}$%. It is made to the Yugoslav Investment Bank and guaranteed by the Government of Yugoslavia. The proceeds of the World Bank loan all go towards the Federal Government's share and are being passed on to the Railway Transportation Enterprises of Belgrade and Titograd which are carrying out the project.[21]

In order to integrate the Belgrade–Bar line more fully into the country's rail network, it will have a junction at Požega with a 23 km line to Čačak and continuing on a line connecting with the Kraljevo–Niš–Skopje and the Kraljevo–Kosovska Mitrovica–Skopje lines. The Požega–Čačak link is being converted from narrow to standard gauge.

At Priboj, the Belgrade–Bar line will have a junction with the narrow-gauge line to Sarajevo which is expected to be maintained for some time.

THE RAILROAD ROUTE

At its extremities, the Belgrade–Bar railroad will connect the Belgrade area of expanding population (some one million in 1969) and industry, as well as the adjacent and prosperous Vojvodinia and Šumadijan agricultural districts with the seaport at Bar and the Montenegrin coast where tourism is developing.

In between, the railroad will traverse four physiographic regions: the Pannonian hills and valleys, the high mountains, the zone of karst, and finally the Adriatic littoral. The landscape en route is mostly mountainous. For a considerable part of the way, the railroad will cling to the sides of the steep and winding Lim, Tara, and Morača valleys.

Climate along the route is mainly continental: warm summers, cold winters, and generally adequate precipitation except during the summer in the south. From Titograd to the coast, a Mediterranean climate prevails. In the zone of continental climate, measurable snowfall is frequent. At Titovo Užice (elevation 440 m), snow fell on an average of 50 days per year between 1955 and 1964. At Sjenica (elevation 1,015 m), in southern Serbia slightly east of the railroad, over the same period, snow fell on an average of 96 days per year.[22]

The railroad's zone of influence includes roughly one million hectares of forest which contain an estimated 8·4 million cubic metres of timber.[23]

Population is generally light south of Titovo Užice. Lumber mills, woodworking factories, and small slaughtering establishments are characteristic of industrial activity in much of the zone of influence. At Nikšić, however, which lies some 50 km northwest of Titograd and which is connected via the Zeta valley with Titograd by standard-gauge rail, there is a steel mill. The mill's requirements for delivery of raw materials and transportation of its finished products provide the port at Bar with much of its current freight volume. In 1968, Montenegrin ports handled some 700,000 metric tons of cargo.[24] Of this, Bar can be estimated to have handled some 600,000 tons.

Mineral resources are considerable in the zone of influence. Plevlje and Ivangrad have coal mines. Bauxite exists near Titograd and Nikšić. Lead and zinc deposits are reported near

Mojkovac, Plevlje and Podvinj. Nickel, chromium, copper and antimony deposits are also reported near the railroad.[25]

Within most of the railroad's immediate zone of influence, both lateral and longitudinal roads are inadequately developed. A bus trip from Belgrade to Titograd, for instance, can take over 10 hours.

Soils tend to be better along the railroad's northern portion than along its southern, with agriculture by far the dominant activity along the route. The railroad's immediate zone of influence will include areas representative of four of the Yugoslav agricultural regions distinguished by Professor Milojević.[26] These coincide with the physiographic regions of traverse.

The Pannonian hill and valley agricultural region extends from Belgrade along the line through Vreoci, Lazarevac, Valjevo, Požega, and Titovo Užice on the Detinja river. This is a region of mixed agriculture. Fruits and livestock are the main cash products. Valley bottoms host cornfields and meadowland. Slopes, where not too steep, bear corn and wheat. Steeper slopes support orchards and vineyards. Apples and grapes are typical to the north, plums to the south. In areas of mild climate, tobacco is sometimes cultivated. Cattle predominate to the north, sheep to the south. Hog-raising is also important. As indicated by much field fragmentation, this is largely a zone of small-scale private farming, a good deal of which is of subsistence character.

The high mountain agricultural region, as distinguished by Professor Milojević, extends along the route roughly from Titovo Užice (440 m) to Kolašin (950 m). With aid of tunnelling, the line crosses the Zlatibor range between Titovo Užice and Priboj (394 m) on the Lim river. Thence, the railroad follows the Lim upstream past Prije Polje (448 m) to Bijelo Polje (586 m). From Bijelo Polje, with aid of tunnelling, the line crosses a saddle in the Kamena Gora to the Tara river at Mojkovac (800 m). From Mojkovac, it follows the Tara upstream to Kolašin. The sparsely settled region of traverse between Titovo Užice and Kolašin is characterized by sheep farming and some cattle raising. It averages well over 1,000 m in altitude, with mountains rising to over 2,000 m.

From Kolašin, the railroad route edges slightly westward to the Morača Valley. As it does so, it enters the agricultural region of the zone of karst. It also leaves the Danubian for the Adriatic watershed. The railroad follows the Morača downstream to Titograd (52 m), where the landscape becomes increasingly karstic. In the zone of karst, only about 20% of the land is arable. Agriculture is concentrated on the floors and sides of basin-shaped fields, called *poljes*. At lower elevations, mostly corn is planted. High elevations are used for grazing and the cultivation of mountain grains, potatoes and cabbage. Sheep, goats and some cows are found in this agricultural region. Transhumance is practised, with livestock herded to higher elevations for summer grazing.

Near Titograd, the railroad enters the agricultural region of the Adriatic littoral. Here, grapes, olives and figs are typical crops. Market gardening is also important, particularly where irrigation is feasible. Some livestock are kept and transhumance is practised. Hay is sometimes grown for winter feed.

THE POTENTIAL SIGNIFICANCE OF THE RAILROAD

Yugoslavia has considerable spatial contrast in regional development. The Belgrade–Bar railroad will traverse a landscape which is largely underdeveloped. While a railroad's zone of influence is not easily measured,[27] the Belgrade–Bar railroad's predictable zone of primary influence is one of low *per capita* income[28] and has all the physical characteristics attributed to Yugoslavia's underdeveloped regions.

> The main characteristics of Yugoslavia's underdeveloped regions are their hilly to mountainous characters, with few exceptions unsuited to modern agrarian technologies, large mineral reserves, and a predominant primary production of minerals, important timber resources and energy reserves...[29]

Yugoslavia's leadership has emphasized that regional developmental gaps should be narrowed. In this direction, Yugoslavia's programme of economic reform, which stresses economic decentralization and self-management, is aimed at stimulating progress in underdeveloped regions and at welding the country into a viable economic whole. In this regard, promotion of capital flow from business enterprise to business enterprise within the whole of the Federation is a significant goal of the reform. At the same time, however, the programme of economic reform puts great stress on factors such as profitability, modernization, productivity and the internationally competitive quality of Yugoslav products. Another significant factor in the Yugoslav scene is political reform which also emphasizes decentralization and self-management in political affairs.[30]

As a result of policies emphasizing decentralization and self-management, there has been considerable articulation of regional interest in the process of pursuing economic advance.[31] Debate has arisen as to the relative efficiency of areas of potential investment and, on the other hand, on the importance of promoting development in underdeveloped regions. Proponents of investment projects in developed areas have reasoned that investment there would result in more immediate profits and, thereby, the Federation's economy and capability to promote development elsewhere could best be improved. On the other hand, spokesmen for investment in underdeveloped regions have pointed out the economic and political undesirability of failure and the need to take immediate steps to narrow developmental gaps.

Prime Minister Mika Spiljak addressed himself to these questions at the end of 1968 in discussing measures taken during the year to assist underdeveloped areas. The net effect, he said, was to accelerate the flow of funds to those areas so that by the beginning of 1969 the 'slackening' in flow caused by adjustments brought about by the programme of economic reform since 1966 would be fully made up. By 1970, he predicted, the original 8·2 billion in new dinars (12·5 new dinars equal U.S. $1·00) obligated by the 'Federal Fund for the Developing Areas' would be exceeded by about 237 million new dinars. The remaining problems were 'qualitative', he said, since the efficiency of investment in less developed areas was still only 76% of the Yugoslav average.[32]

President Tito addressed himself to the subject of the developmental gap at length in his report dated March 11, 1969, which he submitted to the Ninth Congress of the League of Communists of Yugoslavia:

> By promoting the system of self-management, we were able to achieve rapid development and undertake deep economic-

social transformations in our country. However, the level of economic development reached is still low compared with developed countries of Europe and the world in general. Its particular characteristic is the disparity between the developed and underdeveloped regions, the lagging technical structure, and still comparatively insufficient incorporation into the international division of labour. This situation makes it impossible to meet numerous growing needs in a satisfactory way, and so becomes a source of certain contradictions and difficulties in our society.... Under conditions of stagnating economic trends, this led to mounting difficulties, in the course of last year, and even to protests and revolt, because some important problems were being dealt with slowly....[33]

We must devote careful attention in our development policy to the more rapid progress of the economically less developed areas. The general principles and policy in this respect have been laid down in the Programme of the League of Communists of Yugoslavia and the Constitution of the Socialist Federal Republic of Yugoslavia....[34] We have noted that under the present conditions we are already in a position to accelerate the overall progress of the economically less developed republics and regions not only at a more rapid rate than they themselves would be able to achieve by their own efforts and resources, but at a rate which will exceed the all-Yugoslav average. We have, therefore, decided that appropriate measures be taken which will bring about a gradual reduction of the disparities between development levels of the individual republics....

... Investments in the economy of these areas ... during the past few years through the Development Fund for the crediting of the insufficiently developed republics and regions have already begun yielding the positive effects desired.

Effective methods of accelerating the development of these regions should be worked out with a view to achieving the optimum results in this respect. We must retain the Development Fund ... and provide for its new and regular sources of income.[35]

However, in stressing the importance of decreasing gaps in development, President Tito also emphasized the significance to underdeveloped regions of continued advance in developed areas:

> Needless to say, the fact that rapid progress of the advanced regions is also in the interest of the less developed should be evident to one and all, as this would enable the former to raise accumulation and allocate part of these resources to foster the further progress of the insufficiently developed areas. Consequently, this is in the mutual interest.[36]

While the foregoing makes it clear that investment projects in Yugoslavia's underdeveloped regions need not be decided on relatively short-term economic considerations, the World Bank decision to help finance the railroad from Belgrade to Bar was reportedly reached only after officials of the Bank had made 'most extensive enquiries into the undertaking and had become convinced that it was an economic proposition'.[37]

In forecasting probable benefits to agriculture in the railroad's zone of influence, movement towards more efficient land use should be encouraged by completion of the railroad. There should be a trend away from subsistence farming. With cheaper and more efficient transport available, vegetable, fruit and livestock production should intensify. Cheaper transportation should encourage greater use of inputs such as chemical fertilizer. Availability of rapid, reliable and inexpensive transport should prompt market gardening in the region of the Adriatic littoral to concentrate on early crops for consumption in urban centres such as Belgrade and Zagreb. Furthermore, anticipated development of tourism, on the coast in particular, should provide a growing market for agricultural produce from within the railroad's immediate zone of influence as well as adjacent agricultural areas such as the Vojvodina.

The timber potential in the immediate zone of influence was underlined by the inauguration in January 1969 of a $2·3 million United Nations Development Programme project to advise and assist the Government of Yugoslavia in planning rationalization and further development of forest industries in the Republics of Bosnia–Herzegovina and Montenegro.[38] In Montenegro, forests cover 576,000 hectares of the total area of 1·4 million hectares.[39] In the Serbian portion of the zone of immediate influence there also is much forest land. Completion of the railroad should facilitate and encourage development of rational lumbering practices and profitable woodworking industries. Possibilities for profitable mineral extraction within the zone of immediate influence also should improve when the railroad comes into service.

Tourism has much opportunity for expansion in the zone of immediate influence, especially on the Montenegrin coast. A $2·9 million United Nations Development Programme project was initiated in January 1967 to assist the Government of Yugoslavia in drawing up physical plans for the South Adriatic region from the Island of Hvar, 30 km south of Split, to the Albanian border, including master plans for selected areas and tourist facilities.[40] Experts on the project are reportedly enthusiastic over the possibilities for tourism on the coast. During the 20-year period beginning in 1970, a huge development project costing over $1 billion is foreseen for the South Adriatic region. It is understood to include provision for 480,000 new beds, as well as for roads, recreational facilities, airports, harbours, and agricultural and urban development. On the Montenegrin coast, Budva and Ulcinj are foreseen within the project as major centres of tourism.

In his report to the Ninth Congress of the League of Communists of Yugoslavia, President Tito emphasized the importance of tourism and noted an interrelationship between agriculture and tourism:

> We must devote far greater attention to the development of tourism, which is an exceptionally important branch for our country, both because it opens new employment opportunities and brings in substantial foreign exchange earnings. Development of tourism also expands the market for livestock and agricultural products. This is one of the most profitable forms of 'export' of our products, especially agricultural products.[41]

The railroad will represent a major addition to the infrastructure of its immediate zone of influence. Predictably, it will stimulate further development of infrastructure such as lateral roads to provide surrounding areas with ready access to the railroad's services. This should enhance prospects for both agriculture and tourism as well as facilitate their interrelationship.

The questions may well be asked, 'Why build a railroad in this era of emphasis on highway transport? Doesn't the railroad represent an expensive and essentially political concession to satisfy long-standing aspirations for a railroad linking Belgrade as directly as possible with the Adriatic? Wouldn't a highway be less extravagant and more efficient?' The details of a recent World Bank loan to Yugoslavia for highway construction help to answer this. On May 28, 1969, the World Bank announced that it had approved a loan of $30 million for highway construction in Yugoslavia.[42] This loan would assist in financing three sections of highway at a total cost estimated to be the equivalent of $75 million. One section was a 44·8 km, four-lane stretch between Zagreb and Karlovac in Croatia; another was a 32 km, four-lane stretch between Vrhnika and Postojna in Slovenia; and the third was a 46·4 km, two-lane stretch between Gostivar and Kičevo in Macedonia. The total distance was 123·2 km, giving an average cost per km of approximately $608,000. On the Belgrade–Bar railroad 371·4 km are being built at a cost of approximately $211 million, giving an average cost of $568,000 per km over terrain which is generally more difficult than that traversed by the above-mentioned highway sections.

Since a Belgrade–Bar highway matching the railroad in carrying capacity and rate of travel would require four lanes, a highway would need much broader tunnels and viaducts as well as a far wider roadbed and right of way. Costs for a Belgrade–Bar highway of that sort may be estimated as at least twice those of the railroad.

Although existing roads in the immediate zone of influence are being gradually improved, very substantial economic benefits are expected to stem from diversion of passengers and freight from road to rail transport. The steep grades and sharp curves on the highway cause high operating costs and extract a heavy toll of wear and tear on vehicles. As far as the railroad is concerned, tunnelling will make its path relatively straight and will enable elimination of very steep grades.

Furthermore, cargoes well-suited to rail transport are expected to predominate: coal, coke, petroleum products, metallurgical products, ores, timber, cement, building materials, phosphates, cereals, livestock and livestock products, vegetables and fruit. Diversion of heavy cargoes from the main roads to the railroad should help to reduce costs of highway maintenance. Use of electric traction employing locally generated power will reduce the country's requirements for imports of petroleum and petroleum products. In winter, when highways are often blocked by snow, the railroad will provide reliable transportation. The railroad's estimated capacity is 8·5 million tons of freight and 15 million passengers annually.[43]

By filling a void in Yugoslavia's railroad network, the Belgrade–Bar line will substantially reduce rail distances between eastern Yugoslavia and the Adriatic. Belgrade, for instance, will be 153 rail kilometres closer to Bar than to Ploče, the nearest alternate port. Skopje will be 431 rail kilometres closer to Bar than to Ploče.[44] Transport economies will thus be made possible. The Belgrade–Bar line will make possible other economies such as those resulting from the abandonment of some narrow-gauge lines in its immediate zone of influence which have very high operating costs.

Development in recent years of ports at Koper, Ploče and Bar has intensified port rivalries in Yugoslavia. These new ports have supplemented previously existing facilities at Rijeka, Split, Šibenik and Gruz (Dubrovnik) (see Fig. 9.1). In the heat of discussion, Koper, Ploče and Bar have been described as 'political ports', constructed without real economic justification mainly in satisfaction of regional aspirations.[45] In this connection, however, while the port at Bar can not be said to have the fundamental economic and strategic importance that it would have had for the landlocked Serbian state, it already is enabling substantial economies of transport for the Nikšić steel plant which formerly relied on the port of Gruz, connected by narrow-gauge line with Nikšič. Once linked by standard-gauge line with the interior beyond Titograd, Bar promises to enable economies of transport for other major industrial establishments within the zone of influence. Near Belgrade, for instance, the Pančevo fertilizer industry requires major tonnages of imported phosphates.

Neither can it be said that the port at Bar is without strategic importance, for it is the only port on the 170 km of Yugoslav coast south of Ploče which is connected with the interior by standard-gauge rail.

In considering the long-term significance of the Belgrade–Bar project, it is also worth while noting that, as is true in most countries, the well-developed regions of Yugoslavia are those best served by good transportation facilities, and they include mountainous Slovenia. Looking well into the future, F. E. Ian Hamilton has written:

> When complete and linked by new or improved railways with Tuzla from Valjevo, Prahovo from Titovo Užice, Sarajevo from Priboj, and the Kosmet (Kosovo) from Bijelo Polje, the railway (Belgrade–Bar) will become a trunk route, a 'zonal' multiplier for the economy of the eastern two-fifths of the Federation.[46]

In conclusion, while the Belgrade–Bar railroad lacks the immediate and dramatic significance that it would have had for the landlocked Serbian state, it promises to benefit the estimated one million residents of its zone of primary influence and the other 20 million residents of Yugoslavia. By providing critical infrastructure within its zone of primary influence, it should serve as a catalyst encouraging and enabling economic advance. This will help to reduce the developmental gap which exists between most of its zone of primary influence and much of the rest of the country. Furthermore, by linking a hitherto relatively isolated area more directly with the rest of the country, the railroad should promote political and economic integration within Yugoslavia. As Professor Hoffman has observed, advancement of the position of Yugoslavia's underdeveloped regions can best promote unity and co-operation within the Federation.[47]

NOTES

[1] International Bank for Reconstruction and Development, (IBRD), *Bank Press Release* No. 68/15, March 22, 1968.

[2] Stanley H. Beaver, 'Railways in the Balkan Peninsula', *The Geographical Journal*, Vol. XCVII (May, 1941), p. 280.

[3] Beaver, *op. cit.*, p. 281.

[4] Dragomir Arnaoutovič, *Histoire des Chemins de fer Yugoslaves* (Paris: Dunod, 1937), pp. 23–8; Beaver, *op. cit.*, p 282.

[5] Kenneth S. Patton, *Kingdom of Serbs, Croats and Slovenes (Yugoslavia), A Commercial and Industrial Handbook* (Washington, D.C.: Department of Commerce, 1928), p. 99.
[6] Beaver, *op. cit.*, p. 288.
[7] Beaver, *op. cit.*, pp. 288-9.
[8] International Bank for Reconstruction and Development, *Bank Press Release* No. 63/48, October 28, 1963.
[9] International Bank for Reconstruction and Development, *Bank Press Release* No. 64/42, December 11, 1964.
[10] Dennison I. Rusinow, 'Ports and Politics in Yugoslavia', *American Universities Field Staff, Reports Service, Southeast Europe Series,* Yugoslavia, Vol. XI, No. 3 (1964), p. 18.
[11] Brana Vučković, 'The Belgrade-Bar Railway', *Review, Yugoslav Monthly Magazine,* IX (1968), p. 9.
[12] Bernadotte E. Schmitt, 'Serbia, Yugoslavia and the Hapsburg Empire', Chapter IV of *Yugoslavia,* Robert J. Kerner, editor (Berkeley, Cal.: University of California Press, 1949), pp. 41-2.
[13] Arnaoutovič, *op. cit.*, p. 201.
[14] Arnaoutovič, *op. cit.*, p. 202
[15] Arnaoutovič, *op. cit.*, p. 202.
[16] Patton, *op. cit.*, p. 103.
[17] Vučković, *op. cit.*, p. 9.
[18] *Borba,* Belgrade, September 25, 1959.
[19] Vučković, *op. cit.*, p. 8.
[20] Vučković, *op. cit.*, p. 8.
[21] *Bank Press Release* No. 68/15, March 22, 1968.
[22] *Statistički Godišnjak Jugoslavijc, 1968.*
[23] Vučković, *op. cit.*, p. 7.
[24] *Statistički Godišnjak Jugoslavijcs 1969.* A report in *Politika* of February 12, 1970, speaks of 'an annual turnover of about a million tons of goods' at Bar and after completion of 3 million tons.
[25] Vučković, *op. cit.*, p. 7.
[26] Borivoje Ž. Milojević, *La Yugoslavie, Aperçu Géographique* (Belgrade Commission pour les relations culturelles avec etranger, 1956), pp. 42-9.
[27] Rusinow, *op. cit.*, p. 20.
[28] Jack C. Fisher, *Yugoslavia, A Multinational State* (San Francisco: Chandler Publishing Co., 1966), p. 71.
[29] George W. Hoffman, 'The Problems of the Underdeveloped Regions in Southeast Europe: A Comparative Analysis of Romania, Yugoslavia, and Greece', *Annals of the Association of American Geographers,* Vol. 57 (December, 1967), p. 650.
[30] Benjamin Ward, 'Political Power and Economic Change in

Yugoslavia', *The American Economic Review*, Vol. LVIII (May, 1968), pp. 568–79.

[31] On July 31, 1969, for instance, the Slovenian Executive Council met and asked the Federal Executive Council to reconsider the decision to omit Slovenia from sharing in a forthcoming World Bank loan for highway development; see *Delo*, Ljubljana, August 1, 1969.

[32] *Politika*, Belgrade, December 31, 1968.

[33] In June 1968 there were student demonstrations in Belgrade which were related to lack of job opportunities in Yugoslavia. In November 1968, there were disorders in Kosovo, an underdeveloped region of the country.

[34] Part One, Chapter II, Article 27 of the 1963 Constitution of the Socialist Federal Republic of Yugoslavia states in part: 'The social community shall provide the inadequately developed republics and regions with material and other commodities necessary for more rapid economic development, and for the creation of the material bases of social activities.'

[35] Text as distributed at Belgrade during the Congress.

[36] Text as distributed at Belgrade during the Congress.

[37] Vučković, *op. cit.*, p. 7.

[38] United Nations Development Programme Document DP/SF/R.3/Add. 114, October 4, 1968, p. 1.

[39] United Nations Development Programme Document DP/SF/R.3/Add. 114, October 4, 1968, p. 2.

[40] United Nations Development Programme Document DP/SF/R.3/Add. 61, November 18, 1966, p. 1.

[41] Text as distributed at Belgrade during the Congress.

[42] *Bank Press Release* No. 69/33, May 28, 1969; subject: $30 million Loan for Highways in Yugoslavia.

[43] Vučković, *op. cit.*, p. 7.

[44] Although Skopje is closer yet by rail to the Yugoslav Free Zone at Thessaloniki, the Free Zone has never been fully utilized in as much as Yugoslavia has evidently preferred to use its own port facilities. See George W. Hoffman, 'Thessaloniki: The Impact of a Changing Hinterland', *East European Quarterly*, Vol. 2 (March, 1968), p. 25.

[45] Rusinow, *op. cit.*, p. 10.

[46] F. E. Ian Hamilton, *Yugoslavia, Patterns of Economic Activity* (New York, Washington: Frederick A. Praeger, 1968), p. 277.

[47] Hoffman, 'The Problem of Underdeveloped Regions', *op. cit.*, p. 659.

Comments   GUIDO G. WEIGEND

We are living in an era when railroad lines are being abandoned rapidly not only in the United States but also in parts of Europe. In fact, in an address at the American Geographical Society in the spring of 1969, Professor Harold M. Mayer predicted the demise of all railroads in the United States within the next century. It is, therefore, interesting to know that Yugoslavia is carrying out a new railroad construction project of such magnitude. Total cost of the project is $211 million, which means more than $0·5 million per km.

It must be assumed that the World Bank and Yugoslavia have made careful cost analyses, and that the alternative of an expansion and improvement of the highway network was taken into consideration. It is true, of course, that a rail line can operate the year round in all kinds of weather conditions and that the upkeep of railroad equipment is less expensive than the maintenance of a highway network and the vehicles using it. Moreover, the railroad will use electricity for propulsion, while petroleum products for vehicular traffic would have to be imported into the country and the region at considerable cost.

However, from the economic standpoint, the contention[1] that the construction of the railroad line will enable Montenegro to develop its natural resources for domestic and overseas markets is open to serious doubt. The assumptions made in Mr Wilson's paper, that the area along the line will be a zone of primary influence, that a profound change of land-use patterns will result from construction of the line, that investment opportunities will be enhanced and tourism promoted, and that development of mineral and timber resources will be speeded up, may prove to be incorrect unless a feeder line network of rail lines and/or highways can be tied in with the main line. There are no plans at this time to build such a complementary network, and without it the railroad line can not become an energizer, or 'zonal multiplier'[2] as visualized because it can serve – and give access to – only a narrow belt of the countryside which it traverses. In this connection, a study by André

Blanc[3] of the economic impact of railroads in Croatia has shown that the greatest impetus to development was given to the termini of lines with a minimal influence on the route itself.

Another economic argument for the construction of the Belgrade to Bar line has been the development of Bar as one of Yugoslavia's ocean ports, which would thereby give Serbia, Macedonia and Montenegro a more direct access to the sea than they now have. This must be viewed in terms of the total development of Yugoslavia's maritime traffic. This traffic has multiplied since 1939 when less than $2\frac{1}{2}$ million tons of merchandise were handled in Yugoslav seaports.[4] After a post-war stagnation of several years, total maritime cargo tonnage reached close to 7 million in 1959 and some $12\frac{1}{2}$ million in 1966.[5] This increase not only represents cargo with destination or origin in Yugoslavia, but it also includes a rapidly increasing transit traffic, notably to and from Czechoslovakia, Hungary and Austria.

The bulk of the internal and transit traffic – nearly two-thirds or some 8 million cargo tons in 1966[6] – was handled by the port of Rijeka which has been operating at or near capacity. Much of the remainder went through the Croatian ports of Split, Šibenik, and Ploče, all of which are operating considerably below capacity. Current Yugoslav policy with respect to seaports has been to focus the transit trade upon Rijeka and, in as much as this port's capacity is limited, to develop and expand other Yugoslav ports for the national maritime traffic. This has created a rivalry among the Republics for port development. Included among them in Croatia are the older, well-established and reconstructed ports of Split, Šibenik and Zadar, all of which have had standard-gauge rail connections with the interior, and Ploče whose rail link with Sarajevo and the northern interior was recently standardized, modernized and electrified. While Ploče is a Croatian port its main hinterland is Bosnia–Herzegovina and Serbia, and the construction of port facilities and rail line was promoted by these Republics. Slovenia has pursued a very aggressive and energetic policy in developing its only direct window to the sea, the port of Koper, which is located immediately to the south of Italian Trieste.[7] By laying 26 km of track, Koper was tied into the northwestern railnet. The port began to operate in 1959, and by 1970 its annual handling capacity was to have reached some 3 million tons.

Koper's rich and diversified national hinterland is in great contrast to that of Montenegro's maritime outlet, the port of Bar. Except for the Adriatic littoral this Republic lies within the high mountains and the karst of the Dinaric system, an area which is not likely to generate much traffic for the port of Bar. From the standpoint of distance to the interior, it has been argued that with the completion of the Belgrade to Bar rail line, Skopje will be 431 rail kilometres closer to Bar than to Ploče. It is also true that Belgrade will be 150 rail kilometres closer to Bar than to Ploče, but that too represents no economic justification for building a rail line at enormous expense, whose economic functions will duplicate those of the already existing Belgrade to Ploče line which connects the national capital and the large northern interior with a seaport which as yet is operating at a fraction of its capacity. Already large sums of money have been expended in Bar port construction, and the rail connection to Titograd is completed. But the port will remain little used until the railroad has been extended into the interior. Even then it is problematical as to how much maritime traffic will be generated and how rapid the increase will be.

It seems then that the economic significance of the Belgrade to Bar railroad may well be subordinate to political, psychological and strategic considerations.

The development of the Yugoslav rail network has had strong political and strategic overtones in the past. The importance of Austria-Hungary in this development, and its desire to link Vienna and Budapest by rail with north Adriatic ports cannot be over-emphasized, but it was also important that some rail links were not constructed before 1918 because it was Austrian and Turkish policy to keep certain areas in isolation. Thus there was no rail connection between Croatia and Bosnia, and none between Bosnia and Serbia. Even the Banat was separated from Serbia because no bridges had been built across the Danube river.

In connection with the evolution of the railnet,[8] one should mention a grandiose plan which in fact can be considered to be a precursor of the Belgrade to Bar railroad. This was the socalled Trans-Balkan line, which was to lead in a northeast–southwest direction from Romania through Serbia to the Dalmatian coast at Kotor or in northern Albania. Serbia did begin construction of the line, but the Balkan Wars interrupted the project. The idea was revived after World War I. The line from Prahovo, south of Turnu Severin, to Niš was completed by 1922, and the section Priština to Peć, further southwest, by 1935. But Peć was and still is the end of this particular line because to the west of it lies the great mountain wall of the Dinaric system and a further distance of some 320 km in rugged terrain to the coast.

There is, then, a long record of Serbia's desire to gain access to the Adriatic Coast, a goal which was reached with the modernization of the Belgrade–Sarajevo–Ploče rail line. But what about Montenegro? It has remained one of the most primitive and underdeveloped sections in Yugoslavia. Yet, Montenegrins gave great strength to the partisan movement during World War II, and they were promised in turn that their country would be tied more closely to the newly emerging state, and that economic conditions would be improved. Construction of the Belgrade to Bar rail line in that sense has become a symbol for the fulfilment of past promises and future development. With completion of the line the psychological distances to the capital on the one hand and the sea on the other will have been reduced, and the flow of goods and people will in fact reduce the centuries-old isolation, an isolation which may have been desirable and beneficial at one time but which spells doom for the present and future. Finally, from the strategic standpoint, the Belgrade to Bar rail line will give Belgrade additional access to the Adriatic Sea through a modern, electrified line which will be an all-weather route and the shortest connection between the capital and the sea.

Construction of the Belgrade to Bar rail line is but a beginning in the long process of raising hopes in this underdeveloped region of Yugoslavia towards eventual equalization in its political and economic status among the various Republics. Completion of the line certainly will have immediate political and psychological impact upon the population. In order to work towards economic development and greater national cohesion, however, construction of a feeder line network will be essential. The main line *per se* will be only one energizer in a complicated framework. And it might be that the national traffic in Yugoslavia eventually would increase enough to strengthen the economic justification for the Belgrade to Bar railroad line from the national point of view.

NOTES

[1] International Bank for Reconstruction and Development, *Bank Press Release*, No. 68/15, March 22, 1968.

[2] F. E. Ian Hamilton, *Yugoslavia, Patterns of Economic Activity* (New York: Frederick A. Praeger, Inc., 1968), p. 277.

[3] André Blanc, *La Croatie Occidentale. Étude de géographie humaine* (Paris: Institut d'Études Slaves de l'Université de Paris, 1957).

[4] Dennison I. Rusinow, 'Ports and Politics in Yugoslavia', *American Universities Field Staff Report Service, Southeast Europe Series*, Vol. XI, 3 (1964), p. 5.

[5] *United Nations Statistical Yearbook*, 1968, pp. 438–39.

[6] U.S. Department of Commerce, Overseas Business Reports, *Basic Data on the Economy of Yugoslavia*, OBR 67-80, p. 10.

[7] Rusinow, *op. cit.*, pp. 10–14.

[8] Anton Melik, *The Development of the Yugoslav Railways and Their Gravitation Toward Trieste* (Belgrade, 1945), reprint in English, French and Russian of 'Razvoj železnic na ozemlju Jugoslavije', *Geografski Vestnik*, Vol. 14 (1938), pp. 118–34.

Remarks

Further discussion of Mr Wilson's paper began with the statement by Professor Fisher that he fully agreed with Professor Weigend, but that one point could have been made more strongly, the utility of the new port and the money needed to develop Bar when there are already the ports of Ploče and Rijeka and Koper and Šibenik. These bring into question the economic efficiency studies . . . were there any?

As for tourism, Professor Fisher stated that the railroad will have no influence except for those few resorts along the railroad. Road development would be a more important stimulus to tourists and probably could enhance the type of development suggested in Mr Wilson's paper.

The Belgrade to Bar Railroad is the last Federal investment in this type of project. It was a debated question whether it is in the spirit of the new economic reforms or a vestige of an old reform. It was unwise for the World Bank to invest in it.

Professor Velikonja then stated that he had published a brief article in *Bulgarian Geography* some time ago thinking that the railroad would be finished in 1960. He also travelled and saw the construction work. Ploče, he indicated, has been greatly under-used and there has been discussion about whether the great volume of traffic would really materialize. The Sarajevo to Ploče rail line is used, but the railroad station at Titograd is inhabited by gypsies.

As for tourism, he said that prior to 1961 the road was in bad repair, but that even with improvements the road is still unused. The road will provide little increase in tourist traffic and can handle it. There is also a point to be made concerning the high cost of the railroad in the mountainous areas.

Professor Hoffman stated that geographers do not give sufficient attention to the psychological and political aspects of economic development in underdeveloped areas; to be specific, railroad development in Yugoslavia's most underdeveloped regions. Economically, he doubts if this proposed railroad from Belgrade to Bar can compete with the railroad to Ploče. It is a basic fact that the partisan movement in World War II had its great and enduring strength in the underdeveloped parts of the country, in part due to promises of a better life and also the establishment of closer ties with the more developed parts of Yugoslavia. A railroad is a symbol and a chance for the future for the people of this region. For this reason, the political aspects of this railroad are unquestionable. He felt that this railroad is a definite symbol for the underdeveloped regions of 'moving up the ladder'. Professor Hoffman indicated that he was more sceptical about the economic value of the railroad. The pressures for the construction of this railroad are certainly a good example of the power of local pressure groups in a socialist society. On the other hand, every society, capitalist, socialist, or a mixed one, has its pressure groups. They have been important in the past and will be of importance in the future. They play an important role in the decision-making processes of every society.

Finally, Professor Hoffman continued, a word should be said about the economic aspects of this railroad and the region it will serve. Grave doubts have been expressed by numerous Yugoslav economists and geographers in their writings over the years. For example, he said, he had serious doubts about the quality of the timber resources along the railroad's zone of influence. Most of the timber resources must be classified as mediocre, so much so that the paper factory at Ivangrad in eastern Montenegro, on the Lim river, must import its raw materials – and this is not due to the absence of road connections. Also the southern ports of the Adriatic are not fully utilized and it would take considerable artificial restructuring of the traffic flow to increase the freight traffic in this direction. In conclusion, Professor Hoffman said, the justification for this railroad must be sought in both politics and economics.

Dr Alexander then stated that there is generally very little criticism in Slovenia about another railroad built with Slovenian money. Koper and Rijeka have their railroads, so Bar has its railroad. All national groups have their 'political' railroads.

Dr Hamilton pointed out that this is a political and a national issue. He felt that the railroad has much economic justification, but the results cannot be judged until completion. There is an 'historical legacy of gross underprivileged' in this area. In consideration of input versus output he felt that it is more economical than has been assumed. This project will have a lot of social and economic repercussions on the rest of the country. The welfare economics and economic reform complicates the issue. Dr Hamilton questions whether the social principles are the real ones. You have a past in Montenegro, he said, where past decisions combined with other factors may well have been in response to political pressure. Nevertheless, industrial establishments in the area operate at a loss due to the atrocious condition of the roads. These are complex issues. In his personal experience, Dr Hamilton said that he could not think of isolated, projected social political and economic factors since they cannot be separated or quantified.

Dr Crkvenčić commented that now roads have to play the leading role. Dr Romanowski remarked that if the railroad is uneconomical then should there not be plans for roads with the same amount of money?

Dr Hadžić then indicated that he was convinced that the railroad complements others; that it has economical justification, otherwise why would the World Bank give money. There is no politics involved, he said. The railroad relies on economic justification. All other investments are in developed centres. In his opinion Dr Hadžić felt that it should help the whole area; one can not view this in the narrow economical way. There had been a severe discussion about this project and that the positive and negative factors had been balanced. Professor Hadžić agreed with Professor Hoffman in saying 'what chance can we give to this population?' Roads carry cargo to the ports, but the winters are bad and the roads are unusable. We will see if the project is successful in the next ten years.

Dr Rath pointed out that this illustrates that all types of economical plans have social and political implications. The political pressures compare to Western railroad development ... or the example of the Trinity river project – should Dallas have a harbour? One should view the problem in an historical perspective. The Hapsburgs had their local pressures which included national factors.

That the project under discussion is in a strategic region was expressed by Dr Velikonja. He agreed with Professor Hoffman who pointed out the political aspects and indicated that a great deal of attention should be given to them. The idea of a railroad versus a road was studied by the World Bank and the Yugoslav planners. Both felt that the railroad transportation was cheaper than road, especially when covered with snow. Smaller tunnels were required and the wear and tear on road vehicles would be less. Also, the upkeep on railroad equipment is cheaper and electrical power for railroads is available, whereas fuel for trucks must be imported since Yugoslavia is not self-sufficient in petroleum.

Dr Pounds then brought the discussion to a close with the suggestion that we all assemble in 2019, fifty years from now, to view the outcome.

10

Significant Demographic Trends in Yugoslavia, Greece and Bulgaria

HUEY LOUIS KOSTANICK

Southeastern Europe has traditionally been a difficult area for demographic study because of the numerous catastrophic alterations of population structure due to incessant wars and massive population migrations, both voluntary and involuntary. In this century, the Balkan Wars of 1912–13, World War I and World War II all created violent loss of lives totalling millions of people. Similarly, there were large-scale population movements, such as the Greek–Turkish–Yugoslav–Bulgarian population exchanges in 1923 and the expulsion of some 150,000 Turks from Bulgaria to Turkey in 1950–1.

But in the period since the end of World War II in 1945, sufficient time has elapsed to provide some sound evidence of present-day basic trends in population structure and in internal migration patterns. Some of these trends are in line with previous population projections and follow major demographic trends in the world today. Chief among these would be the movement from the countryside to the city and a decline in infant mortality.

In sharp contrast, some of the current trends are directly opposite to the pre-World War II predictions. Before World War II, Eastern Europe had a much higher birth rate than Western Europe. It was assumed, therefore, that Eastern Europe would continue to have a faster population increase than Western Europe. The rationale offered was that the more industrialized nations of Western Europe would have a declining population as people chose to have smaller families and a higher standard of living.

But this rationale must be re-examined in view of the fact that, in the 1960–6 period, the countries of Western Europe had a higher annual rate of increase, 1·2%, than the other sections (Table 10.1). Southern Europe had a rate of only 0·8%, Northern Europe, 0·7%, and East-Central and Southeast Europe, lowest of all with only 0·6%.

These changes in annual rate of increase are the result of complex shifts not only in birth, death, marriage and divorce rates, but also in migration to foreign countries. In addition, there are less easily documented factors, such as standard of living, legal and illegal abortions and the relatively new possibility of mass use of contraceptive pills and devices. Without doubt, political conditions are also an inherent determinant of people's decisions to have children or not to have children. The political factor is not an easy one to evaluate, especially in reference to the effects of ideology, because both Greece and Hungary have low rates of population increase, although one is Western-oriented and the other Communist-oriented. In like ideological orientation, Turkey and Albania demonstrate high rates of increase. Therefore it is easier to show what the results are than to prove their specific causes.

Major regional generalizations can also be misleading in terms of what is happening to specific countries. To show the wide range possible within a geographic region, in East-Central and Southeastern Europe in the period 1963–7 Hungary had the surprisingly small annual increase of 0·3%, whereas Albania had an increase of 2·8% and the European portion of Turkey had an even higher rate of 2·9%. The other countries occupied an intermediate position, with Romania at 0·6%, Greece at 0·7%, Bulgaria at 0·8% and Yugoslavia somewhat higher with 1·2%.

For comparative purposes it should be noted that in the same period Ireland had a lower rate of 0·1% and East Germany, West Berlin, East Berlin and Malta all showed migration losses with a minus 0·2%. Yet West Germany, Poland and France all had the same rate as Yugoslavia, and Switzerland was even higher with a 1·9% annual increase.[1]

The same general pattern is evidenced in crude birth rates and in projected fertility of women. Again, Hungary had the lowest, Albania the highest. Hungary combined the lowest birth rate of 13·6 per thousand with the highest death rate, 10·0 per thousand to produce the lowest natural increase of 3·6 per

TABLE 10.1 – *Europe: Annual Rate of Increase, Birth and Death Rates*

	Annual Rate of Increase per cent			Birth Rate per 1,000		Death Rate per 1,000	
	1958–66[3]	1960–6[3]	1963–7[4]	1960–6[3]	1963–7[4]	1960–6[3]	1963–7[4]
EUROPE[1]							
Western Europe	0.9	0.9	0.8	18	19	10	10
Southern Europe	1.2[2]	1.2[2]	1.0	18	18	11	11
Eastern Europe	0.8[2]	0.8[2]	0.9	21	21	9	9
Northern Europe	0.7	0.6	0.6	17	17	9	9
	0.7	0.7	0.7	16	18	11	11

Western Europe: Austria, Belgium, France, West Germany, Liechtenstein, Luxembourg, Monaco, Netherlands, Switzerland, West Berlin.
Southern Europe: Albania, Andorra, Gibraltar, Greece, Vatican, Italy, Malta, Portugal, San Marino, Spain, Yugoslavia.
Eastern Europe: Bulgaria, Czechoslovakia, East Berlin, East Germany, Hungary, Poland, Romania.
Northern Europe: Channel Islands, Denmark, Faeroe Islands, Finland, Iceland, Ireland, Isle of Man, Norway, Svalbard and Jan Mayen Islands, United Kingdom.

[1] Excluding U.S.S.R and European Turkey.
[2] Rate reflects combined effect of natural increase and migration.
[3] United Nations. Department of Economic and Social Affairs. *United Nations Demographic Yearbook, 1966.* (New York: 1967), p. 95.
[4] *Ibid., 1967* p. 97.

thousand. Albania combined the highest birth rate, 34·0 per thousand, with the lower death rate of 8·6 per thousand to produce the highest natural increase in the region of 25·5 per thousand.

Again, for comparative purposes, France had a birth rate of 17·5 per thousand and Spain, 20·9. In terms of natural increase, Luxembourg was lower than Hungary with 3·4 but France and West Germany were only slightly higher than Bulgaria, and Spain's rate of natural increase was very close to that of Yugoslavia.[2]

Yet life expectation patterns do not conform to the previously indicated Albania-Hungary high–low orientation. Although statistics are not available for precisely the same period, statistics of the early 1960's indicate that, for males, Bulgaria has the highest age expectancy at birth, 67·82 years, and Yugoslavia the lowest, 62·39 years. But for females, Hungary had the highest life expectation, 71·83 years, and Yugoslavia the lowest, 65·58 years. These were neither the highest nor the lowest for Europe.

These differing factorial relationships highlight the basic reality that each country of Southeastern Europe exhibits different internal circumstances, hence it is necessary to analyse each country separately. Furthermore, a word of caution should be clearly expressed that each country uses its own definitions and has its own methods of gathering and classifying statistics. Therefore not only is it difficult to make comparisons between countries, but in cases where definitions or methods of enumeration have changed, it may be impossible to make comparisons between different censuses. Tolerance is, therefore, a necessary quality for the geographer studying Southeastern Europe, especially when combined with a sense of humour. In the words of an old Bulgarian peasant of a remote village in the Rhodope, his view of census takers was expressed as follows: 'Every so often bureaucrats come from Sofia to ask all kinds of foolish questions, such as who was born and who died. Then they go back to Sofia and must be very careless because they lose all the information we gave them and then they come back and ask the same questions all over again.'

Because of space and time limitations, this study focuses on only three of the countries in Southeastern Europe – Yugoslavia, Greece and Bulgaria. These countries were selected because, although they have many physical, cultural and historical similarities, they show significant variations in both demographic structure and settlement patterns.

YUGOSLAVIA

Yugoslavia has the largest population in Southeastern Europe. In mid 1970, it had a population of 20,500,000.[3] In the next decade, this is expected to increase to 22,132,000, or about 0·6% per year. This is the same rate as during the 1960–6 period, but is considerably lower than the average rate of increase of 1·41% per year over the 45-year period 1921–66.

World War II produced a catastrophic loss of life in Yugoslavia, not only because of enemy action on the part of the Germans, Italians, Hungarians, Romanians and Bulgarians, but also through the tragic internal situation of civil war, wherein Yugoslavs destroyed each other. In 1945 there was a statistical decline in population of 380,000 over 1939. If one adds to this loss what would have been the increase of population during the war years, the true loss was over one and a half million people and the effects of this have continued in terms of what would have been the present population.

Neither emigration to foreign countries nor immigration from foreign countries has made much change in the over-all post-war growth pattern. After World War II, emigration was not permitted, but was resumed in 1953. In the few years following, it reached some 50,000 a year. But in recent years it has been more like 10,000 a year with immigration at some 1,000 per year.

Since the end of World War II in 1945, there have been three censuses taken in Yugoslavia, in 1948, 1953 and 1961. A number of volumes have appeared focusing on different aspects of the 1961 census, but the complete results are still not available. More recent data and estimates have appeared in the *Statistical Yearbook* (*Statistički Godišnjak SFRJ*), in the bulletin *Demografska Statistika*, and in a few selected statistical publications, such as *Jugoslavija, 1945–1964*, published in 1965. These are the basic sources of the discussion which follows.

VITAL STATISTICS

Since 1921 there has been a consistent decline in the birth rate in Yugoslavia from 36·7 per thousand to 20·2 per thousand in 1966 (Table 10.2). The death rate declined from 20·9 per thousand to 8·0 per thousand. Although both have shown a decline over the period of time, their ratios have fluctuated from year to year producing yearly variations in the natural increase. The end result has been a drop in natural increase from 15·8 per thousand in 1921 to 12·2 in 1966. This drop in natural increase is the critical factor in the decreasing annual rate of growth. Of particular interest has been the decline in infant mortality from a high of 164·5 per thousand births in 1931 to 61·3 in 1966. This reflects better prenatal care and childbirth facilities, but also indicates that more improvement is possible and should take place.

It is probable that a lower growth of population will continue, because there has been a decrease in marriages from 13·1 per thousand in 1947 to 8·5 in 1966. This detrimental effect on potential childbirth is compounded by the rise in divorce. In 1966, the figures indicated that there were about 168,000 marriages and 22,000 divorces, or 13% as many divorces as marriages in that year.

Regional contrasts

Yugoslavia is a land of great physical and cultural diversity. As might be expected there is marked regional variation in population characteristics, both in terms of physical regions and in ethnic patterns. Standard of living, ethnic tradition and historical association are significant individual keys to demographic trends in the different areas. Because the republics are primarily based on ethnic differences, they serve as adequate regional units for comparative purposes, although they mask topographic differences in great part.

There is a strong regional gradient expressed from north to south. The northern areas have the highest standard of living, lower birth rates, higher death rates and lower natural increase than the southern lands. One might expect them to also have lower marriage rates, yet the opposite is true in that more marriages take place in the north than in the poorer area of the

TABLE 10.2 – *Yugoslavia: Vital Statistics (per 1,000 inhabitants)*

Year	Population (mid-year) in 000's	Live Births	Deaths	Natural Increase	Infant Deaths	Immigration	Emigration	Marriages
1921	12,059	36.7	20.9	15.8	—	—	—	13.0
1931	13,982	33.6	19.8	13.8	164.5	18,135	15,368	9.0
1939	15,596	25.9	14.9	11.0	132.3	23,763	27,402	7.9
1947	15,679	26.6	12.7	13.9	—	—	—	13.1
1953	17,048	28.4	12.4	16.0	116.1	350	13,450	9.8
1961	18,612	22.7	9.0	13.7	82.0	739	11,418	9.1
1966[1]	19,735	20.2	8.0	12.2	61.3	907	10,583	8.5

Source: Yugoslavia. Federal Institute for Statistics. *Statistical Yearbook, 1967* (Belgrade, 1967), p. 82.

[1] Provisional data.

south. This may be explained, in part, by the migration of eligible males to the north so that there is less chance for marriages in the south (Tables 10.3 and 10.4).

Of the republics, Croatia shows the lowest birth rate and the lowest natural increase. These were, in 1966, respectively 16·7 and 7·8 per thousand. Croatia also has the lowest fertility rate of the republics, 64·9 per thousand females. But the incidence of both marriages and divorces was relatively high. It also has the highest proportion of females to males.

Slovenia shows similar demographic patterns. A birth rate of 19·0 in 1966 and a death rate of 9·3 produced a natural increase of 9·7 per thousand. It has the smallest average family, 3·47 members, and, by far, the lowest infant death rate, 24·5 per thousand.

Bosnia and Herzegovina shows a transition to the south with a higher birth rate, 26·2 per thousand. This, combined with the lower death rate of 6·6 per thousand, produced the highest natural increase of 19·6 per thousand. A high marriage rate is linked with a low divorce rate.

Montenegro has the lowest death rate, 6·0 per thousand, but still has a high natural increase due to a high birth rate, 23·1 per thousand. It also has the lowest marriage rate of the republics, 6·5 per thousand.

Macedonia, the southernmost republic, has the highest birth rate, 27·1 per thousand. It also has the distinction of having the lowest divorce rate of the republics, 43·2 per thousand marriages, and the largest family, averaging 5·02 members per family. Macedonia is the only republic in which males predominate; in 1966 there were 980 females to every 1,000 males.

The Republic of Serbia reflects these same north to south patterns in terms of the units that make up Serbia. Taken as a whole, Serbia would hold an intermediate position in these various demographic factors, but the individual units show extreme contrasts in almost every feature. Lowland Vojvodina, a fertile area inhabited by Serbs, Croats and neighbouring minority groups, has the lowest birth rate of 14·5 per thousand, a death rate of 8·5 and a resulting natural increase of only 6·0 per thousand, again the lowest unit increase. But it had the highest marriage rate and, to confound the statisticians, the highest divorce rate of 202·0 per thousand marriages – 20%!

TABLE 10.3 – *Yugoslavia: Population by Republics (present-day territory)*

	Yugoslavia	Bosnia and Herzegovina	Montenegro	Croatia	Macedonia	Slovenia	Serbia (Total)
Area sq kms	255,804	51,129	13,812	56,538	25,713	20,251	88,361
Population							
1953	16,991,449	2,847,549	419,873	3,936,022	1,304,514	1,504,427	6,979,154
1961	18,549,291	3,277,948	471,894	4,159,696	1,406,003	1,591,523	7,642,227
Number in Family							
1953	4·29	5·04	4·55	3·81	5·24	3·66	4·32
1961	3·99	4·64	4·43	3·56	5·02	3·47	3·96

Republic of Serbia

	Serbia proper	Vojvodina	Kosovo (Kosmet)
Area sq kms	55,968	21,506	10,887
Population			
1953	1,004,731	484,936	127,004
1961	1,215,899	560,737	152,598
Number in Family			
1953	4·44	3·50	6·42
1961	3·97	3·31	6·32

Statistically, it also had the smallest families, 3·31 members per family. Another characteristic is that the Vojvodina has a high number of females, 1,054 to 1,000 males.

Serbia proper conforms to much the same patterns as Slovenia and Croatia. But Kosovo (formerly Kosovo–Metojia), mountainous and populated primarily by Albanians, provides a study in contrasts. A very high birth rate, 37·6 per thousand, and a death rate of 8·9 create a net increase of 28·7 per thousand per year. Families are large, 6·32 members, even though infant deaths are high, 102·0 per thousand live births, and were previously even higher. There is both a low rate of marriage and a low rate of divorce.

Within republics, there is again a great variety of individual conditions within the smaller administrative divisions. Just a few examples can suffice to show the tremendous range of differences, as indicated in the 1961 census. In Vrbovec *opčina* in Croatia, there was a birth rate of 11·0 per thousand, a death rate of 14·0 and an infant mortality rate of 41·0.[4] In Dragas *opčina* of Kosovo, the birth rate was 50·0 per thousand, death rate, 16·8, and infant mortality rate, 194·1 per thousand livebirths![5]

Demographic trends

The basic pattern which emerges from these assorted factors is that population growth is highest in the plains instead of the mountains, highest in the southern areas rather than the northern, and highest among the Albanians rather than among the Serbs, Croats or Slovenes.

This raises some basic issues of capital investment procedures, in that the standard of living and standard of productivity are higher in the regions of smaller natural population increase thereby creating a natural attraction for migration from the poorer regions, and, by the higher productivity, creating a source of investment funds for the poorer regions of the south. Thus Slovenia and Croatia argue that it is not fair to make them expend their capital in the southern republics, when they could more productively use their capital in their own areas.

These economic rivalries have produced severe political crises, compounded by long-standing ethnic animosities among the

TABLE 10.4 – *Yugoslavia: Vital Statistics by Republics (per 1,000 inhabitants)*

					Natural Increase				
	Yugoslavia	Bosnia and Herzegovina	Montenegro	Croatia	Macedonia	Slovenia	Serbia Total	Serbia proper	Vojvodina	Kosovo (Kosmet)
1950-4	16·4	24·3	22·1	11·5	23·9	11·9	15·0	14·8	10·9	25·5
1955-9	14·3	23·7	21·6	10·2	22·3	9·7	11·7	10·3	8·1	26·6
1966[1]	12·2	19·6	17·1	7·8	19·0	9·7	10·0	7·4	6·0	28·7

Source: Yugoslavia. Federal Institute for Statistics. *Statistical Yearbook, 1967*, pp. 326-9.

[1] Provisional data.

various groups. The greater rate of increase of the Albanians will not reduce the ethnic tensions, especially in view of local riots and demonstrations by the Albanians in Kosovo in 1968 and 1969.

MIGRATION TRENDS

It is obvious that there is a migration from village to city and from the poorer republics to the more prosperous republics. There is what might realistically be called a 'flight from the villages'. It is estimated that perhaps 80% of the internal migrations consists of people leaving villages for the city.[6] The potential for such large migration lies in the fact that even though there was a numerical change in the agricultural population from 10,606,000 in 1948 to 9,197,597 in 1961, this still gave Yugoslavia the high agricultural proportion of 49·7%.[7]

And it is true that there is a constant migration from the mountain areas to the plains and inland basins. But there is also a great deal of migration within the republics from village to small town to capital city and from republic to republic. Actually the only republic to show a sizeable net increase in immigration is Serbia. This is associated with considerable migration to Belgrade and to the fertile Vojvodina.

One way to evaluate such movement is in terms of 'native' population, that is the percentage of people who remained in their place of birth. In 1961, such 'native' population included 62·4% of the total population in Yugoslavia, undoubtedly a low rate of migration compared to Western Europe and the United States.

The figures for the republics were: Bosnia and Herzegovina, 72·4%; Montenegro, 65·7%; Croatia, 60·0%; Macedonia, 65·9%; Slovenia, 46·9%; and Serbia, 61·9% (Serbia proper, 62·8%; Vojvodina, 55·0%; and Kosovo, 70·6%).[8]

This shows that actually the greatest migration is not necessarily from the rural, mountain areas but can instead consist of the better skilled, more educated people from the urban areas who can be more mobile in seeking jobs elsewhere. This question of mobility often means that an area loses its more energetic and more highly qualified people who move to better areas instead of helping to upgrade their home areas. Such a case is evident in the migration from many of the offshore islands of Dalmatia, where there is a movement not only to the immediate Dalmatian mainland, but also to the inland regions.[9]

A map of population density would show that population is concentrated in a crescent-shaped zone stretching from the upper Dalmatian coast northeastward through the Slovene and Croatian republics across the Sava and Danube plains to the borders of Hungary and Romania, then southwards through the plains and hills of northeastern Bosnia and Serbia ending in the Kosovo area. In contrast to this populated zone the Dinaric Mountains project as poorly settled territories, as do the mountain zones bordering on Bulgaria, Italy and Austria.

Within this broad zone there are specific nuclei of population. Probably the greatest concentration is in the Zagreb–Zagorje area, but the Vojvodina–Belgrade–Kragujevac nucleus poses a question as to which of the two is the greater. To the west, Ljubljana constitutes another major nucleus.

These cities are all in the major population zone, but there are also two local, isolated nuclei. These are Sarajevo, the capital of Bosnia-Herzegovina, and Skopje, the capital of Macedonia.

Although a multiplicity of statistics has been issued in the post-war period covering the 1948, 1953 and 1961 census, there is a surprising difficulty in obtaining statistics for towns and cities. Part of the difficulty is the recent practice of dividing a city into *opčine*. This tends to mask the precise area of the municipality, especially when such municipalities are undergoing change. But, even beyond this, there is the problem that statistics in different official and unofficial publications simply do not agree (Tables 10.4 and 10.5).

In the major population area, Belgrade has shown the greatest growth, 60·0% over 1948, while Zagreb and Ljubljana have both increased at a surprisingly similar rate, 37·1% to 36·9%. Rijeka has grown slightly faster than these two, with 46·8%.

But, outside this zone, Skopje has had the high increase of 88·8% in the same period, and Sarajevo and Split have enlarged by 145·2 and 155·0% respectively.

Thus, Belgrade, the largest city, has grown at the fastest rate in the major population area. It certainly indicates the attraction of a country's capital city. Yet its population of 585,234 in

1961 constituted less than 4% of the country's total population. This contrasts tremendously with the situation of Athens in Greece and Sofia in Bulgaria, which dominate the urban structures of their individual countries. Furthermore, Zagreb is a growing competitor, only some 25% smaller, and if the conurbations are compared there is even less difference. In

TABLE 10.5 – *Yugoslavia: Population in Cities subdivided into communes*

City	Population			Growth Index		
	1948	1953	1961	1953–48	1961–48	1961–53
Belgrade	365,766	437,641	585,234	119·7	160·0	133·7
Ljubljana	97,845	112,728	134,169	115·2	137·1	119·0
Maribor	62,677	70,815	82,560	113·0	131·7	116·6
Rijeka	68,780	75,238	100,989	109·4	146·8	134·2
Sarajevo	98,555	109,585	143,117	111·2	145·2	130·6
Skopje	87,654	119,134	165,529	135·9	188·8	138·9
Split	64,262	75,695	99,614	117·8	155·0	131·6
Zagreb	314,669	350,829	430,802	111·5	136·9	122·8

Source: Yugoslavia. Federal Institute for Statistics. *Census of 1961* (*Popis Stanovnistva 1961*) (Belgrade: 1965), Vol. X, p. 19.

addition, there are a number of other significant centres in the regional capitals of Ljubljana, Skopje and Sarajevo. Novi-Sad, Rijeka-Sušak and Split are also in the 100,000 category and if one considers nearby areas as a total conurbation complex, Niš, Subotica, Osijek, Maribor and Banja Luka might also be considered. This makes a total of 13 major cities and a considerable number of smaller towns.

Why this widespread urban development? It is to be expected that sharp differences in physical features, such as mountains and plains and continental versus Mediterranean climates, as well as in littoral and inland relationships would create varying patterns. And some of these features do explain particular site characteristics such as Belgrade at the confluence of the Sava and the Danube or Skopje in a mountain basin.

Yet, cultural factors may have even greater relevance to this decentralized urban structure. The first of these is the long period of occupance by different empires and states, the Turkish empire, the Austro-Hungarian empire and, for a shorter time, the Italian empire. These regimes precipitated development of a number of specific cities as administrative centres and military centres. The boundaries between these empires served as considerable barriers to migration so that groups tended to be separated again into nuclei. The historical influences of these empires are still visible on the present landscape in the form of railroads, roads, architecture and farmsteads as well as in the traditions and even languages of the people.

A second factor is undoubtedly the matter of ethnic cohesion, and Yugoslavia has a considerable number of ethnic groups indeed. In the past ethnic groups have tended to migrate to a particular regional city, such as Slovenes to Ljubljana, Croats to Zagreb and Macedonians to Skopje. This helped create a number of regional centres, in contrast to Greece where there has been a traditional 'urge to Athens'.

This sense of decentralization is now being augmented, perhaps even escalated under the present administrative system of autonomous republics, where a sense of competitive pride can be expressed in regional urban growth, and where both republic and Federal funds can be procured for settlement, industrialization, and in case of tragedy, such as the Skopje earthquake, for rapid reconstruction. Because of the governmental focus on rapid industrialization and on upgrading the 'less-developed' areas, capital investment is creating special, and perhaps premature, urban growth in the poorer areas of the south. This, as previously noted, has already caused economic and political friction between the northern republics and the southern ones.

All of these features – regional discrepancies, accelerating urbanization, intra-republic and inter-republic migration, a flight from village to town amid rapid technological changes – pose a growing dilemma as to which direction population growth will take.

For Yugoslavia certain trends and features merit study and analysis of possible new policies. Salient among these would be the movement from the villages to nearby cities. Would it be worth while to consider the upgrading of life in the villages and small towns? Should agriculture receive greater emphasis in the national economy or would other aspects of the economy, such as tourist and recreation facilities be utilized as a population 'sponge'? And on a different level the basic question must be faced that there is inevitable competition for growth between Zagreb and Belgrade. Which, if either, should receive the greatest support?

Yet for Yugoslavia these questions can be considered in a fairly hopeful natural environment where nature still offers much to be exploited. But in Greece, the outlook is quite different.

GREECE

Greece is one of the smaller and poorer countries of Southeastern Europe. In mid 1970 it had a population of 8,800,000.[10] In 1921 the population was 5,049,500. In the 45-year period, 1921–66, the average annual growth statistically was 1·6%. But this average includes the addition of nearly a million immigrants from Turkey in 1922–3. To be sure, there were large war-time losses in World War II, but this huge increment in the early 1920's seriously masked the true rate of growth as expressed in other years. In actuality, Greece has had traditionally a low rate of growth, and, in the past decade, this has dropped to a fantastic 'low-low' of 0·37% in 1963, 0·36% in 1964 and 0·47% in 1965 due to the combination of low natural increase and an annual emigration total of over 100,000 yearly (Table 10.6).

In future years the outlook may be somewhat more optimistic. Greek population projections are based on 0·7% per year for the 1970–80 period, which is the same rate as the increase in 1966. This would produce a population of 9,627,500 in 1980.[11]

VITAL STATISTICS

In the period between World War I and World War II, Greece had a rather high birth rate, the rate fluctuating from some 20 to over 30 births per thousand inhabitants. But, since World War II, there has been a steady drop to just under 18 per thousand in 1966. Death rates were also high in the inter-war period, but an even more marked drop occurred after World War II to less than 8 per thousand. The result has been a small decline in natural increase to 10 per thousand in 1966. But, in the past decade, there has been a slight increase in marriages, which could produce a corresponding increase in births

TABLE 10.6 – *Greece: Vital Statistics (per 1,000 inhabitants)*

Year	Population (mid-year)	Live-Births	Deaths	Still-Births	Marriages	Natural Increase	Emigration	Population Increase
1921	5,049,500	21·18	13·63	—	5·61	7·55	—	—
1931	6,462,772	30·83	17·70	11·47	7·04	13·13	—	1·50
1939	7,221,896	24·77	13·91	10·08	6·59	10·86	—	1·41
1949	7,482,748	18·59	7·94	9·94	5·63	10·65	—	—
1955	7,965,538	19·35	6·87	12·02	8·31	12·48	29,787	0·91
1961	8,398,050	17·94	7·61	13·55	8·44	10·33	58,837	0·85
1966	8,613,651	17·94	7·88	16·03	8·32	10·06	86,896	0·74

Source: Greece. National Statistical Service. Various Statistical Yearbooks (Athens).

(see Table 10.6). In contrast to the usual pattern of post-war decrease in the number of stillbirths, Greece shows an increase in the recent period.

The net result of this is that Greece shows a smaller rate of natural increase than Yugoslavia and this difference is compounded by the more significant incidence of Greek immigration, a pattern which is likely to continue.

REGIONAL CONTRASTS

In both Greece and Yugoslavia there are strong regional contrasts which are reflected in marked demographic contrasts. One of the most marked of these is the area of concentration of high birth rates. In the west and north of continental Greece, stretching from the Gulf of Patras to Thrace and the Turkish border, there is a zone of higher birth rate exceeding 22 per thousand. This zone consists primarily of mountains in the west, but in the northeast the land consists mainly of plains and foothills. Culturally, these lands are inhabited in great part by Macedonians and Turks, both of whom have had traditionally higher birth rates. Thus in Greece as well as in Yugoslavia, high birth rates are associated with more isolated terrain and with specific ethnic groups.[12] And in Greece as in Yugoslavia the islands usually have a lower birth rate, though certain islands have high rates, such as the Dodecanese Islands in Greece that have birth rates of over 25 per thousand inhabitants. Hence they too are a source area for emigration whether into other regions of Greece or to other countries.

MIGRATIONS

As in most other countries of the world, there has been a growing urban trend in Greece. In the census of 1961, a new category was introduced, 'semi-urban', which has made it more difficult to differentiate between the usual 'urban' and 'rural' categories. Nevertheless, the trend towards urbanization as evidenced by the past four censuses is clear. The rural category declined from 54·4% in 1928 to 43·8% in 1961 (Table 10.7). The urban group increased from 31·1% in 1928 to 43·3% in 1961. Thus if the semi-urban group, which has been about 14% rather consistently, are included with the urban group, then the urban percentage would be 56·2%, clearly a majority. The basic motivation of this 'flight from the village' trend is rather easy to perceive. The generally low standard of living in the villages and the daily rigours of ordinary agricultural life on impoverished lands makes even the worst conditions of the city seem unduly attractive. But the movement from village to town is usually but part of a larger regional movement, in that

TABLE 10.7 – *Greece: Urban, Semi-Urban and Rural Population*

Census Year	Urban	Population Semi-Urban	Rural	Urban	Per cent Semi-Urban	Rural
1928[1]	1,931,939	899,466	3,373,281	31·1	14·5	54·4
1940[1]	2,411,647	1,086,079	3,847,134	32·8	14·8	52·4
1951[1]	2,879,994	1,130,188	3,622,619	37·7	14·8	47·5
1961	3,628,105	1,085,856	3,674,592	43·3	12·9	43·8

Source: Greece. National Statistical Service. *Statistical Yearbook, 1967* (Athens: 1967), p. 21.

[1] According to the criteria of 1961.

migration is ordinarily first to a local city or regional centre, rather than to either the Athens or Thessoloniki conurbations. Thus a degree of centralization is evidenced which contrasts sharply with the more decentralized patterns of Yugoslavia.

The period between the 1940 and 1951 censuses marked the tragic loss of population in World War II. In the 1951 census, of the geographic regions of Greece only Thessaly and Crete showed an increase in population in addition to Greater Athens, which increased a remarkable 22·6%. All the other geographic regions showed a net loss of population, ranging as high as a −8·8% in the Ionian Islands (Table 10.8).

The subsequent census in 1961 showed quite a different pattern – one which is more indicative of the trends today. Countrywide, there was a 9·9% increase, with the greatest increase again shown by Athens. This was a rise of 34·4% over 1951. All the other geographical regions gained somewhat, except for the Peloponnesus and the Ionian and Aegean Islands.

This insular loss was also characteristic within the geographical regions as well as in the country in general. The continental portion increased 12·96%, but the insular part lost by 3%

TABLE 10.8 – Greece: Percentage of Population Change, 1940–1951, 1951–1961

Geographic Regions	Number of Inhabitants			Change %		1951–1961[1]	
	1940	1951	1961	1940–51	1951–61	Continental	Insular
1. Greece	7,460,203	7,632,801	8,388,553	2·3	9·9	12·96	3·03
2. Greater Athens	1,124,109	1,378,586	1,852,709	22·6	34·4	34·39	8·11
3. Rest of Central Greece and Euboea	908,511	908,433	970,949	–0·0	6·9	8·50	1·44
4. Peloponnesus	1,156,189	1,129,022	1,096,390	–2·4	–2·9	–2·88	–12·50
5. Ionian Islands	250,626	228,597	212,573	–8·8	–7·0	—	–7·01
6. Epirus	332,132	330,543	352,604	–0·5	6·7	6·68	—
7. Thessaly	585,430	624,342	689,927	6·6	10·5	10·94	–10·57
8. Macedonia	1,705,664	1,756,434	1,896,112	2·9	11·2	11·23	3·95
9. Thrace	359,923	336,954	356,555	–6·4	5·8	6·02	–10·05
10. Aegean Islands	548,380	528,766	477,476	–3·6	–9·7	—	–9·70
11. Crete	438,239	462,124	483,258	5·5	4·6	—	4·57

Source: Greece. National Statistical Service. *Statistical Yearbook, 1967* (Athens: 1967), pp. 32–40.

[1] Greece, National Statistical Service. *Statistical Yearbook, 1964*, pp. 24–31.

(see Table 10.8). In Thessaly, the continental gain was 10·94% but the insular loss was about the same proportion, 10·57%. While the continental part of the Peloponnesus decreased by 2·88%, the insular area has a much higher loss of 12·5%. Thrace, too, had a disproportionate loss with a 6% gain on the continental area but a 10% loss in the adjacent islands. But the magnetic attraction of Greater Athens, again asserted itself even in the off-shore islands, which had a population increase of 8%.

The basic trend of population distribution and of migration is easily evident in Greece. The eastern littoral of the mainland has the largest concentration of people. The Attic Plain, site of Athens and the classical heart of Greece, is the key area of concentration. But it has a rival in the Macedonian Plain, which has a larger total area and, according to the 1961 census, had a larger population, 1,986,112, compared to 1,852,709 in Greater Athens. Actually the Attic Plain includes more than just Greater Athens, so that if the rest of the Departments of Attica and Piraeus were added, the Attic Plain still has the largest total, but not by much.

Between these two population poles is the lowland of Thessaly which forms another possible nucleus of population, but it forms a poor third because the two large plains include some 80% of the country's urban population, and, since 1940, have had a rate of growth 60% higher than the total for all Greece.

The urban pattern is quite different from this regional pattern. Athens is undisputed king with 22% of the nation's population concentrated in Greater Athens, formed of Athens, its port, Piraeus, and the newer satellite communities. While Piraeus has had the anomalous role of the only municipality in the area to lose population, the newer suburbs have increased astronomically in the decade 1951–61. A few examples suffice to show the pattern, Aghios Dimitrios had a 362·3% increase; Ilioupolis, 243·2%; Peristerion, 122·0%; and Nea Liosia, a soaring 666·7%, showing the change from a rural village to a new municipality of 31,000 people.

In 1961, Athens with its population of 1,852,709, was some five times greater than that of Thessaloniki, 378,444. The next largest conurbations were Patras, 102,244, Iraklion on Crete,

69,983 and Volos, 67,424. All the other cities were substantially smaller, again pointing up the predominance of Athens.

The smaller cities of 25,000 inhabitants or less show quite sharp contrasts in the census gains and losses. Some showed gains as high as 40%, but many showed losses of population as high as 10%, in the case of Corinth.

This urban contrast poses the basic problem that in much of the country's villages and small towns life is unattractive at best and extremely difficult at worst. The low birth rate and high outflow of emigration appear most likely to continue with the inevitable result that Greece will gain but slowly in population over the next decade. It is perhaps paradoxical that, in a world that speaks of a population 'explosion' in saddened tones, a country which does not have a constantly escalating population finds itself on the defensive. This is precisely the case in Greece when questions are raised as to why people are having such small families. It raises not only economic issues, such as poor agricultural conditions, but also industrial queries as to why industry has been permitted to increase so much in the Athens area and why it was Thessaloniki that was chosen as the site of a new industrial park, which could have been placed in one of the deficit regions. No less does it raise some political questions as to why the standard of living has remained so low. But whatever the reasons, it is inescapable that in the post-war period internal conditions have not been attractive to major population increase.

BULGARIA

In mid 1970, Bulgaria had an estimated population of 8,500,000.[13] A comparison of growth rates between censuses shows a decline in annual average increase to about half of what it was at the beginning of this century. This was from about 1·5% per year to 0·8% during the decade from 1947 to 1956 and 0·9% in the more recent 1957–65 period.

At the beginning decades of the century, the increase in rural population was somewhat greater than in the urban category. But, in the inter-war period, urban increase began to surpass the rural gain. In the post-World War II decades, urban population in Bulgaria has risen at a rate of some 4% per year, while the rural population has actually declined to −1·5% in the 1957–65 period. This change reflects the strong migration to cities and towns and points up the growing problem throughout the Balkans of holding people on the farms.

VITAL STATISTICS

The birth rate in Bulgaria has made a drastic plunge from some 40 per thousand in the early 1920's to 14·9 in 1966. There was also a decrease in the death rate from about 21 per thousand to 8·3 in 1966. The corresponding drop in the rate of natural increase was from 18·5 per thousand inhabitants to only 6·6 per thousand in 1966 (Table 10.9).

The decrease in natural increase is the cause of the present low annual increase because both emigration and immigration are so small in recent years as to cause little effect. But, during 1950–1, there was a forced emigration from Bulgaria of some 150,000 Turks to Turkey. The emigration did have some effect on the subsequent rural decline because almost all of the Turks came from the countryside, and the Turks have had a higher rate of birth than the Bulgarians. Thus ethnic traditions are significant in terms of natural increase, just as they are in both Yugoslavia and Greece.

And again as in both Yugoslavia and Greece, there was a spectacular decline in infant mortality. The average mortality rate for 1926–30 was 147·1 per thousand. By 1961–5 this had gone down to 35·0. In 1966, there was reportedly a further drop to 32·2 per thousand. There is still a sizeable range in the greater incidence of child mortality in the country than in the city. Similarly, it is interesting to note that the number of male child deaths is greater than that of female children, which may account for the gradual change from the former preponderance of males in the population to the present slight preponderance of females (Table 10.10).

MIGRATIONS

A basic change in population composition is indicated by growth of urban population from some 16·7% in 1880 to nearly 46·5% of the population in 1965 (see Table 10.10). This latter figure, however, is expressed differently in a three-fold classification of 43·9% as urban, 2·5% as 'settlements of an urban type',

TABLE 10.9 – *Bulgaria: Vital Statistics (per 1,000 inhabitants)*

Year	Population mid-year	Live-Births	Deaths	Marriages	Natural Increase	Immigration	Emigration
1921	4,897,000	40·2	21·7	12·3	18·5	—	—
1931	5,808,200	29·5	17·0	9·6	12·5	—	—
1939	6,294,600	21·4	13·4	9·1	8·0	—	—
1946	7,000,200	25·6	13·7	11·0	11·9	—	—
1956	7,575,800	19·5	9·4	8·9	10·1	36	370
1966	8,256,800	14·9	8·3	8·2	6·6	5	141

Source: Bulgaria. Central Statistical Office. *Statistical Yearbook, 1967* (Sofia, 1967), pp. 14–62.

and the remainder, 53·6% as rural.[14] Therefore rural population still predominates in Bulgaria, traditionally a peasant country, but at the present growth of urbanization this may not continue for very long.

In the cases of Yugoslavia and Greece, administrative boundaries conformed in general with both physical and

TABLE 10.10 – *Bulgaria: Population, Male–Female Ratio and Urbanization*

Year	Population	Urban Population %	Females to 1,000 Males
1880[1]	2,007,919	16·7	954
1884[2]	942,680	23·8	978
1887	3,154,375	18·8	965
1892	3,310,713	19·7	958
1900	3,744,283	19·8	961
1905	4,035,575	19·6	962
1910	4,337,513	19·1	966
1920	4,846,971	19·9	1,002
1926	5,478,741	20·6	997
1934	6,077,939	21·4	990
1946	7,029,349	24·7	999
1956	7,613,709	33·6	1,004
1965[3]	8,226,564	46·5	1,001

Source: Bulgaria. Central Statistical Office. *Statistical Yearbook*, 1967 (Sofia, 1967), p. 13.

[1] Northern Bulgaria.
[2] Eastern Rumelia.
[3] Provisional Data.

cultural regions, so that statistical generalizations could be made with some relevance to regional conditions. But this is not the situation in Bulgaria where in the post-war decades administrative divisions have tended to be rather arbitrary. In 1949, 14 provinces (*okrugi*) were created, replacing the previous nine counties (*oblasts*). In 1951 two were abolished, leaving twelve. Then in 1959 the country was divided into 30 provinces. But in 1964, two of these, Varna and Plovdiv, were amalgamated into their provinces, leaving Sofia as the only separate city province plus a rural province also called Sofia.

These provinces, again called *okrugi*, show great disparities in total population ranging in 1966 from the 167,400 in Smolyan province on the Greek border with a density per square kilometre of 47·6 to the large population of the town of Sofia, 915,000, with a density of 881·4 persons per square kilometre (Table 10.11). A study of population changes in the provinces from 1946 to 1965 shows some curious changes. Sofia (town) province gained 71%, but the surrounding rural province of Sofia lost 2%. And Smolyan, the smallest province in population, was second to Sofia in its growth, 44%. Kyustendil, also a

TABLE 10.11 – *Bulgaria: Population by Administrative Divisions (End of Year)*

Okrug	*1962*	*1966*
Bulgaria	*8,045,200*	*8,284,000*
Blagoevgrad	301,400	305,700
Burgas	372,300	384,600
Varna (town)	159,300	
Varna	188,900	373,800
Vidin	187,600	179,000
Vratsa	311,700	307,300
Gabrovo	158,800	173,000
Shumen (Kolarovgrad)	238,900	245,100
Kŭrdzhali	276,500	286,100
Kyustendil	197,600	197,400
Lovech	214,500	217,500
Mikhaylovgrad	245,300	239,300
Pazardzhik	299,800	298,800
Pernik (Dimitrovo)	190,200	180,900
Pleven	356,000	351,000
Plovdiv (city)	309,500	
Plovdiv	305,800	649,500
Razgrad	194,400	198,200
Ruse	263,000	275,900
Silistra	170,300	171,400
Sliven	216,100	228,600
Smolyan	150,300	167,400
Sofia (city)	784,600	915,000
Sofia	330,200	316,800
Stara Zagora	338,100	356,200
Tolbukhin	240,500	236,100
Tŭrgovishte	173,500	177,300
Tŭrnovo	344,300	337,500
Khaskovo	295,500	291,700
Yambol	230,300	222,900

Source: Bulgaria. Central Statistical Office. *Statistical Yearbook, 1963, 1964, 1965* and *1967*.

mountain province, lost 2%, while Vidin province on the Danube lost a record 8%.

What these statistics really show is that there is no major area of population concentration in Bulgaria as there is in Yugoslavia and Greece. The basic situation in Bulgaria is that the greatest concentration of population is in the Sofia basin, with

TABLE 10.12 – *Bulgaria: Population of Cities, by Censuses (50,000 and above in 1965)*

City	1926	1934	1946	1956	1965
Burgas	31,157	36,230	44,449	72,526	106,127
Varna	60,563	69,944	76,954	120,345	180,062
Gabrovo	10,473	13,668	21,180	37,912	57,758
Pazardzhik	21,578	23,228	30,376	39,499	55,410
Pernik	12,296	15,844	28,545	59,930	75,844
Pleven	28,775	31,520	39,058	57,555	79,234
Plovdiv	84,655	99,883	126,563	161,836	222,737
Ruse	45,788	49,447	53,523	83,453	128,384
Sliven	29,263	30,571	34,291	46,175	68,331
Sofia	213,002	287,095	366,801	644,727	800,953
Stara Zagora	28,957	29,825	37,220	55,094	88,522
Tolbukhin	—	—	30,522	42,661	55,111
Khaskovo	26,256	26,516	27,435	38,812	57,672
Shumen	25,137	25,486	31,327	41,546	59,362
Yambol	23,037	24,920	30,576	42,333	58,405

Source: 'Broi na Naselenieto na Gradovete v B'lgariya Spored Prebroyavniyata prez Proslednite Chetirideset Godini' (Population of cities in Bulgaria according to censuses during the last 40 years), *Geografiya* (Knizhka I, 1966), pp. 21–3.

a secondary nucleus at Plovdiv. For the rest of the country there are numerous nuclei scattered in different localities.

This pattern is clearly evidenced by the distribution of towns. Sofia, like Athens, is undisputed king with a population of 800,953 in 1965 (Table 10.12).

Plovdiv is the next largest city, only about a fourth as large with 222,727. But whereas Thessaloniki is centred on a large, fertile plain area with a large rural and small town population, Plovdiv is in a smaller valley without such a large hinterland. Of the other three largest cities, two are seaports on the Black Sea, Varna with 180,062, and Burgas with 106,127. The third largest of these is Ruse, 128,384, on the Danube river in the north. There are 33 additional cities with populations from 20,000 to 100,000 inhabitants (see Table 10.12).

This demonstrates a third great pattern, different from those of Yugoslavia and Greece, where a capital predominates, but where the rest of the population is scattered in the form of dispersed nuclei.

SUMMARY, CONCLUSIONS AND QUERIES

These three countries show contrasting demographic processes and differential areas of population growth and concentration. Each country is different from the others, just as in the entire region of Eastern Europe. Therefore it is not possible at present to predicate a basic, universal 'model' of population structure and change, except for the two basic truisms of worldwide population increase and growing urbanization.

In terms of ethnic composition, these three countries show greater homogeneity since World War II, but the remaining ethnic minority groups generally have a higher birth rate than the majority population. This again reflects the other countries of Eastern Europe.

But if drastic changes have occurred in population structure and processes, little change is indicated in terms of the specific areas occupied in each country. This statement must, of course, be accepted as a generality with exceptions possible, but it is clearly evident that the centres of concentration of today are the same as those a century ago, only as time goes on the urban focal areas show an increasingly higher density in contrast to rural areas. Thus each country has to concern itself not only with demographic policies, but also with regional policies of development and change.

Countries can thus forge population policies and seek to implement them. But this does not mean that the people will necessarily conform to governmental policies. What is instead quite evident is that people do what they want to do regardless of governmental policies. In short, people respond to the 'family climate' rather than political exhortation. One of the indices of 'family climate' is the average age of marriage. In Bulgaria in recent years the highest statistical age bracket of marriage for women is under nineteen years, and 20–4 years for the men.[15] In Yugoslavia the largest number of marriages is, for both men and women, the 20–4 year old category.[16] But, in Greece, although the 20–4 year old category still applies to women, for men the age bracket is 25–9 years.[17] This significantly older marriage age for men mirrors the harsher 'family climate' of Greece and can be compared with the situation in Ireland where the older age of marriage has a significant effect on subsequent lower birth rates.

The statistics of the demographic factors of birth rate, death rate, marriage rate, internal and external migrations, etc., all bear witness to the changes occurring in the 'family climate'. But, although they document the results, they do not in themselves convey the basic causes. Economic, social, political, military and perhaps even global forces have direct bearing on the 'family climate' and hence on demographic trends in Eastern Europe. In much of current demographic literature the popular focus is upon new technological devices as outside creators of demographic processes. But in East-Central and Southeast Europe, the quite individual national trends indicate that perhaps we could as profitably seek population growth answers by studying the individual 'family climates' to see what processes are at work on the assumption that individuals themselves exert a great deal of voluntary decision as to when to get married and how many children to have in the family.

NOTES

[1] United Nations. Department of Economic and Social Affairs. *United Nations Demographic Yearbook, 1966* (New York, 1967), pp. 109–11.

[2] *United Nations Demographic Yearbook, op. cit.*, pp. 118–19.

[3] Official estimate.

[4] Savezni Zavod Za Statistiku, *Statistički Godišnjak SFRJ – 1967* (Belgrade, 1967), p. 512.

[5] *Statistički Godišnjak SFRJ – 1967, op. cit.*, p. 519.

[6] Josip Roglić, 'Neke Značajke Kretanja Stanovnistva Jugoslavije' (Some characteristics of Yugoslav population movements), *Matica Iseljenicki Kalendar* (Zagreb, 1966).

[7] Savezni Zavod Za Statistiku, *Jugoslavija 1945–1964* (Belgrade, 1965), p. 46.

[8] Roglić, *op. cit.*, p. 41.

[9] Roglić, *op. cit.*, p. 45.
[10] Official estimate.
[11] National Statistical Service of Greece. *Statistical Yearbook of Greece – 1967* (Athens: 1967), p. 21.
[12] Bernard Kayser, *Géographie Humaine de la Grèce* (Paris: Presses Universitaires de France, 1964), pp. 47–51.
[13] Official estimate.
[14] Ljubomir Dinev, 'Niakoi Izmeneniya v Razpredelenieto na Naselenieto na NRB B'gariya (The study of changes in the distribution of Population in Bulgaria)', *Geografiya* (Knizhka 1, 1966), p. 3.
[15] Tsentralno Statistiches Ko Upravlene Pri Ministerskiya Suvet, Sofiya, *Statisticheski Godishnik na Narodna Republika Bulgariya, 1967* (Statistical Yearbook of Bulgaria, 1967), p. 17.
[16] *Statistički Godišnjak SFRJ – 1967*, *op. cit.*, p. 91.
[17] National Statistical Service of Greece, *Statistical Yearbook of Greece – 1968* (Athens: 1968), p. 29.

Comments
LESZEK A. KOSIŃSKI

1. An interesting theoretical question that has to be discussed in a study of any area is to what extent the experience of that area is unique and to what extent it is just a variation of general trends. Also in this discussion on East-Central and Southeast Europe the problem of uniqueness versus generality offers an interesting point of departure.

(a) The experience of West European countries was generalized in a theory of demographic transition that tries to explain changes in the levels of fertility and mortality and their interrelationship in terms of social and economic changes. At present, there is discussion among experts of population problems regarding the extent to which the transition theory (i) has universal validity, and (ii) can be used as a predictive tool for areas of different cultures. In most cases the feeling about this question is sceptical; it is argued that the theory was based on the unique experience of Western Europe and cannot be applied to other areas.

The question arises as to whether the transition theory applies to East European countries and how the present changes in those countries comply with the theoretical patterns. It seems that the experience of Eastern Europe does fit the theoretical expectations quite remarkably. The drop in mortality was followed by a delayed but equally dramatic drop in the level of fertility and the rates of natural increase declined very drastically in almost all the countries except Albania. Consequently, it seems that the experience is not unique although the rate of change might be higher than it was in different Western European countries.

(b) Social and economic development leads to a concentration of population. This seems to be a general rule which all the countries of East-Central and Southeast Europe certainly confirm. In the post-war period, in particular, increasing industrialization led towards greater concentration of population. I can quote here the results of some studies on population concentration for Poland. If one compares the area occupied by 20% of the population, where the density is highest, in the present territory of Poland this area decreased from 4·2% in 1910 to 3·2% in 1930 and 2·5% in 1960. At the same time, the area occupied by 20% of the population with lowest density increased from 38·8% in 1910 to 42·5% in 1960. In the 1960's, the respective proportion of this area in the various countries of East-Central and Southeast Europe varied a great deal according to the settlement pattern and level of industrialization. For example, in Bulgaria the area occupied by 20% of the population living in the highest and lowest density areas was 9% and 30% respectively. In the German Democratic Republic 3% and 47%; in Hungary 2% and 35%.

It seems that the process of concentration of population in East-Central and Southeast Europe shows similar characteristics to other countries of Europe that have gone through the later stages of industrial revolution earlier.

(c) The patterns of migration had very unique characteristics in East-Central and Southeast Europe as a result of war and post-war transfers of population. However, later more normal migration from rural to urban areas, reflecting the process of industrialization, conformed to the theoretical expectations that one might have for any country that is going through the process of modernization. In details it varied depending on the policy adopted by an individual country; generally it was a flow of people from rural to urban areas.

To answer the first question, to what extent is the experience of East-Central and Southeast Europe unique and to what extent it reflects the general trends, one might qualify the answer depending on the level of the generalization. Basically, it is a variation of general trends, though in detail the uniqueness is much more pronounced.

2. What are the motivations and causes of population

changes? During the Conference at Austin the influence of ideology was emphasized and discussed in detail. There is no doubt that the fact that the countries under consideration have a socialist or communist type of government is reflected in the policies adopted and consequent changes. However, it seems that ideology has a direct impact only in a very few cases. On the other hand, its indirect influence can be far-reaching.

(a) The first and most important cause of population changes seems to be modernization of society. One can argue that the modernization was initiated and directed by the Communist theoretical concepts. On the other hand, the process of modernization shows remarkable similarities all over the world, and in particular, all over Europe. Industrialization brings people to towns; the pattern of family life changes; the consequences for lowering fertility are pronounced. These are developments independent of the system of government, and certainly Eastern Europe experiences the process of modernization at a very rapid rate.

(b) The policy of the government can influence population changes. Generally speaking, in the post-war period the governments had pro-natalist policies. This was justified by the desire to make up for the tremendous war losses, especially high in Poland and Yugoslavia. Later, this policy changed, although large families were encouraged and family allowances paid by the government. Yet, after 1955 abortions were permitted and birth control methods and means made available and encouraged. This, together with the effect of modernization, led to a sudden decline of fertility and of natural increase, and consequently in some countries, most notably Romania, the policy was very drastically reversed. It seems that the population policy reflects much more the national interest of individual countries than the ideological attitude. Besides, the discussion that goes on among population experts in Eastern Europe and in the Soviet Union, reflects a confusion as to the basic Marxist theory of population. The feeling is growing that such a theory does not exist and should be developed. If there is no overall theory, how could one expect an overall unified policy based on the theory. Therefore, the policy is much more pragmatic in each case, although lip-service is being paid to the theoretical justifications.

(c) If (i) the process of modernization that has certain general qualities, and (ii) the policies that reflect pragmatic national interests are two basic factors of population changes, there is not much room left for ideology. It seems, however, that the latter has some place although not a very pronounced one, and its impact is only indirectly felt in the population trends.

3. During the discussion in Austin the question was asked; is East-Central and Southeast Europe a homogeneous demographic region, or is it just a set of countries and should be regarded as such? In any discussion of the past demographic trends, Eastern Europe was considered as a more or less homogeneous region. Its characteristics were low degree of industrialization, economic and social backwardness, very high fertility and a low level of urbanization. The peasant societies of Eastern Europe were considered to be similar. Admittedly, countries in the northern part of the area such as Germany or Poland, Bohemia, part of Hungary were much more industrialized and did not quite fit the picture. Generally, however, this definition was true. The area then was a homogeneous demographic region. The common feature for all of the area now is that it is undergoing drastic and far-reaching changes. In that sense, it is homogeneous.

The answer to the question depends really on the level of generalization. It seems that one can talk about Eastern Europe as a demographic region defined by the changes. On the other hand, certain groups of countries, individual countries and even regions show remarkable differences.

Remarks

Professor Chauncy Harris began the discussion which followed the paper given by Professor Kostanick, suggesting that the rate of natural increase in East-Central and Southeast Europe is higher than in the West. They are the results of cultural diffusion in two innovation waves. The first wave was a reduction in the death rate due to modern medicine. The second was brought about by a reduction in the birth rates. The classic fluctuations in demographic characteristics of Eastern Europe are a result of these two waves. Perhaps abortions may be a factor in the reduced birth rate.

Professor Pounds offered a different point of view, stating that historically population growth has fluctuated up and down without the innovation waves suggested by Professor Harris. For example, population decline during the late Roman Empire in Europe occurred for no apparent reason. The decline in Europe came long before the plague in the Middle Ages. One reason for this type of fluctuation might be the land inheritance systems as was the case in East Anglia.

Professor Pounds also commented on the changes in density over the last 100 years. By analogy with studies for Western Europe where densities have not changed, it is then suggested that the same holds true for Eastern Europe. Migrations have been short distance movements, he said. Cities grew in dense rural areas.

Dr Crkvenčić used Croatia as an example in citing his reasons for low population increase. He said that post-war population grew 10%, but this is smaller than the national rate due to out-migrations. The migrants were young, unmarried males who leave for better work, not other areas.

Dr Baxevanis then pointed out that there is a problem in reading population data for East-Central and Southeast Europe because of the two different methods of census taking which may cause confusion. He indicated that perhaps there is a distortion because certain Greek towns are good for 'dying' and many old people, including Greeks and American Greeks, go there to spend their last years.

11

Regional Development Processes in Southeast Europe. A Comparative Analysis of Bulgaria and Greece*

GEORGE W. HOFFMAN

THE CONCEPTUAL FRAMEWORK

Comparative studies of the Regional Development Processes of countries of differing ideologies are extremely rare in the social sciences, and geographers have scarcely touched on this subject.[1] This paper attempts to contribute to the meagre literature of comparative studies, especially on those countries with different ideologies, by analysing the regional development processes in two countries of Southeast Europe, Bulgaria and Greece. Emphasis will be given to a comparative analysis of the type and impact of the most important regional development processes during the post-war period, their institutions and tools, their decision-making constraints and the methods of planning for the purpose of reducing spatial inequities within Bulgaria and Greece. The selection of Bulgaria and Greece as a case study will permit a comparative assessment of two political systems which operate under very different social and economic

* Research for this study is closely connected with the author's long-term interest in the changing economic and political geography of Southeast Europe. This study is part of a larger one entitled 'Regional Differences and Economic Development in Southeast Europe' which was supported by the National Science Foundation between 1963 and 1969. The author also gratefully acknowledges the assistance of his colleagues, both in the United States and in various European countries, whose extensive knowledge of various problems of the area has been extremely valuable in pursuing this project. Appreciation is also extended to the publishers for permission to use material first published in the author's studies 'Eastern Europe', Chapter 8, in *A Geography of Europe*, edited by George W. Hoffman, (New York: The Ronald Press 3rd edition, 1969), pp. 431–524, 'The Problem of the Underdeveloped Regions in Southeast Europe', *Annals of the Association of American Geographers*, 57 (December, 1967), 637–66, *The Balkans in Transition*, Searchlight book No. 20, Princeton, New Jersey: D. Van Nostrand Co., 1963 and (with Fred W. Neal), *Yugoslavia and the New Communism* (New York: The Twentieth Century Fund, 1962).

conditions in their objectives and the approach used to find a long-term solution to ameliorate their spatial inequities. The discussions here will provide a conceptual framework on which a regional synthesis emphasizing the post-war development strategy can be presented. It is also hoped that this analytical comparison of development processes in two countries of different ideologies will have both methodological and practical value and will encourage other similar studies, especially since Bulgaria and Greece are in many ways representative of the broad spectrum of developing countries.

Regional development processes in all countries of Southeast Europe have been conditioned by a long history of foreign domination which in turn resulted in an extremely complex cultural landscape. The formation of this cultural landscape and its impact on the social structure and economic activities was a process which affected all the countries of Southeast Europe, often in a similar way. To understand the interaction of the spatial variations in the physical and cultural environment, and their relationship to the development processes of the two countries under discussion it is necessary briefly to cite the background of the various problems of the area as a whole.

Few areas in the world show as great spatial contrasts as the five countries of Southeast Europe. The complex relief of the peninsula, with its large fragmentation which developed as a result of the rugged relief, encouraged particularism and isolationism and was largely responsible for the absence of political unity. The crossroad position of the peninsula, with its easy accessibility by both land and sea, subjected various parts of the region to cultural and political influences from Central Europe, the Mediterranean region and the Orient, thus also contributing to the absence of political unity. These influences, as well as the interaction of natural and social factors have left a deep impact on every one of the countries of Southeast Europe, complicating relationships with neighbouring countries as well as with the various regions of each national territory. There is no country in Southeast Europe in which internal spatial inequities and contrasts brought about by past history have not been expressed in cultural and economic diversity. This diversity has played a decisive role in the developmental processes and is shown especially in its regional application between and within the individual countries of Southeast Europe. In addition, any region under the domination of the Ottoman Empire was not only undeveloped, but was 'also left a legacy of maladministration and corruption in public life that influenced and burdened the new governments.'[2] These past associations created great spatial differences, including greatly differing cultural and economic levels which have not been overcome in the relatively short period since independence was won.[3] In addition, due to the location of the newly independent states within various Great Power spheres, the economic and political initiative of these countries was, and still is, severely limited, even after their liberation from direct foreign control.

In many ways the five countries of Southeast Europe had reached a similar social and economic development level at the outbreak of World War II. Resistance of many regions to the spread of innovations and improvements very often retarded change. World War II and the changes in the political geography of the region left a deep impact on each of the five countries. Bulgaria (also Albania, Romania and Yugoslavia) went through a political revolution which brought about basic social changes. Old institutions were transformed, the static society with its traditional social structure was radically changed, Soviet institutions and practices were copied and Soviet Russia made financial contributions to the economic transformation, especially for heavy industry and raw materials. Greece, taking the path of evolution (after a bloody civil war), began to modernize some of its institutions along traditionally democratic lines and was assisted by large-scale aid from the United States. However, the long-term impact of the 1967 military take-over is still difficult to assess.

From these brief introductory remarks it can already be seen that the reason for the great social and economic regional contrasts in every one of the countries of Southeast Europe is longstanding. These contrasts are especially marked on the regional level, and for this reason the respective governments have lately given attention in varying degrees to regional developmental planning in their respective countries.

In presenting a comparative analysis of regional development processes between and within Bulgaria and Greece, which includes the distribution and impact of the location of economic activities one is reminded of the geographer's special concern with the whole problem of regional differences. This was clearly stated by N. Ginsburg:

> the geographer is interested in those differences in resource endowment and potentials that favour one such region or country over another; with the type of ecological relationship to bear upon the use of natural resources; with the functions of distance and transport costs that bear upon regional development and retardation.[4]

An analysis of the

> Character and dimensions of these regional entities, the ways in which they have evolved, the commonalities that they display in different situations, the discovery of principles ordering their evolution, their functioning as dynamic systems and their relationship with one another[5]

as well as the fluidity and flexibility within the two societies as to interaction among social and economic forces, but also as to the spatial mobility, such as the relationship of internal regions, including the interaction between urban and rural areas, are essential prerequisites for a better understanding of the development processes. Obviously a detailed analysis of all these factors is certainly beyond the reach of this limited study.

The whole question of regional development processes is closely tied to a regional planning policy for individual countries. Bulgaria has had such policies all through the post-war era. They are highly centralized and follow, generally, the Soviet model. Until the early 1960's rational development planning was completely lacking in Greece, and even the 1968–72 Economic Development Plan for Greece 'is essentially an indicative Plan for the private sector of the economy', with the State initiating projects only 'when the projects are of fundamental importance for the industrial development of the country, and provided that private initiative takes no interest in the matter'.[6]

Every one of the five Southeast Europe countries has emphasized in various degrees the reduction of spatial inequities, and the achievement of inter-regional equality as a long-term goal, but as J. Friedman pointed out, problems of 'spatial organization' must be faced by all countries which have 'reached a certain stage in the development of their economies'.[7] Spatial inequities or contrasts in many of the countries of Southeast Europe, in spite of much emphasis by some countries on regional economic planning since 1945, are unfortunately as strong as ever, though certain 'broad growth zones', as well as 'growth centres', are clearly visible throughout the area. Those of Bulgaria and Greece will be detailed later in this paper. The problem of finding a solution to the inequities in regional contrasts is closely associated with the question of a balanced v. unbalanced growth in each country,[8] but the solution is made more difficult by the need for assistance to both the underdeveloped areas and the crowded urban agglomerations.

REGIONAL DEVELOPMENT PROCESSES

The focus in this section will be on the development processes of two countries of Southeast Europe, Bulgaria and Greece, since their formation, with special emphasis on their economic and social transformation which cannot be understood without an analysis of the distribution and changes in economic activities since the time independence was reached (Table 11.1 and Figs 11.1 and 11.2).[9]

THE FORMATIVE YEARS

Bulgaria and Greece were typical 'Balkan states' whose nominal political independence was relatively new (Bulgaria, 1878, and Greece, 1830) and whose autonomous economic policies date from the decay of the Ottoman Empire. With the collapse of the three Empires – Austro–Hungarian, Russian and Ottoman – all influential in the political and economic life of the Southeast European countries up to the end of World War I, Bulgaria and Greece for the first time could pursue essentially their own economic and political interests. The task of building a viable economy and developing the agricultural and manufacturing resources of newly independent Greece and Bulgaria was a difficult one in the face of the backwardness of the area and other obstacles. The limited productive soil, especially in Greece, the structure of land ownership, the deplorable transportation

TABLE 11.1 – *Comparative Economic Growth: Bulgaria and Greece*

	Bulgaria Pre-war	Bulgaria 1950s	Bulgaria 1960s	Pre-war	Greece 1950s	Greece 1960s
Area (sq. ml)	39,824	42,822		50,147	50,547	
Population (000s)	6,292	7,667	8,400	7,222	8,100	8,800
Population density (per sq. ml. agric. land)	285	358	370	240	241	245
Per cent of total population in towns of 2,000 or more inhabitants	21·4	33·6	46·5	47·6	52·5[a]	56·2[b]
Annual rate of population increase (per 1,000)	13·8[c]	10·1	6·6	13·0[c]	11·7	10·2
Proportion of active population:						
In manufacturing, mining, construction	8[d]	19	33	16[e]	19	23
In agriculture, forestry, fishing	80	64	45	50	54	47
In trade, services and other activities	12	17	22	34	27	30
Cultivated land (% of total land)	43·2	40·9	41·1	19·9	26·6	29·9
Fertilizer used (kg. per ha. of arable land)	NA	16	79	8	33	65
Tractors, average for agricultural use	3,200	17,000	42,000	1,500	13,000	49,000
Physicians	3,127	9,271	13,593	NA	10,423[b]	12,573
Death rate per 1,000	13·4	8·6	8·1	14	7·6	7·9
Birth rate per 1,000	21·4	18·4	15·3	25	19·3	17·9
Infant mortality (deaths per 1,000 live births)	138·9	66·3	32·2	118·2	44·2	34
Illiteracy (over 10 yrs old)	31·4	13·5		40·8[d]	23·6[d]	17·8[b]
Students graduated from universities and colleges	1,223	5,860	7,781	NA	3,727	5,337
Road density (length of all roads in km.)	19,554	25,719	28,914	15,760	32,365	50,000
Railroad density (length of all railways in km.)	4,426	4,926	5,657	2,557	2,664	2,585
Length of railway lines (ml. per sq. ml. of territory)	0·07	0·07	0·08	0·03	0·03	0·03
Estimated *per capita* income (in U.S. dollars)	NA	NA	407	NA	326	600
Energy consumption in coal equivalent (metric tons *per capita*)	0·14	1·0	2·7	0·18	0·4	0·8
Percentage contribution to GNP:						
Of agriculture, forestry, fishing	51	26	21	40	34[f]	25
Of manufacturing, mining construction	31	63	70	21	22	25
Of trade, services and other activities	18	11	9	39	44	50

Source: Official Statistics (including those prepared by banks).
In general the pre-war period is based on data from the years 1933–40, for the 1950s, from 1956–8, and for the 1960s, from 1965–6.
NA – data not available; [a] 1951 census; [b] 1961 census; [c] average rate, 1931–5; [d] 1934 census; [e] 1928 census; [f] 1952.

FIGURE 11.1 – TERRITORIAL GROWTH OF BULGARIA, based on map in George W. Hoffman, *The Balkans in Transition* (Princeton, N.J.: Van Nostrand Co., Searchlight Book 20, 1963), p. 55

network, the absence of easily obtainable resources, and the few urban centres made the economic development and integration of these new, highly independent nationalist countries into the broad currents of the Western and Central European development difficult.

Basically, both countries after becoming independent had similar economic and political ambitions:

to create modern armies and administrations, to fulfil their 'national aspirations' – which, alas often meant the acquisition of the same pieces of territory – and to become integrated into the broad currents of capitalist growth and industrialization embracing western and central Europe ... these countries encouraged industrialization by resorting to tariffs, subsidies, and state purchases of the products of domestic industry, and moreover, by facilitating the growth of an integrated banking system and creating favourable conditions for foreign investments.[10]

Expansion of the economic activities was made possible by heavy borrowing from abroad, and this in turn resulted in an increasing public indebtedness which necessitated an increasing

FIGURE 11.2 – TERRITORIAL GROWTH OF GREECE, based on Homer Price, 'Southern Europe', in *A Geography of Europe*, Edited by George W. Hoffman (New York: The Ronald Press Co., 3rd edition, 1969), p. 415

taxation burden. This forced an increased reliance on exports of agricultural products, resulting in further depressing the rural standard of living. Due to insufficient diversification in economic production, the exports were carried by a very limited number of products, mainly tobacco in Greece and wheat in Bulgaria. Protectionist measures such as laws 'encouraging domestic industries' and a variety of tariff measures were enacted in Bulgaria in 1894 and in Greece in 1910.[11]

Foreign debts increased rapidly,[12] and while some of the money was used for the construction of railroads, bridges, ports and other essentials for the development of their infrastructure, unfortunately much too large a proportion was spent in building complicated and cumbersome administrative machinery, and in the creation of modern armies.[13] Still, considering the insurmountable backwardness, especially of the rural areas, the series of wars in which both countries had been involved since their independence, and the archaic economic and social structure, considerable economic growth was accomplished, though neither of these countries either before or following World War I, reached the standard of living and broad economic development of the Western and Central European countries.

INTER-WAR PERIOD

World War I left a profound political and economic impact on Bulgaria and Greece, as well as on the whole region of Southeast Europe. Bulgaria lost valuable territory to Greece and Romania. Greece soon became involved in another war with Turkey which caused a great population upheaval – 1,350,000 refugees entered Greece between 1922 and 1928, and 434,000 Turks departed. Also 54,000 Bulgarians were exchanged for 46,000 Greeks. The whole region was economically disorganized, but it was hoped that the various population transfers would at least bring some reduction in the size of the various minority groups and thus contribute to greater political tranquillity in the area. Unfortunately, this hope remained unfilled throughout the inter-war years.

The socio-economic structure of Greece and Bulgaria was quite similar. Both countries were typically agricultural countries with low and variable crop yields,[14] lacked modern technology, including improved farming and livestock practices, and relied on only a few grains and tobacco. In 1920 over 82% of Bulgaria's labour force was engaged in agriculture, while only 50% of the population of Greece was so occupied,[15] but in the latter country an important part of the labour force was occupied by trade and shipping. In 1939 Bulgaria's labour force engaged in agriculture had declined only by 2% (to 80%). Increased farm fragmentation was a serious problem in both countries. In 1897 an average Bulgarian farm occupied 7·2 hectares; in the 1930's 14% of all farm holdings had less than one hectare and 43% had less than three hectares, and in addition each farm consisted on an average of ten strips, often some distance apart. In 1939 in Bulgaria 112 peasants were dependent upon 100 hectares of cultivated land for their livelihood in comparison to 98 in Greece and 75 in Germany. When many refugees returned to Bulgaria from Greece, Turkey and Yugoslavia after World War I and land became scarce, the government intervened passing an agrarian reform law (1920) and voluntary co-operative associations were organized.[16] This was the second land reform law within a short time, and a third land consolidation law was enacted in 1933, all aimed at helping the rapidly increasing rural population to obtain their own, though highly fragmented land holdings. This, obviously, gave rise to increased subsistence farming. The rapidly increasing rural population had little opportunity to seek other employment. In spite of considerable effort, *per capita* production of the main crops hardly increased in Bulgaria during the inter-war period.

The picture was somewhat better in Greece. The great population pressures brought about by the influx of 1·2 million refugees from Turkey gave tremendous attention to land expansion through reclamation and land improvement. Cropland was more than doubled by major reclamation works between the two wars[17] while the supply of farm labour increased by only one-third. In both Bulgaria and Greece the government was forced to intervene, though foreign assistance in the form of capital and technical aid greatly helped the latter. Farm incomes in both countries were extremely low. According to a study by A. Pepelasis 'the reclamation program was not dictated entirely by economic considerations, but was inspired by a policy of autarky. This intensified the high cost structure of Greek agriculture.'[18]

All through the inter-war period protectionist policies encouraging autarky and industrialization were an important facet for all countries in Southeast Europe, and Bulgaria and Greece were no exceptions. Stress in both countries was placed on the development of light industry (domestic industry) largely producing for home consumption. The great depression of the early 1930's, with its flight of foreign capital, the catastrophic fall of agricultural prices and numerous other factors brought a momentary halt to the economic (industrial) expansion of the 1920's in both countries. Due to systematic encouragement for the building of a domestic industry by the state in both Bulgaria and Greece, the impact of the depression on the development of those relatively protected industries was more easily overcome. As a matter of fact, it must be made clear that before World War II the state, more in Bulgaria than in Greece, played an important role in the control and even outright ownership of numerous enterprises, including all important mines, railroads, public utilities, and, in Bulgaria only, river and maritime shipping. But in spite of the emphasis on industrialization, the industrial labour force, including handicraft, hardly changed in the twenty years of the inter-war period in both countries, 9% in Bulgaria and 16% in Greece.[19] At the outbreak of World War II, industry and handicraft contributed 16% of the national product. The earlier-mentioned lack of domestic capital in Bulgaria necessitated an intensive effort for foreign investments which contributed 43% of all co-operate capital, 60% in power generation, 46% in the paper industries, 43% in food and tobacco, 24% in metal and chemical industries. Twenty-four per cent of the foreign capital was invested in mining industries.

Industrial development in Bulgaria and Greece was heavily protected by tariffs and this had an adverse long-term effect on the economic development of the countries. High tariffs encouraging domestic production produced inefficiency, and in both countries acted as barriers to greater regional distribution of industrial production. As a result a relatively heavy concentration of industrial production developed in the two capitals, Sofia and Athens. Industry in both countries relied heavily on the domestic resources produced by agriculture, and this, too, was protected. In part this development was shown in the composition of foreign trade which was typical of an underdeveloped economy. In Greece all through the inter-war years, foreign trade emphasized three commodities: tobacco, raisins and currant products, which accounted for 61% of the total value of commodity exports in 1937. Receipt of invisibles (emigrant remittances and shipping income), accounting for two-thirds of the value of commodity exports, covered the trade balance.[20]

Bulgaria also relied heavily on agricultural exports during the inter-war years. Nearly 54% of the total export income was derived from tobacco, wheat and egg exports. Imports were largely industrial goods, consisting mainly of machinery and capital equipment. The large balance of payments deficit was covered in part by foreign loans.

Throughout the inter-war years the two countries made great effort in their economic development and this period was also the high time (at least between 1924 and 1936) of their independence.[21] With the enactment of land reforms and a policy of autarky and industrialization, the social and economic structure of both Bulgaria and Greece had become quite similar by the outbreak of World War II.

The developments in these two countries before World War II can be briefly summarized to serve as a focus for a better understanding of the post-war changes: (1) a backward agriculture based on subsistence farming with a complete lack of technical progress and outmoded methods of animal husbandry; (2) heavy demographic pressures which resulted in land fragmentation, and some attention to land expansion by reclamation, with greater attention in Greece due to the larger influx of refugees, than in Bulgaria; (3) a domestic industrial expansion largely of light industry heavily dependent upon agricultural resources, based on protectionist policies, and a nearly complete lack of heavy industries; the majority of the labour force employed in extremely small-scale, low-productivity plants;[22] (4) a heavy reliance on foreign capital for the development of industries and an infrastructure, which resulted in large indebtedness; (5) heavy military outlays absorbing between one-fourth and one-third of the budget;[23] (6) state involvement in various spheres of the economy, including ownership of most public transportation and utilities and industries for the development of national military capabilities; (7) concentration of industrial expansion in a few poles located in a few urban centres, which resulted in their rapid growth, but hardly any changes in the traditional countryside; Athens and to a much lesser extent Thessaloniki in Greece, and Sofia and to a lesser extent Plovdiv and Varna in Bulgaria.[24]

Greece's whole economic development was also much affected, more than that of Bulgaria, by the permanent or temporary emigration of its citizens. Close to 2·5 million people had left between the late nineteenth century and 1940, some of the most active age groups thus were removed from the economic development processes of the country. On the other hand, the remittances in foreign currencies by its emigrants were of importance to Greece's economy. Unfortunately their value was largely lost by the poor management of Greece's economic and political affairs.

AFTERMATH OF WORLD WAR II

Greece emerged from World War II with a devastated economy and serious social and political problems.[25] The civil war, shortly after the end of World War II, further destroyed a great amount of human and material resources. The period 1945 to the middle of the 1950's is characterized first, by a struggle for survival and, once this was established, by a tremendous foreign and domestic effort for recovery.[26] While American aid was most impressive, it covered 68% of Greek imports and invisibles in 1949, excluding military aid, it also created a real inflationary problem. Only after thorough monetary and economic reforms, including drastic currency devaluation, was a more stable basis for future economic development created. Therefore our analysis of comparative regional development takes the development in the mid-1950's as its starting point. During the recovery period Greece had already shown some of the political instability, in part caused by the breakdown of the multi-party system, for which she was known from the pre-war period and from which she suffered almost continuously during the post-war period. Its impact was felt in every major economic decision, and unfortunately, much too often, in the absence of much needed decisions for economic and political development.

With the exception of Greece, the Communist party gained support in all countries of Southeast Europe. In Bulgaria it swept away quickly most of the past social and economic structure, sharply limiting private ownership in agriculture and completely eliminating it by nationalization in transport, banking, trade and industry by 1948. Bulgaria escaped, with few exceptions, serious war damages and losses in manpower, which played such havoc with the early Greek post-war economic development. This, in turn, made it possible for the Communist government, shortly after the end of World War II, to institute long-term economic planning within the framework of a centralized socio-economic political order. The country's whole economic policy is based almost exclusively on the experience of the Soviet Union. This, in turn, resulted in the reproduction in miniature of the Soviet pattern of economic development, with all its centralized, bureaucratic and microplanning apparatus,[27] channelling a major part of Bulgaria's investments into industry (roughly five times the pre-war rate) and especially heavy industry (Table 11.2). The impact of this decision affected every aspect of her economy, including her foreign economic relations. Bulgaria's post-war economic development policies vary only insignificantly from those of most other socialist countries, including the Soviet Union, inasmuch as she has devoted a greater percentage of her investment to agriculture. Bulgaria carried out her plan of economic development generally without foreign aid, while Greece received substantial United States aid which represented, in some years, up to 10% of her gross yearly product. Still, the two ideologically differently-oriented countries, with widely differing planning tools, achieved throughout the 1950's very similar growth rates as evidenced in industrial income and total product – between 6·1 and 7·5%.[28]

Development processes in Greece were greatly affected by the civil war and her dependence upon foreign advisers (U.S.) who, in turn, were able to re-evaluate the focus of capital expenditures which at their height in the period 1949–55 was equal to 9% of the GNP.[29] As a result of World War II and the civil war, the level of the pre-war GNP per head was again

TABLE 11.2 – *Bulgaria and Greece: Percentage Distribution of Gross Fixed Investment*

	Bulgaria	Greece
Gross fixed investment as percentage of GNP		
Total Investment		
1950–4	23·7	15·9
1955–9	27·7	19·2
1960–3	41·5	28·9
Investment in Industry and Construction		
1950–4	10·6	4·3
1955–9	12·1	3·5
1960–3	19·3	NA
Investment in Agriculture and Forestry		
1950–4	4·1	1·6
1955–9	7·6	2
1960–3	11·2	NA
Investment in Services		
1950–4	9	10
1955–9	8	13·7
1960–3	11	NA
Gross fixed investment as a percentage of total		
Investment in Industry		
1950–4	45	27
1955–9	44	19
1960–3	47	NA
Investment in Agriculture		
1950–4	17	10
1955–9	27	10
1960–3	27	NA
Investment in Services		
1950–4	38	63
1955–9	29	71
1960–3	26	NA

Source: Maurice Ernst, 'Post-war Economic Growth in Eastern Europe', in *New Directions in the Soviet Economy*, a study prepared for the Joint Economic Committee of the United States Congress, Washington, D.C., 1966.

NA – data not available.

reached only in 1956. It was realized in Bulgaria by 1950. The problem of the great spatial differentials in productivity and *per capita* income was left untouched with the result that *per capita* income for the leading region, Central Greece in 1958, was 33% above the national average.

Earlier we spoke of the period of the mid-1950's which can be counted as the one on which 'normal' economic development can be based. It is therefore of interest to see the conclusions in N. Ginsburg's *Atlas of Economic Development*.[30] Data shown measures various aspects for each of the countries relative to each other as well as to the rest of the world. It appears that in 1955 Greece was much more typical of the world's nations than was Bulgaria, judging by the number of properties in which Greece was at or very close to the median of all countries measured. Bulgaria ranked higher in all these areas except for energy potential, which relates the obvious fact that Bulgaria was a more developed country at the beginning of the period of 'normal' development. Benjamin Ward came to very similar conclusions in his study comparing Greece and Yugoslavia.[31] In summary, it appears from the relative rankings of these two countries that both countries in the mid-1950's presented a rather typical picture of the economic development for some developing countries and, quoting Ward, 'of the not-too-distant future for others'.[32]

THE POST-WAR YEARS

The preceding discussion makes it clear that economic development in Bulgaria and Greece based on quite similar human and economic resources and using strikingly similar tools had reached a very similar level of economic growth by the end of the inter-war period. World War II and its aftermath left a major impact on both countries, more so in Bulgaria where a new political philosophy exerted its influence on the whole social and economic structure of the country and brought about a complete restructuring of the economy.

Greece gave renewed attention to its long-term economic development once the immediate problem of recovery from the war, including the civil war, had been accomplished, but a national development plan was not initiated until the early 1960's,[33] and even then it was expressed in very general terms. One of the main aims of this programme was the gradual elimination of disparities among the various regions of the country, though 'regional development was hardly considered in the plan'.[34] Greece all through the post-war years has pursued a policy of economic development based on market demand and supply factors, with monetary policy as a major weapon of economic control. It uses governmental policy to provide assistance for economic growth, especially industrial development, in most cases without controlling it. Funds for a fairly large-sized public investment programme have been channelled through various semi-government institutions. Generally, the private sector has had little state interference and Greece's fiscal structure is based on providing incentives for private and foreign investment. The only direct controls in use affect agriculture and the Agricultural Bank organized in 1929 is the main Government vehicle. Government plans until recently had little direct influence on economic development, and only during the last year, with the promulgation of the 'Economic Development Plan for Greece 1968–72',[35] has a more elaborate programme with stated goals been provided.

To illustrate some of these general points it should be noted that between 1950 and 1963 the gross national income in real terms increased at an average rate of 6·3% with agricultural and industrial output more than doubling – one of the highest increases among OECD countries. Gross investment rose from 10% of the gross domestic product (excluding ships) from the end of the civil war to about 25% in 1965. While Greece's overall economic policies in this period were strikingly successful, numerous fundamental, long-term problems remain. A solution of these problems is imperative if Greek products are to compete successfully in the international market, something essential in view of the forthcoming termination of the transitional period of tariff protection in connection with Greece's membership of the Common Market.[36]

One of the most pressing problems remaining for the Greek economy is the unchanged basic structure of the economy.[37] Agriculture still accounts for close to 47% of the labour force. In the origin of GDP (at factor cost in 1967) agriculture ranks 23% (29% in 1956), and industry at 26% (27% in 1956). Tobacco and currants still constituted 41% of the total exports in 1967, 55·5% of the total agricultural exports. No less distressing is the basically unchanged structure of industry. In spite of considerable growth, Greece's industry is one of a small closed family enterprise with a few large units dominating a particular activity.[38] The devastating effect of the protectionist policies of all pre-war governments which resulted in an absence of initiative, technical skills and lack of experience in international competition is still felt in the whole economic development of Greece.[39] Certainly alternative sources of income, a greater diversity of agricultural output and a greatly increased regional spread of economic activities together with a greater social awareness are essentials if Greece is to take its place among the developed societies. Finally, of concern to Greece is her dependence upon income from tourism and shipping, as well as remittances by Greek workers from abroad. The total income from these sources amounted to $573 million a year in foreign exchange in 1967. Partly as a result of the military coup, a considerable reduction in these foreign incomes on which the Greek economy was and is so heavily dependent occurred after mid-1967. Economic experts disagree in their evaluation of the economic development in Greece and the reliability of official reports since mid-1967.

The major characteristics of Bulgaria's economic development processes since World War II were indicated earlier in the discussions under 'Aftermath of World War II'. Bulgaria is unique among socialist countries in having recognized the need for continuous investments in the agricultural sector, though it can be argued that they have been insufficient, considering the importance of this sector in the national picture. Bulgaria was the first East European country to achieve full collectivization (1958). The post-war economic development of all socialist countries benefited from a rather rapid and widespread nationalization and the diverting of savings according to carefully devised plans resulting in powerful resources being in the hands of these states. 'It is possible in these conditions to increase capital formation in a rapid pace and to achieve economic growth according to the priorities set by the planners.'[40] In Bulgaria this policy has meant a rapid expansion of state ownership and a sharp limit on personal consumption. With the rapid expansion of production, the problem of allocation of capital investment, both as to branches and in terms of its spatial distribution has come to the forefront of the discussions of the political leadership, as well as the academic community, with the latter often working on the methodological problems only, with little regard to political and even economic realities. The best possible distribution of new productive forces, both in terms of efficiency of production and greatest benefit to the underdeveloped parts of the country, without doubt, has been a major problem for the Bulgarian planners all through the post-war period.

Bulgaria has greatly re-oriented its whole economy to the quick development of industry, using its resources of minerals in addition to depending on heavy imports largely from the Soviet Union.[41] Production of crude oil jumped from 4,700 tons in 1954 to 499,000 tons in 1967, and recent discoveries should bring production to close to one million by 1971. Before the war there were no integrated iron and steel works in the country; today there are two, one at Pernik started in 1949, and a second built at great cost during the 1960's at Kremikovci. Entire new industrial branches were started and these took a large part of available investment funds. Engineering industries, all built since 1950, account today for close to 30% of the total industrial growth, and employ more workers than the total Bulgarian industry in 1939. Bulgaria produces a variety of machinery with concentration in heavy capital goods, and this production is concentrated in several of the larger cities (Sofia, Pernik, Varna, Ruse, Burgas, Plovdiv), with 35% of all production in Sofia. Generally, the machine-building industry developed very rapidly in the post-war years due, in large part, to favourable sales possibilities to the Soviet Union. As a result of the one-sided industrial building programme great structural disproportions have been formed and in future years will undoubtedly have a negative influence on the economic growth of the country. Though agriculture received some investment funds during the whole post-war period (with the exception of the 1948–52 period) it was more a starvation diet than one of abundance (see Table 11.2). While agriculture was already rather intensive in the inter-war years, quick emphasis on collectivization required additional investment funds for the greatly increased needs of mechanization. In addition, Bulgarian agriculture was assigned a special role by COMECON and its specialized products, such as vegetables, especially tomatoes, fruit, tobacco and certain specialized crops, find a ready market both in the COMECON and West European countries. Agriculture still furnishes roughly one-third of Bulgaria's exports.

A considerable part of its agricultural surplus population in the last 20 years has found employment in the newly established industries with the result that in 1966 the percentage of population engaged in agriculture had declined to 45% thus creating at times a shortage of labour in the countryside.[42] In part this was accomplished by large-scale consolidation of the many small farm holdings into less than 1,000 co-operatives averaging about 11,000 acres, thus releasing a considerable labour force to work in the newly built industries, and also making possible the use of machinery.

A certain slow-down in the Bulgarian economy became obvious during the 1960's, but no decrease in the investments of heavy industries accompanied this, although increased investments in agriculture were evidenced, largely as a result of Comecon's needs. Bulgaria's whole post-war development processes were directed towards the transformation of her economic base from that of an agrarian, underdeveloped country with industries oriented largely towards the processing of agrarian products into a modern, broad based, diversified industrial economy with heavy emphasis on engineering and lately on chemicals. This basic change in the structure of her economy resulted only in a slow progress in the standard of living for her people and still left large parts of the country basically backward and underdeveloped.

From these brief remarks it can already be seen that economic development and policies have basically differed in the two countries, though the ultimate goals of achieving a better-balanced economy within the framework of long-term regional development, and ameliorating the negative characteristics of underdevelopment and backwardness in the two countries are much alike. If one were to make a general statement as to the overall progress of Bulgaria and Greece in the post-war years, it must be recognized that in spite of a much later start in its post-war economic development, Greece, by a policy of less interference by government in the market demands and supply factors of its economy, in the long run has been eminently more successful in improving the level of economic growth than Bulgaria, though in its regional aspects of economic development, little progress has been accomplished. Both countries have few very large industrial complexes (Pernik and Kremikovci metallurgical combine, Northern Greece–Thessaloniki industrial complex) and a large number of small-scale industries more numerous in Bulgaria, mainly family owned in Greece, and an underdeveloped and often backward agricultural sector, though less serious in Bulgaria. Bulgaria's large number of small-scale industries, and specifically 'the complex of farming, food and light industries', according to a recent study by Enčo Dragnov, will remain for the near future the 'pillars of industrialization'.[43] Largely by its centralized planning policies, Bulgaria has been able to bring about a wider distribution of economic activity, especially in the field of industrialization, drawing on its agricultural overpopulation as can be seen by the drastic reduction in the numbers deriving their income from agriculture.

Greece's progress of economic development is closely tied to the much-needed structural reorganization and a greatly increased investment effort. Her generally unplanned economic effort has resulted in further concentration, largely in Athens and perhaps one or two other urban places, of the means of production, capital accumulation and labour force.

PLANNING INSTITUTIONS AND TOOLS

BULGARIA

The new post-war government in Bulgaria introduced a highly centralized, authoritarian micro-plan, a system of planning known as 'administrative mechanism',[44] but by the late 1950's certain changes and deficiencies in this system of management already had become obvious. The essence of this system was a policy of administratively determined prices, with high prices for industrial products and low prices for agricultural products. The main planning organ was the State Committee for Planning, and it established numerous compulsory indices in the economic plan. Often these indices were inconsistent and enterprises were, 'practically speaking, prevented from displaying greater initiative in setting the total volume of production and in matters concerning production costs, productivity of labour, range of goods, price-formation procedure, etc.'[45] Starting in the early 1960's a trend towards greater liberalization and the use of market forces in the planning mechanism became evident. This manifested itself in the decreased influence of central planning, in decentralization and an emphasis on regional planning, a greater concern for the producers, a gradual expansion of responsibilities and rights of enterprises in planning and management, in greater responsibilities for expenditure of funds, etc. Generally market elements in the planning mechanism were stressed and this tendency was closely related to the increasing role of modern technology in the more advanced production of the Bulgarian economy.

Planning practices presently in force, with some exceptions, date from the decision of the plenum of the Bulgarian Communist Party Central Committee of May, 1963, which reviewed the structure and functioning of economic management in the country. A number of basic changes in the way in which national economic plans were to be drawn up in the future were initiated, and later approved by the Council of Ministers.[46] The main reason for these changes was reported in a lengthy speech by Party Chief Zhivkov, but basically they were obvious throughout the 1950's: poor long-term planning, excessive centralization, failure to link and observe planning with actual production, insufficient scientific research for individual planning goals, unreal plans, etc. As part of the reorganization and changes proposed, new organs were added to the State Committee for Planning.[47] Plans are still drawn up by the State Committee for Planning which allocates targets among various Ministries, State Committees and other State Agencies and among various economic branches. Plans pertaining to agricultural procurement and local industry must be allocated among *okrugi* and approved by the Council on Agriculture and the Council of Ministers. Greater responsibilites on the *okrug* level and the establishment of *okrug* planning commissions were an innovation, but economic plans were still centrally approved and *okrug* authorities and individual enterprises had little initiative in planning.

Ever since planning of all socio-economic aspects of the State received such high priorities, Party, government officials and academicians have been concerned with the establishment of proper economic territorial units, so-called administrative-economic regions, as functional economic regions, to serve as regional groupings for integrated economic activities following the Soviet example (territorial-productional complex),[48] which reflect the economic and political aims of the system. These units (regions) had to have the capabilities to develop all aspects of its economic potential, yet they had to fit into the overall national plan. Special emphasis in the planning was given, as mentioned earlier, to industrialization with each unit having ultimately to show a balanced development between industrialization and agriculture, preferably self-sufficient in fuels with a strong industrial centre. In addition, each unit had to assure adequate labour pools and transportation. All in all, these units according to the Soviet model had to be well-balanced in all their economic activities.

In 1959 Bulgaria introduced basic administrative changes, administrative-economic regions, functional regions which were all-encompassing, and with precisely prescribed boundaries.[49] These 30 new administrative economic districts (*okrugs*), now reduced to 28, are strictly administrative divisions, serving also planning purposes, and replaced the earlier 13 districts and 117 *okolias*. The capitals of these administrative-economic regions usually are of importance to the region, for the most part centrally located cities (Sofia, having the most populous *okrug*, has the rank of a region).

Geographers and economists were evidently little consulted on the most appropriate regional division of the country, and there is a strong argument that Bulgaria is too small and interdependent to have such a large number of regional divisions.[50] The argument for unity of the functions of administrative and economic regions is in many ways a continuation of the discussions on regionalization which filled so many pages of Soviet journals, though the fact remains that these regions have an important administrative-economic purpose. Decisions from the State Committee for Planning go to the *okrug* planning commissions and are profoundly influencing spatial interactions, and for this reason the *okrugs* are rapidly assuming the characteristics of 'economic regions'. In addition, implementation of important sections of the plan are now left to the *okrug* planning commissions and even individual enterprises. In the process new relationships of individual enterprises and collective farms for procurement and marketing are established, crossing customary administrative and planning regions and forming new links. This is of special importance for the establishment of close links between raw material processing enterprises.[51]

Weaknesses in the decision making processes related to planning activity were obvious and further economic reforms therefore were introduced in 1966 leading to further liberalization and decentralization in planning activities, including increased stress on market mechanism. It was admitted that inefficient planning largely cancelled the effect of high growth rate of industry (an average of 11% over the five-year period 1961–5). Cost factors and accurate pricing were thus far completely neglected and in spite of earlier reforms the decision-making power of the central authorities had hardly changed.

Differences of opinion about the correct way to plan the economy brought disagreements between those advocating further economic reforms in terms of liberalization (meaning decentralization) of planning activities, the increased use of market mechanism, the establishment of proper commodity-money relations,[52] the transfer of real responsibilities to enterprises, and those advocating a more cautious policy and a continuously dominant role of central planning and even decreased use of the market mechanism. These matters were again discussed in the Plenum of the Bulgarian Communist Party in July, 1968. According to some observers, there appeared a retreat from the previous 1966 commitments to broader economic reforms, including decentralization of the decision making processes. These expressions reflect similar discussions in the Soviet Union and previous discussions among Czech and Slovak and Hungarian economists, and even earlier in Yugoslavia. One of the main arguments in the 1968 discussion was related to a new system of price-forming procedures, the improvement of centralized planning, the so-called economic reform and a certain division of responsibility between the 'organs which implement centralized planning and those of local planning'.[53]

It is obvious from this discussion and from the 1969 Economic Plan passed by the Bulgarian National Assembly in December 1968 that decentralization and economic reform was largely a question of degree. Basically all important decisions affecting economic development processes are political decisions of the Party[54] (closely following Soviet policies), and they in turn are executed by the State Committee of Planning and the *okrug* planning commissions. The main feature of the economic reform, started on a trial basis in 1967 and almost uniformly adopted in 1969, deals with a reduction of planned indices from the centre and a greater authority of the enterprises in the preparation of their plans. The introduction of commodity-money relations, the long-term planning for production needs, the establishment of a basis for measuring productivity of the individual and the enterprise, as well as profitability of the production of individual goods, are also included. The application of the new economic system is reflected in a different title for the plan from previous years, the 'unified plan on social and economic development', a one-year State Budget for 1969, and in modifications in the implementation and fulfilment of the plan, though individual targets set differ little from previous years. At the same time, new factory prices of industrial production, new transport tariffs and a new amortization base introduces important changes in the Bulgarian economy.[55]

GREECE

Contrary to economic planning in Bulgaria with its centralized micro-plan and cumbersome planning mechanism, Greece, until the 1960's, completely lacked a national planning policy, and even now the aim of its Plan 'is to provide an institutional and organizational framework for tackling the country's vital long-term problems on a rational basis'. In its initial economic growth after World War II the Greek economy was largely financed by capital from abroad, mainly from the United States. The Greek government since the conclusion of the civil war has had a public investment programme and funds available were and still are channelled through a number of semi-government special institutes such as the Industrial Development Bank, Agricultural Bank and the Productivity Centre. Basically, however, the initiative comes from private enterprise and the government simply assists.

The first official commitment by the Greek government was expressed in a very general *Five Year Programme for the Economic*

Development of Greece *1960–64,* with projections to 1969. One of the main aims of this programme was the gradual removal of disparities among the various regions. This document was published by the Ministry of Coordination[56] with the collaboration of a South Italian consulting team. In it the hope was expressed that 'a coherent policy aiming at the structural obstacles which prevent the country from developing economically',[57] could be one of its major accomplishments.

The need for a detailed overall plan, including regional studies, became quite obvious in the early 1960's. Various foreign organizations were invited by The Ministry of Coordination to study individual regions, e.g., Crete and Peloponnesus. The government created some organizational measures; the Regional Development Service as a section of the Planning Directorate within the Ministry of Coordination as well as an advisory body to the Ministry. However, the conclusion of the OECD Case Study evaluation in 1964 was by no means optimistic about planning activities in Greece.

A new draft *Five-Year Economic Development Plan for Greece (1966–1970)* was completed in late 1965. It is a detailed factual document and contains a general discussion on regional development policy. A special committee to examine the draft of the Five-Year Plan was established. It analysed and praised the draft plan and suggested ways of implementation,[58] but in the absence of a stable Greek government, implementation had to wait until a somewhat revised and up-dated *Economic Development Plan for Greece 1968–1972* was officially published by the new government. Between the end of the first Plan and the new plan for 1968–72, the Ministry of Coordination created five planning regions with regional development centres. They are: Epirus with the islands of Corfu and Levkas in Ioannina; Thessaly in Larissa; Northern Greece (Macedonia and Thrace) in Thessaloniki; Crete in Iraklion[59] and Peloponnesus, with the islands of Ithaki, Kefallinia and Zakinthos; and two neighbouring *nomes* of Central Greece in Patras. The main task of these regional development services was the preparation of detailed regional plans using as much as possible of the existing studies. In addition to these plans, contracts were let during 1965 for city plans to include eight to ten of the major cities. These are primarily physical planning schemes and take little account of economic aspects of urban development. Detailed feasibility studies for industrial investment are still completely lacking, and even basic studies which could assist in the modernization of agriculture or open up new tourist areas were only initiated in 1968.

The 1968–72 plan is essentially in its own words 'an indicative plan for the private sector of the economy. This is completely justified by past experience, which indicates that the effective operation of the free market is the best means of promoting the country's economic development.'[60] The main emphasis of the Plan is much more detailed than the two earlier documents. Again and again the authors make it clear that 'the State will initiate business ventures, through organizations operating on a strictly entrepreneurial basis (Hellenic Industrial Development Bank, Agricultural Bank, etc.), only when the projects are of fundamental importance for the industrial development of the country, and provided that private initiative takes no interest in the matter.'[61]

This Plan varies from earlier studies inasmuch as it gives greatly increased attention to effective implementation and follow up. Greek economic planning as now expressed in official policy is extremely flexible and proposes to operate simultaneously on four levels: (1) flexibility of goals on a national and regional basis relating to changing economic and social conditions; (2) elaboration of specific programmes in a number of basic branches of the economy; (3) policy measures which must be put into specific legislation and administrative order, and (4) the establishment of close relations to the budget and the annual public investment programmes. Greece has an old tradition of a very thorough administrative control over its provinces. The new Plan gives greatly increased emphasis to effective regional development policies and this is indicated by the new office of 'Central Services for Regional Development' within the Ministry of Coordination.

In terms of organization, the Ministry of Coordination is the responsible organ for all planning activities with the 'Central Committee for the Economic Development Plan', a semi-government advisory organ working under the Ministry, but it is too early to evaluate the actual workings of this and other planning organs. Attached to the 1968–72 Plan is a rather detailed and unique list of proposed measures and institutional reforms 'to implement the plan'. Perhaps the weakness of the Plan, like all plans, is the fact that little is said of alternatives, though Greek planning, contrary to its Bulgarian counterpart, possesses sufficient flexibility for changes and adjustments. But again, it is an indicative plan on the macro level without setting parameters of action for individual enterprises with the exception of a few large and very specific projects, e.g. farmers' vocational and technical training programmes, increased social services for regional urban centres and small towns, allocation of funds to the Agricultural Bank for the specific purpose of land consolidation of the many scattered plots (the newly organized Regional Development Councils will be charged with the preparation of detailed plans) and allocation of funds to achieve a more efficient structure of production of Greek manufacturing industries.[62] This is of special importance in view of the transition period (to be completed in 1983) of tariff protection *vis-à-vis* the EEC countries in connection with the proposed entry of Greece into the Common Market. The existing Industrial Development Councils are charged to propagate the advantages offered by certain locations and co-ordinating local efforts, but the responsibility for drafting regional programmes rests with the Regional Development Services. The plan sets out in general statements possible initiatives, especially for the creation of 'industrial poles', industrial zones (industrial estates) without going into detail, leaving this to joint planning of private and government initiators, though it clearly states the responsibility of the State for providing the necessary basic infrastructure.

LONG-TERM DEVELOPMENT STRATEGY

Both Bulgaria and Greece, in their post-war development strategy, gave major attention to reduce socio-economic spatial inequities among the various regions of their respective countries. Two, usually simultaneous strategies were emphasized in these countries. One is the planned emphasis on a more equitable distribution of production throughout the country.

This generally meant a more even dispersal of investments among the underdeveloped areas of the country and was planned, though by no means perfectly, in Bulgaria and consisted of encouragement of private investors and/or direct regional or central government initiative in Greece. Two less controllable factors were more pronounced in Greece than in Bulgaria. These included various aspects of demographic changes, such as population mobility within the country (agriculture to industry, rural to urban, low to high productivity occupations), and emigration to foreign employment opportunities, mainly, but not exclusively, from underdeveloped regions. For Greece this emigration, especially pronounced between 1960 and 1966, was mainly to the Federal Republic of Germany (West Germany).

The problem of producing greater regional equality, besides the large investments needed, raises some very fundamental strategic and tactical questions, which in turn are closely tied to the question of limiting and thus stabilizing regional population mobility, and the decision-making objectives such as the allocation of scarce resources. In Greece, as will be shown later in some detail, one result of the great regional inequities has been the unplanned population movements towards two urban concentrations, Thessaloniki and Athens, especially the latter,[63] and large-scale permanent and temporary emigration.

Closely connected with the availability of funds for investment and a source needing relatively little input in comparison with output expected, is tourist income. Such income has been on the increase in most European countries, including the socialist countries.[64] The development of the tourist potential has equal importance in both countries. The emphasis lies in increasing facilities, improved services, both in tourist places and on the highways leading to the resorts (network of gas and repair stations), and the building of transport facilities, including modernization of the highway network. Greece has an old tradition in this area. The many beauty spots and places of interest due to their archaeological importance, have for many years attracted thousands of foreign tourists, but the potential has by no means been exhausted. Only after the mid-1950's was governmental initiative responsible for greater emphasis on attracting foreign tourists. Funds have been set aside for the development of an improved and greatly enlarged first-class road system, the building of new, and the modernization of old, hotels, etc. The 1968–72 Economic Development Plan gives considerable attention to the achievement of an accelerated growth rate in the tourist sector. The Hellenic Tourist Organization (HTO) in 1968 received the additional responsibility of planning and designing the necessary investments, as well as advertising the attractions. Even though foreign arrivals tripled between 1960 and 1966 – 344,000 to about one million – (an average annual rate of increase of 20%) and foreign exchange receipts rose from $49 million in 1960 to $144 million in 1966 (they declined to $127 million in 1967), Greece in comparison to some other Mediterranean countries (Italy, Spain) is behind in both input (investments) and output (foreign exchange income). That there exists a great tourist potential in Greece is without question. Constraints of a political nature (political stability) will undoubtedly have a greater impact on the future development than domestic economic decisions.

Bulgaria too has given greatly increased attention to this branch of the economy, especially to its potential foreign exchange earnings and the impact such earnings have on the whole development process, including providing the government with greater freedom in its foreign exchange allocations. A central tourist agency, Balkantourist, is both charged with planning and allocation, including advertising, and serves as the main link in the country's overall planning objectives. Tourist spots, especially the Black Sea Coast, received considerable investment, and today offer over half of the facilities available for foreign tourists. Several new towns have been established. The number of foreign tourists increased from 200,602 in 1960 to 1,752,214 in 1967. Figures of actual earnings are not available.

BULGARIAN DEVELOPMENT STRATEGY

Emphasis during the 1950's in its drive for rapid industrialization was largely directed towards existing industrial and handicraft centres, with a few new developments (Dimitrovgrad, the large extension of the iron and steel industries in Pernik and later Kremikovci) receiving the remaining investment funds. This resulted in 'spatially inelastic plans'.[65] Starting in the early 1960's the emphasis was shifted to industrial dispersion, to the erection of numerous smaller plants located in smaller towns and central market places. A number of industrial complexes with specialized production have been formed in the past twenty years and new poles of activity are now being linked to various existing territorial concentrations of production. Obviously, Bulgaria's better raw material endowment attracts new industries. Cities such as Sofia, Plovdiv, Varna and Dimitrovgrad, a new city specializing in chemicals, cement and some light industry, have important industrial concentrations. New industrial locations such as Kremikovci with its newly integrated iron and steel works, Reka Devnja a new chemical complex, and others, act as focal points for priority growth zones or complementary growth zones.

These poles of industrial activity left an important impact on the traditional pattern of rural life with large-scale population migrations and the resultant rapid growth of many cities and rural market towns (Figs. 11.3 and 11.4).[66] While urbanization

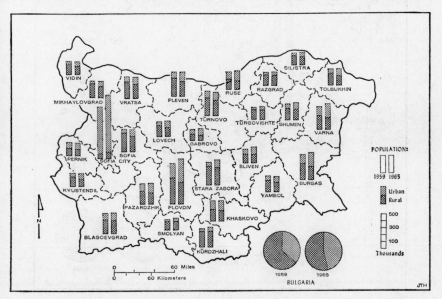

FIGURE 11.3 – DISTRIBUTION OF URBAN AND RURAL POPULATION IN BULGARIA, 1959 and 1965

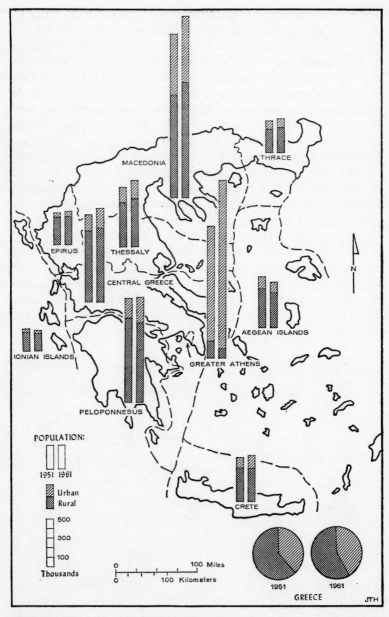

FIGURE 11.4 – DISTRIBUTION OF URBAN AND RURAL POPULATION IN GREECE, 1951 and 1961

and concentration is less pronounced in Bulgaria than in Greece, Sofia had one-eighth of the total population of the country in 1967 and 20% of its industrial production (33% in 1939!). In 1966, 47·6% of Bulgaria's population was classified as urban, as against 22·6% in 1939, but included under urban classification are a number of so-called 'urban type localities'. There were 41 with a total population of 206,990 in 1965. Between 1948 and 1966 roughly 100,000 people per annum (1·77 million) mostly from agriculture moved to industries, construction and services. Cities added, between 1946 and 1966, 1,657,000 people from the countryside and 560,000 in natural growth. This process of urbanization continues with emphasis of growth on smaller cities and towns, though the large agrarian population reserves will in the near future be a thing of the past.[67]

The formation of economic regions,[68] oriented towards a pole of development, has been rapid in the post-war period in Bulgaria. It was encouraged by the investment policies of the government giving top priority to new industries and, to a lesser degree, to the expansion of some already existing ones. According to Bulgarian geographers, economists and planners,

> the rational territorial distribution of the productive forces in accordance with socialist principles is one of the factors for raising labor productivity in industry and agriculture, while economic regionalization is one of the means for improving the planning process and the territorial organization of social problems.[69]

It is generally agreed that regional divisions to be of value for integrated regional planning should define (1) areas having a growth potential but insufficient local labour forces, (2) areas where the labour forces exceed growth potentials, and (3) areas with unbalanced economic activity, in part due to failure to utilize available local resources.[70] Today spatial differences in the level of economic activity must be carefully studied for planning purposes in order to give maximum advantage to planning authorities in determining possible areas of future growth, so-called underdeveloped or backward regions. Bulgaria has no programme which specifically stresses regional planning strategy, and the only reference in the very detailed

Resolution of the Plenum of the Central Committee of the Bulgarian Communist Party[71] is found in item VI, 'Directions for Raising the Living Standard of the People', indicating the intention

> to introduce modifications in the present policy of shifting and productive forces, by giving more attention to the line of building appropriate industrial enterprises in the small towns and bigger villages. To launch on a course to building agrarian-industrial complexes in the peasant regions... to work out the fundamental lines for the development of the national economy by *okruga* and regions for a period of 10–15 years...

Earlier in the Resolution, the State Committee for Planning is advised to assist local People's Councils to develop initiative in using budgets of the People's Councils to create conditions for the efficient development of the productive powers of the territory of the *okrug*.

Academicians in their writings have expressed great concern for 'rational distribution of production and consumption' in the country. Numerous publications indicate concern for regional distribution of production in less-developed regions and the impact of new economic connections and territorial combinations. The Marxian regional development theory is of recent origin and many of the principles and problems therefore are debatable. The whole discussion about territorial-production complexes and their 'regionalization' is still open for argument and in many ways is purely academic, controversial, and at times borders on 'nit-picking'.[72] Iordanov and Marinov, in their discussion, cite the creation of the industrial complex 'Maritsa-Iztok' which produced its own hinterland for the recruitment of its labour force and supply needs. Other similar developments can be cited. In some cases economic connections are formed independent of existing *okrug* borders.[73] Even Iordanov and Marinov come to the conclusion that 'socialist industrialization has not reached the stage of complexity and combination' to make a correct assessment for determining the scope of economic regions.[74]

While not specifically stressing regional developments and spatial differences in the country in its Plans, Bulgaria gives special attention to the spread of economic activity in certain less developed regions such as the Black Sea Coast, which in this process is deprived of its peripheral character. Two new growth poles have already been identified: Varna-Devnja and Burgas-Yambol, which ultimately will combine with nearby territorial organizations. The Soviet Union is evidently greatly interested in the development of these regions and has made some investment contributions. The earlier-mentioned Burgas refinery and petrochemical plant, which has a capacity of 6,000,000 tons, relies on Russian coal.[75]

Perhaps at the conclusion of this brief analysis mention should be made of Bulgaria's close integration with the Soviet Union and her increased reliance on Russian sources for much-needed raw materials. Many of the new fairly large-sized industries cannot be supplied from Bulgaria's own raw materials. Almost all newly constructed power stations are now designed to run on Soviet coal; the Burgas petrochemical plant and refinery, with an ultimate capacity of 6,000,000 tons, will have to be supplied entirely with Russian oil. The impact of this dependency on essential raw materials will obviously tie the Bulgarian economy even more closely to that of the Soviet Union. It also will affect any progress towards a more realistic pricing system which is at the core of the country's economic problems and is essential for successful economic reform.

GREEK DEVELOPMENT STRATEGY

It has been said, and not without reason, that each region of Greece has its own problems posed by the physical environment, history and tradition. Contributing to its problems are the various periods of liberation and political association of the regions, and the strong centralized administrative system with its concentration of all related activities and functions in Athens. The problem of spatial inequities is one of Greater Athens, Central Greece, and Euboea considered the leading or most advanced regions, containing less than one-fifth of the country's total area and 34% of the population, versus the rest of the country. Greater Athens alone has 22% of the total population of Greece (8,388,533 in 1961). A recent OECD report classified the remainder of the country as 'lagging or underdeveloped regions'.[76]

Table 11.3 cites figures for the national product and *per capita* gross regional product in its regional distribution. It is evident that the position of Central Greece, including Greater Athens, as compared with the rest of Greece, is extremely disproportionate as a source of national income, and that the *per capita* income is two and a half times higher than the average income for the country, perhaps as much as five times that of some northwest mountain communities. Moreover, according to these compilations the purchasing power of the population of the capital was 53% above that of the country as a whole. *Per capita* income in the leading region of the country was more than 33% above the national average in 1965, with all other regions having a *per capita* income below the national average, and much variation between, e.g. *per capita* incomes are 93% of the national average in Macedonia, 72% in Thrace and 57% in Epirus.

The spatial differences are accentuated by a very sharp division between the coastal plains and the mountain regions, very limited power and mineral resources, and by the dominant economic position of Athens in nearly every branch of the economy, followed by Central Greece, and, lagging far behind, Northern Greece. The distance to important sources of industrial raw materials or significant markets, and the small size of her productive potential and markets do not make her attractive to international transport. The unfavourable geographical location of Greece casts a shadow on any large-scale plan for the economic development and the mitigation of spatial inequities. The OECD in a recent study flatly stated that 'structural problems thus lie at the very core of all the future changes and development in Greece'.[77]

Athens, in addition to its heavy population concentration, accounted for 68% of the total urbanization movement in Greece between 1951 and 1961,[78] and also has more than half the manufacturing industries of the country, accounting for roughly 42% of the employment and 57% of the value in manufacturing industries (1961).[79] Nine per cent of the employment is in Thessaloniki with 6% of the manufacturing industries and about 10% of the manufacturing value and 24% of the employment in other towns with a population over 10,000. Forty per cent of the labour force in the secondary

TABLE 11.3 – *Greece: Regional Differences*

	Total Greece av.	Greater Athens (nomos of Attica)	Rest of Central Greece and Euboea	Peloponnesus	Ionian Islands	Epirus	Thessaly	Macedonia including Thessaloniki	Thrace	Aegean Islands	Crete
Percentage of increase in population 1951–1961	9.9	34.4	6.9	−2.9	−7	6.7	10.5	11.2	5.8	−9.7	4.6
Urban population, percentage of total, 1961	43.2	100.0	20.6	24.2	14	18.6	27.6	36.1	26.5	24	26.8
Illiteracy (percentage of total population 10 years old and older), 1961	17.8	10.1	21.8	19.5	26.6	22.4	21.7	16.9	29.5	18.7	18.4
Per capita consumption of electric energy in Kwh, 1966											
(a) Total	579	1,292	960	282	123	92	274	490	74	145	157
(b) In households	146	492	42	50	45	32	44	76	22	50	47
Drinking water installations in homes (percentage of total number of households) 1961	39.2	78.1	24.8	25.9	20	17.8	24.6	32.5	21.5	24.9	27.9
Per capita gross regional produce, 1965	100	152.6	97.7	88.9	73.3	61.7	73.0	86.9	65.1	79.6	75.3

Sources: Statistical Yearbooks of Greece; Reports by Commercial Bank and National Banks of Greece; OECD, *Economic Surveys* – Greece; Ministry of Coordination, *The Five Year Programme of the Economic Development of Greece, 1960–1964*, Athens, 1960; Centre of Planning and Economic Research, *Draft of the Five-Year Economic Development Plan for Greece (1966–1970)*; Ministry of Coordination, *Economic Development Plan for Greece, 1968–1972*, Athens, February, 1968.

sector, two-thirds of the Greek labour force employed in firms with more than 100 wage or salary earners, and 43% of the active population employed in commercial occupations also lived in Greater Athens during 1963-4. The *nomos* of Attica (largely Greater Athens) accounted for 41% of the Gross Domestic Product in 1962. The ports of the Athens area unloaded 87% of the Greek imports in 1965. One-half of all the industrial employment is thus concentrated in the two major cities. The nine departments (*nomes*) located along the Athens–Thessaloniki–Kavala axis together, according to the earlier-cited OECD report, 'account for nearly two-thirds (64·2%) of the persons employed in the secondary sector'. In Greece 31 out of 55 cities (three-fifths of the total) have shown an increase of more than 10% in the period 1951-61; half of the population lives in 5% of the area of the country (see Fig. 11.4). Central Greece and Euboea, with 13% of the country's total area, have one-third of the population. Spatial differences in nearly every index are dramatic, considering the size of the country.

Another problem related to the great spatial inequities and their impact on economic development of the country is the long-term effect of emigration, both temporary and permanent.[80] While in its short-term impact emigration has a favourable effect on the Greek economy in terms of reduction of unemployment and underemployment, the importance of emigrant remittances, in its long-term impact resulted in a permanent or temporary loss of some 250,000 people in manpower over the last 10 years. This was especially serious because 85% of the gross number of the emigrants between 1961 and 1966 were in the most productive age groups, between 15 and 30 years, and between 9·2% and 10·9% of those came from the 3 northern *nomes* Macedonia, Epirus and Thrace. No country's economy can afford the loss of such vital manpower over any length of time, and it is for this reason[81] that serious labour shortages, especially at peak agricultural production periods, have already occurred. It is also for this reason that various Greek governments have encouraged greater specialization in agriculture (by abandoning its support of wheat), including the cultivation of labour-intensive crops. But it must be recognized that important political constraints are placed on any government seeking a broad-based political support, especially among the farmers.

It took a long time before the structural inequities of Greece were considered serious enough to finally merit special attention. In spite of numerous studies, mostly by international organizations, little was done within Greece to deal with this problem and point up its urgency. The 1960–4 Plan mentions the problem without implementing its recommendations, and regional feasibility studies are absent. The earlier-mentioned planning regions, with some exception, were largely paper organizations. Only with the promulgation of the 1968–72 plan by the new government was serious attention given to the regional inequities. A regional strategy with long-term goals is now part of the government's policy.[82] To enable implementation of the Plan a reorganization of the government apparatus was undertaken to allow it to give greater attention to this serious question. The tools are now available, but execution depends on the men and on government initiative. It is too early to say if all the paper efforts will materialize in a real improvement in Greece's regional structure. This is perhaps its last chance. The Plan itself is very specific and consists of 27 chapters. It analyses the underlying factors for the regional inequities which it traces specifically to the low productivity in agriculture (in part due to the great farm fragmentation) and the 'concentration of industrial and other high-productivity activities in Greater Athens'. It proposes both a national and a regional strategy in attacking the basic structural problems.

One thing should be emphasized here concerning Greek development policy. Proposals for implementation of administrative measures are essential for the success of the Plan. Therefore, it is hoped that by establishing and naming specific 'planning regions', an appropriate organization of a regional administration and hierarchy, an expansion of the role and jurisdiction of the newly organized Regional Development Councils (in part brought about by the reorganization of the former Regional Development Services), and the preparation of feasibility studies, ultimately the structural problem can be successfully attacked and ameliorated. Regional strategy also calls for close co-ordination with public investment programmes focusing on greatly increased investments in the infrastructure in carefully selected industrial areas and zones linking isolated, though potentially important, regions. The whole regional plan is based on voluntary co-operation, but for the first time a detailed implementation of the needed steps has been provided. On a successful solution of these steps may well depend the success of the Plan. The association of Greece with the countries of the Common Market could offer the needed incentives for accomplishing these goals.

Among the key programmes are two of vital importance. They concern the broadening of the industrial base of the country. Two new industrial axes are to be planned: (1) the Western Greece axis between Kerkyra and Pylos and Calemai with close connections with Taranto and Bari, Italy. This is considered the EEC connecting link with Western Europe; (2) a series of industrial poles at important urban centres on the main North–South transportation artery along the coast north of Athens via Lamia, Larissa and Katerini to Thessaloniki and Kavalla in Northern Greece, and southwest from Athens to Patras, connecting it with the western industrial zone.[83] Obviously, careful planning is needed to realize such an ambitious project. In addition, Greece's unfavourable geographical location offers a real obstacle to the development of additional large-scale industrial poles. Disagreements among economists and politicians about the need and ultimate success of such greatly increased industrialization of the country versus the need for an expansion of its service industries is also sharp.[84] A base of light industries in the smaller urban centres is available. The recently developed industrial complex near Thessaloniki[58], projects under discussion by various Greek developers, and the project initiated under the sponsorship of the 'Committee for the Development of the NATO countries' in northeast Greece[86] provide the base for a long-term structural re-evaluation of the country.

EPILOGUE

From the preceding discussions it has become quite obvious that the regional development processes of Bulgaria and Greece are of quite different nature and are basically handled differently. The over-concentration of almost all economic activity in Greater Athens and the large foreign emigration accounts for an important part of the present underdevelopment and even backwardness of Greece (due to large-scale emigration and migrations important tracts of the country are virtually unpopulated). A carefully administered and enforced macro-national and regional plan is an absolute necessity to control inequities in Greece. Collaboration between private entrepreneurs, foreign investors and the government is the basic policy of the Greek government, and it is the only way to handle this problem. Greece's 'quiet masses' expect an increased standard of living from the results and successful membership in EEC may very well depend upon it.

The problem is an entirely different one for Bulgaria. Great spatial contrasts and the over-concentration of population in one place does not exist. If credit is due, it must be said that Bulgarian planners, in spite of their authoritarian, centralized and highly inflated planning apparatus, have succeeded in spreading their industrial investments to a number of hitherto underdeveloped regions, and thus increased economic activity in many smaller urban and rural market centres all over the country, thereby creating a number of industrial poles thus laying the basis for a number of growth zones. The whole structure of the Bulgarian economy has changed from a predominant agrarian economy to a mixed agrarian-industrial. With more sophisticated production processes and a much more diverse economy as a whole, national and regional developmental processes in the future call for the introduction of greater local and regional (*okrug*) initiative. The highly centralized micro-plan with a complete disregard of the realistic price system is in the process of being overhauled. In a small way this was recognized in the recent Resolution of the Communist Party dealing with fundamental economic and social problems of the country.[87] When the government of Bulgaria now speaks of the need for economic reform, Bulgarian leaders have at least started to think of modernizing their planning institutions and tools, though the initiative and leadership of the Soviet Union will obviously have to be carefully observed.

Bulgaria and Greece over the years have accomplished a surprising similarity in their overall economic growth. This was true in the inter-war as well as post-war years. Though the institutions, and tools and basic planning mechanism are quite different in the two countries and are a clear indication of their different ideological base, their approach to, and the long-term aims of, regional development in finding solutions to the vital problems of increasing and diversifying economic development, including an attempt to ameliorate their spatial inequities, invariably led to the same end product. The great spatial complexities in the physical and cultural environment, the different natural resource endowment of the two countries, a different demographic pattern and the various levels of economic development which the region inherited, have often resulted in similar spatial variations. These variations demand long-term regional development strategy in every country of similar conditions. In addition, a long and at times common history of foreign domination has left its impact and forced constraints on important aspects of decision-making, e.g. an absence of a regional infrastructure of education, such as vocational and technical training opportunities (a problem more severe in Greece than in Bulgaria), low productivity and underemployment in agriculture and high uneconomical military expenditures (one-third to one-fourth of the entire budget in each of the countries). It is only natural, therefore, that their long-term regional development strategy has many common traits, in spite of their different ideology which makes it necessary to approach spatial inequities and regional development from a basically different methodology.

NOTES

[1] Some of the few studies thus far discussing comparative developments in Southeast Europe are George W. Hoffman, 'The Problem of the Underdeveloped Regions in Southeast Europe', *Annals of the Assoc. of American Geographers*, Vol. 57 (December, 1967), pp. 637-66; Zoran Popov, 'Komparativna analiza privrednog razvoja SFR Jugoslavije i NR Bulgarske' (Comparative analysis of the economic development in Yugoslavia and Bulgaria), *Ekonomist*, Vol. XX (1967), pp. 294-320, and Benjamin Ward, 'Capitalism *v.* Socialism: A Small Country Version', unpublished paper, Department of Economics, University of California, Berkeley, 1967.

[2] Charles and Barbara Jelavich, *The Balkans*. Spectrum Books (Englewood Cliffs, N.J.: Prentice-Hall, Inc., 1965), p. 134.

[3] The first political unit receiving autonomy in Southeast Europe was Serbia in 1830, though full independence was not won until 1878. Greece won independence (though at first only for the Peloponnesus and Athens with Central Greece) in 1832, though it took until 1913 before the present mainland territorial extent was reached. The Dodecanese Islands were acquired only in 1947. Two principalities, those of Walachia and Moldavia, comprised the Romanian state when it became autonomous in 1861 and fully independent in 1878. A large autonomous Bulgaria was created by the Treaty of San Stefano in 1878, though the subsequent Congress of Berlin substantially modified the Treaty's provisions. An autonomous principality of Bulgaria with its capital, Sofia (largely the area of northwest and north Bulgaria) and the province of Eastern Rumelia, including Plovdiv and Burgas remaining under the authority of the Sultan, was created. A few years later, in 1885, the two provinces were united into one political unit, though full independence did not come until 1908. Albania, the fifth political unit of the area, was established in 1913 as a result of the final expulsion of Turkey from Europe after the Balkan Wars, though because of World War I, actual independence was not accomplished until 1920. In mid-1970, Bulgaria had 8·6 million people occupying an area of 103,000 sq. ml., and Greece 8·9 million with an area of 130,000 sq. ml.

[4] Norton S. Ginsburg, 'On Geography and Economic development', Saul B. Cohen, editor, *Problems and Trends in American Geography* (New York: Basic Books, Inc., 1961), p. 186.

[5] Norton S. Ginsburg, 'On Regions and other Geographies', *Introductory Geography, Viewpoints and Themes*. Commission on College Geography, Assoc. of American Geographers, No. 5 (1967), p. 107.

[6] Ministry of Coordination, *Economic Development Plan for Greece 1968–1972*, Athens (February, 1968), pp. 7–8.

[7] For a theoretical analysis see John Friedmann, 'Regional Economic Policy for Developing Areas', *Papers and Proceedings, Regional Science Association*, Vol. 11 (1963), p. 41. Friedmann uses the term 'spatial organization' rather than 'regional organization'.

[8] Niles M. Hansen, *French Regional Planning* (Bloomington, Ind.: Indiana University Press, 1968), p. 7.

[9] For a discussion of the economic development of the countries of Southeast Europe, including Bulgaria and Greece before World War II, see the two valuable chapters on 'The State and Industrialization', and 'The Pace of Change in the Economic Structure of the

Balkans', in Nicolas Spulber, *The State and Economic Development in Eastern Europe* (New York: Random House, 1966), pp. 12–88. I have leaned heavily on this background material in the following discussions. Other important background material for the discussions in this section are from André Blanc, *L'Economie des Balkans* (Paris: Presses Universitaires de France, 1965); Charles and Barbara Jelavich, eds. *The Balkans in Transition* (Berkeley, Cal.: University of California Press, 1963); Leo Pasvolsky, *Economic Nationalism of the Danubian States* (Washington, D.C.: The Brookings Institution, 1928); Doreen Warriner, ed. *Contrasts in Emerging Societies* (Bloomington; Indiana University Press, 1965); Theodor D. Zotschew, 'Spezielle Aspekte der Wirtschafts- und Sozialentwicklung seit der Jahrhundertwende in Bulgarien', Lecture at 10th Internationalen Hochschulwochen der Südosteuropa-Gesellschaft, Hamburg, October, 1968, mimeographed, 18 p., and two papers by the author: 'Die Umwandlung der landwirtschaftlichen Siedlungen und der Landwirtschaft in Bulgarien', *Geographische Rundschau*, 17 (September, 1965), pp. 352–61 and 'Transformation of Rural Settlement in Bulgaria', *Geographical Review*, Vol. LIV (January, 1964), pp. 45–64. Also Adamantios Pepelasis, 'Greece', in A. Pepelasis, Leon Mears *et al.*, eds., *Economic Development* (New York: Harper & Row, 1961), chapter 16.

[10] Spulber, *The State and Economic Development in Eastern Europe*, op. cit., pp. 62–3.

[11] *ibid.*, pp. 63, 69.

[12] Greece as early as 1893 had foreign debts comparable to 35% of her foreign receipts.

[13] A. D. Sismandidis, 'Foreign Capital Investment in Greece', *Balkan Studies*, Vol. 8 (1967), pp. 339–52. The present administrative structure in Greece with its highly centralized decision-making bodies offers a great constraint to a rapid regional economic development. This administrative structure was in many ways copied from the French administrative system.

[14] Crop yields generally were unreliable, modern fertilizing techniques were hardly known and rainfall was insufficient in key grain-growing regions. See S. H. Cousens, 'Changes in Bulgarian Agriculture', *Geography*, Vol. 52 (January, 1967), pp. 12–13; Adamantios Pepelasis and Kenneth Thompson, 'Agriculture in a Restrictive Environment. The Case of Greece', *Economic Geography*, Vol. 36 (April, 1960), pp. 145–57, and Kenneth Thompson, *Farm Fragmentation in Greece*. Research Monograph Series No. 5 (Athens: Center of Economic Research, 1963).

[15] *The Statistical Yearbook of the League of Nations* lists the following

figures for the distribution of the labour force in Bulgaria and Greece in 1920: mining and industry 8·1% and 16·2%, trade 2·7% and 8·1%, other (mostly services) 6·8% and 26·1%.

[16] Dimiter A. Jankoff, *Labor Cooperative Farms in Bulgaria*, Free Europe Committee, Inc., Mid-European Studies Center. Mimeographed Series No. 1, 1953; also Andreas Piperow, *Zemedilskoto stopanstvo Na Bulgarija opit za kratka charakteristika* (Agriculture in Bulgaria, a short characteristic), (Sofia, 1939).

[17] Pepelasis, 'Greece', *op. cit.*, p. 509.

[18] Pepelasis, 'Greece', *op. cit.*, p. 510. For a very valuable background study on Greek agriculture bringing the situation up to the early 1960's see Pepelasis and Thompson, 'Agriculture in a Restrictive Environment', *op. cit.*, pp. 145–57.

[19] Spulber, *The State and Economic Development in Eastern Europe*, *op. cit.*, p. 74.

[20] For detailed figures see Greece, *Annuaire statistique de la grèce, 1939*. For a critical analysis of Greece's inter-war economic policies see Adamantios Pepelasis, 'Socio-Culture Barriers to Economic Development of Greece', unpublished Ph. D. dissertation, Department of Economics, University of California, Berkeley, Cal., 1955, pp. 95–106.

[21] French influence had declined and German influence did not exert its influence until 1936.

[22] In 1939 industrial production of textiles, food and tobacco industries comprised 70% of the industrial production in Bulgaria.

[23] In Greece between 1894 and 1922 a large part of her foreign loans went to support her wars. The situation for Bulgaria was not much different.

[24] In the period 1928–50, Athens population increased by 40%, Sofia's by 20%. See F. W. Carter, 'Population Migration to Greater Athens', *Tijdschrift voor Econ. Soc. Geografie*, Vol. LIX (March–April, 1968), pp. 100–5.

[25] Greece's population losses amounted to approximately 7% of the population in the years 1940–5, while at the same time there was a natural decrease of 240,000 births. In spite of the acquisition of the Dodecanese Islands, the total natural increase between 1940 and 1951 amounted to only 0·2% per annum. Another demographic problem was the large increase in urban population which amounted to close to 20% in the period 1940–51. Much of this was connected with the insecurities in the rural areas. Physical devastation was widespread, e.g. all major ports were destroyed, 95% of the railroads, with industry at a virtual standstill, and agricultural production only at 35% of the pre-war level.

[26] Total American aid amounted to over 2 billion dollars. It was largely used to buy food and materials. In addition, this aid contributed between 1948 and 1954 to the addition of 175,000 acres of new land, the productivity of 1·2 million acres was greatly increased by such projects as flood control, drainage, irrigation. Also emphasis was given to the training of thousands of young people from the rural villages and farms and the retraining of farmers in modern agro-technical methods. The American Farm School in Thessaloniki is making important contributions to these programmes of modernization. See also D. J. Delivanis, 'Marshall Plan in Greece', *Balkan Studies*, 8 (1967), pp. 333–8.

[27] It should be pointed out here that the Soviet model that was blindly copied was not a 'direct product of Marxian ideology', but rather 'a collection of *ad hoc* practices that had emerged over the previous quarter of century of Bolshevik rule', Thomas M. Poulsen, chapter 6.

[28] Spulber, *The State and Economic Development in Eastern Europe*, *op. cit.*, p. 80.

[29] Angus Maddison, Alexander Stavrianopoulos, Benjamin Higgins, *Technical Assistance and Greek Development*. Development Centre Studies, No. 6 (Paris: OECD, October, 1965).

[30] Norton S. Ginsburg, *Atlas of Economic Development* (Chicago, Ill.: University of Chicago Press, 1961).

[31] Ward, 'Capitalism *v.* Socialism', *op. cit.*, p. 4.

[32] *ibid.*, p. 5.

[33] Ministry of Co-ordination, *The Five Year Programme of the Economic Development of Greece, 1960–1964*, Athens, 1960.

[34] Benjamin Ward, *Greek Regional Development*. Research Monograph No. 4 (Athens, Greece: Centre for Economic Research, 1962).

[35] *Economic Development Plan for Greece, 1968–1972*, *op. cit.*

[36] Greece officially became an associate member in December 1961. For specific details see the excellent discussions in S. G. Triantis, *Common Market and Economic Development*. Research Monograph Series No. 14 (Athens, Greece: Centre of Planning and Economic Research, 1965), part B, pp. 63–110.

[37] Bernard Kayser *et al.*, 'Développement régional et régionalisation de l'espace en Grèce', *Tiers Monde*, Vol. VI (October–December, 1965), pp. 1003–25, also by the same author 'Pour un inventaire géographique de l'espace grec', *Sociological Thought*, Vol. 1 (January, 1966), pp. 100–10.

[38] Small-scale industries, including handicraft, small crafts, etc., employ approximately one-half of the labour force engaged in

manufacturing, but contribute only one-third of the manufacturing product.

[39] See the critical analysis of the Greek post-war economic problem in Richard M. Westebbe, 'Greece's Economic Development: Problem and Prospects', *International Development Review*, Vol. VIII (March, 1966), pp. 11–16.

[40] Spulber, *The State and Economic Development in Eastern Europe*, *op. cit.*, p. 79, also Friedrich Slezak, 'Geographische Aspekte der bulgarischen Wirtschaftsplanung', *Zeitschrift für Wirtschaftsgeographie*, Vol. 2 (February, 1962), pp. 48–51.

[41] Bulgaria's iron ore is generally of low quality. There also is an insufficient quantity of coking coal, though chromite and manganese are available. With its increasing industrialization and especially the enlargement of the pre-war Pernik iron works ('Lenin works') and the newly built iron and steel works at Kremikovci near Sofia, adjacent to iron ore supplies, vital mineral raw materials are now imported from the Soviet Union.

[42] Ivan Veltchev, 'Teritorialni i strukturni promeni u selskoto na selenice Na Bulgariya prez perioda 1946–1965' (Territorial and structural changes in Bulgaria's rural population during the period 1946–65), *Bulletin de la societé bulgare de géographie*, Vol. VII–XVII (1967), pp. 109–26.

[43] Enčo Draganov, 'Po nyakoi vuprosi na strouktourata na promishlenostta v NR Bulgariya (On certain questions of the structure of industry in the People's Republic of Bulgaria)', *Plan stopanstvo*, Vol. 23 (1968), pp. 3–13.

[44] For definitions of terms, including the planning instruments see the short but very useful study by the late Rudolf Bicanić, *Problems of Planning – East and West* (The Hague: Mouton and Co., 1967), see also G. Filipov, 'Further Development of the New System of running the national economy', *Novo Vreme*, No. 10 (1968), as excerpted and translated in *Bulgarian Press Survey*, No. 688, November 8, 1968. Research Departments of Radio Free Europe, November 8, 1968, pp. 10–19.

[45] Nikolo Popov, 'On the Character of the Economic Reform in the Socialist Countries', *Novo Vreme*, No. 10 (1968), summarized and translated in *Bulgarian Press Survey*, No. 688, November 8, 1968. Research Departments of Radio Free Europe, November 8, 1968, p. 3.

[46] *Decree* No. 69 of June 13, 1963.

[47] These new organs are (1) Scientific-Research and Planning-Economic Institute, (2) Administration for Regional Delineation and Territorial Distribution of productive capacity, and (3) a

Technical-Economic Council. In addition, a Water Resources Committee and a Department for Economic Evaluation were created within the State Committee for Planning.

[48] Abraham Melezin, 'Soviet Regionalization', *Geographical Review*, Vol. LVIII (1968), pp. 593–621, also Richard E. Lonsdale, 'The Soviet Concept of the Territorial-Production Complex', *Slavic Review*, Vol. XXIV (September, 1965), p. 466 for the term 'territorial production complex' (which) denotes a functionally organized area within which economic activities are sufficiently interrelated as to form a single integrated unit. The unity of the 'territorial production complex' is thus based on economic relationships or 'linkages' rather than on a condition of economic homogeneity.

[49] Anastas Beškov, *Volksrepublik Bulgarien, Natur und Wirtschaft* (Berlin: Verlag die Wirtschaft, 1960), pp. 42–5, also R. H. Osborne, 'Economic Regionalization in Bulgaria', *Geography*, Vol. 45 (November, 1960), pp. 291–4. It is not clear what criteria were used in delimiting the new regions.

[50] The literature on how best to demarcate the country is numerous and each geographer seems to have his own pet theory. Tian Iordanov and Christo Marinov, *Ikonomicheska Geografiya* (Economic Geography) (Varna: State Publishing House, 1967), pp. 315–24, present a good summary. Ivan Zahariev, 'Regionalizacja ekonomicza w Bulgarskiej Republice Ludowej' (Economic regionalization in the Bulgarian People's Republic), *Przegląd geograficzny*, Vol. 38 (1966), pp. 611–18, represents an important point of view. In addition, there exist numerous publications arguing for various divisions as well as publications on individual regions within the general six-, five- or three-fold division of the country. A good example is V. Velev, 'Po vuprosa za formiraneta na Yugoiztočen ikonomičeski raion V Bulgariya' (A discussion of the formation of an economic region in southeast Bulgaria), *Bulgarska akademiia naukite* (Sofia), Geografski institut, *Izvestiia*, Vol. IX (1965), pp. 223–249.

[51] Nikolina Ilieva, 'Planning and Specialization in Bulgarian Agriculture', *Ikonomicheska misul*, No. 6 (1965) as translated in *Eastern European Economics*, (September, 1967), pp. 27–38.

[52] Filipov, 'Further Development of the New System of running the national economy', *op. cit.*

[53] The July plenum never wavered on the compulsory character of the state plan, which it considered 'the central problem of planned management under socialism'. Filipov, 'Further Development of the New System of running the national economy,' *op.cit.*, p. 13, criticizes both the Yugoslav and Czech reformers inasmuch as it was 'based

on the concept that enterprises should not be given compulsory indices from the center. According to them, the state plan should be just a forecast of society's development. As far as enterprise plans are concerned the above theories provide that they must be made freely, and based on the state of the market. Our experience has shown that in working out their plans the enterprises are not able to estimate conditions correctly or to determine the public's needs. These needs and conditions can be determined by the state and in order that the whole society may develop harmoniously, they should be compulsory for all members of the society ... A fundamental innovation of our practice of planning is the working out of programs for the complex development of the specific branches and production processes, regardless of their departmental subordination. With the assistance of these programs it will be possible to discover the actual opportunities for and means of more speedily catching up with the world level in a given branch or production process. The programs for complex development will be the real basis for working out and economically justifying the central state five-year plan. In this context, the state plan will be regarded in the future as a system of mutually-connected, complex programs for various branches and for the national economy as a whole.' See also *Bulgaria: The 1969 Economic Plan*, Research Department of Radio Free Europe, Bulgaria 6–21 (January, 1969), 10 p. and Filipov Grisa, 'Ponatatushno razvitie na novata sistema na rukovodstvo na narodnoto stopanstvo' (Further development of the new system of management of the national economy), *Novo vreme*, Vol. 44 (1968), pp. 42–5.

[54] 'Resolutions of the Plenum of the Central Committee of the Bulgarian Communist Party', July 24–6, 1968, from *Rabotnicheski Delo*, July 27, 1968, translated.

[55] 'Experimente auf Kosten des Planungsprinzip? Reformprobleme der bulgarischen Wirtschaft', *Wissenschaftlicher Dienst Südosteuropa*, Vol. XVI (August, 1967), pp. 131–40; Rumen D. Yanakiev, 'Wirtschaftsplanung und technischer Fortschritt in Bulgarien', *Österreichische Osthefte*, Vol. 10 (March, 1968), pp. 96–104, and *Bulgaria: The 1969 Economic Plan, op. cit.*

[56] The Ministry of Coordination is superior to all other economic ministries, has broad powers and is in overall charge of economic planning. The Ministry is considered one of the key organs of the State.

[57] This programme 'fixes objectives of growth considered mutually compatible, establishes a program of investment, and defines an administrative monetary, financial and social policy which the government has the intention of carrying out in order to

achieve the general objectives'. As translated in Rudolf Bicanić *Problems of Planning – East and West, op. cit.*, p. 34.

[58] Peter O. Steiner, 'Economic Prospects and Planning in Greece: An American's View', *Balkan Studies*, Vol. 8 (1967), pp. 353–63, discusses this plan and is very critical that no alternatives were discussed and evaluated and gave little attention to the long run choices.

[59] Chrysostomos Kosseris and Elizabeth Clutton, 'A Review of the Development Plan of Crete 1965–1975', *The Geographical Journal*, Vol. 134 (March, 1968), pp. 64–9.

[60] *Economic Development Plan for Greece, 1968–1972, op. cit.*, pp. 7–9.

[61] *Economic Development Plan for Greece, 1968–1972, op. cit.*, p. 8.

[62] Funds will be channelled through the various semi-government institutions.

[63] Greece urban population: 1926 – 33%; 1946 – 37%; 1966 – 43%, of which Athens and Thessaloniki contain approximately 66%. For details see Bernard Kayser, *Géographie Humaine de la Grèce* (Paris: Presses Universitaires de France, 1964) and John Campbell and Philip Sherrard, *Modern Greece* (New York: Frederick A. Praeger, Inc., 1968), esp. chapter 11.

[64] Wigand Ritter, *Fremdenverkehr in Europa*. Europäische Aspekte, Schriftenreihe zur Europäischen Integration, Reihe A: Kultur, No. 8 (Leiden: A. W. Sijthoff, 1966), especially pp. 186–91.

[65] F. E. Ian Hamilton, *Yugoslavia Patterns of Economic Activity*, chapter 6, p. 32 (New York: Frederick A. Praeger, 1998).

[66] Ivan Veltchev, 'Osnovni problemi na Geografiya na gradovete v NR Bulgariya' (Basic problems in the geography of cities in Bulgaria), *Bulletin de la Société Bulgare de Géographie*, Vol. VI–XVI (1966), pp. 99–105.

[67] Ljubomir Dinev, 'Changements dans la Géographie de la population et des localités en Bulgarie après la deuxième guerre mondiale', *Mélanges de Géographie*, M. Omer Tulippe (Brussels: editions J. Duculot, S. A. Gembloux, 1968), Vol. 1, pp. 552–9, specifically pp. 553 and 557. Bulgaria's urban population was 21% of the total in 1926, 25% in 1946 and 44% in 1965, of which two cities, Sofia and Plovdiv, contained about 46% of the total urban population of the country. Also Todor D. Zotschew, 'Wandel und Wachstum der Bulgarischen Volkswirtschaft', *Bulgarische Jahrbücher*, Vol. 1 (1968), p. 262.

[68] See the discussions in Hamilton, *op. cit.*, pp. 421–2. Though the author uses Yugoslavia as an example, his theoretical discussions about economic regions, the intentions of regionalization as delimiting 'areas with similarities and linkages in their economic develop-

ment "potentials" rather than existing structure' can also be applied to Bulgaria. Also Bobri Bradistilov, 'Efektivnost na teritorialnata organizatsiyn na froizvodstvoto' (Effectiveness of the territorial organization of production)', *Novo vreme*, Vol. 44 (1968), pp. 56–67.

[69] Ivan Zahariev, Dobri Bradistilov *et al.*, 'Problems of the General Economic Regionalisation of Bulgaria', *Ikonomicheska misul* 6 (1962), as translated in *Eastern European Economics*, III (Fall, 1964), p. 3. This problem has been discussed in numerous articles and books by Bulgarian, Soviet and other socialist planners and academicians. See also Melezin, 'Soviet Regionalization', *op. cit.*, Lonsdale, 'The Soviet Concept of the Territorial-Production Complex', *op. cit.*, and the various contributions in *Geographia Polonica* in connection with the work of the IGU Commission on 'Methods of Economic Regionalization'.

[70] Hamilton, *Yugoslavia*, *op. cit.*, p. 322.

[71] Held in July 24–6, 1968. The resolution was published in full in *Rabotnichesko Delo*, July 27, 1968, and is available in English translation.

[72] Besides the earlier mentioned Zahariev, Bradistilov *et al.*, 'Problems of the General Economic Regionalisation of Bulgaria', *op. cit.*, analytical studies are Iordanov and Marinov, *Ikonomicheska Geografiya*, *op. cit.*, Christo Marinov, *Ikonomicheski Raionirane Na NR Bulgariya* (Economic Regionalization of Bulgaria) (Sofia: Bulgarian Academy of Sciences, 1963), and Ivan Zahariev, Dobri Bradistilov *et al.*, Komisiia za izuchavane na proizvoditalnite sili-*Ikonomichesno raionirane na Bulgariia* (Commission for the study of productive forces and institute for economic science – Economic Regionalization of Bulgaria) (Sofia: Bulgarian Academy of Sciences, 1963).

[73] Iordanov and Marinov, *Ikonomicheska Geografiya*, *op. cit.*, p. 319.

[74] Iordanov and Marinov, *Ikonomicheska Geografiya*, *op. cit.*, p. 318.

[75] Christo Marinov, *Socialisticjeski Mezhdubarodni Kompleksi i raioni* (Socialist International Complexes and Regions) (Varna: D'rzhavno Idatelstvo, 1965), pp. 106–7. The author also discusses among others a Black Sea project with participation by the USSR, Romania and Bulgaria. The proposal is not unlike that by the Soviet geographer, Valev, which received considerable criticism from the Romanian officially inspired press, but also privately from some Bulgarians. Monsieur Pierre Yves-Pechoux from the University of Toulouse in a recent communication to the author points out that during a recent visit to the Black Sea he was told by Bulgarian officials that the Burgas refinery expects to rely heavily on coal found

north of Varna. This coal will also supply the new industries in the Varna–Devnja district.

[76] Maddison, Stavrianopoulos, Higgins, 'Technical Assistance and Greek Development., *op. cit.*, p. 9.

[77] OECD, *Manpower policy and problems in Greece*. Reviews of Manpower and Social Policies, No. 3 (Paris: OECD, 1965).

[78] G. S. Siampos, 'The Trend of Urbanization in Greece', Paper presented at UN World Population Conference, August–September, 1965, Belgrade, Yugoslavia.

[79] George Coutsoumaris, *The Morphology of Greek Industry*. Research Monograph No. 6 (Athens, Centre of Economic Research, 1963) and 'The Location Pattern of Greek Industry', Lecture Series No. 4 (Athens: Centre of Economic Research, 1962).

[80] 36·9% of total invisible receipts in 1966 or $234·9 million, though the 1968–72 Economic Development Plan estimates a considerable reduction in the years to come. See also *Economic Development Plan for Greece, 1968–1972, op. cit.*, pp. 31–3.

[81] *Economic Development Plan for Greece, 1968–1972, op. cit.*, pp. 32–3, Candilis, 'The Economy of Greece', *op. cit.*, pp. 152–60, and Kenneth Thompson, 'Recent Greek Emigration', *Geographical Review*, Vol. 57 (1967), pp. 560–2.

[82] *Economic Development Plan of Greece, 1968–1972, op. cit.*, chapter 6, pp. 37–43. Bernard Kayser, 'Les transformations de la Grèce du Nord', *Bulletin de l'Association de Géographie Français* (May–June, 1964), pp. 59–70.

[83] For a detailed discussion see chapter 9, 'Industrial Policy', pp. 65–81 of the *Economic Development Plan for Greece, 1968–1972, op. cit.*, and a paper submitted by the Greek Government to the International Symposium on Industrial Development, 'Location of national Industry within a wider Economic Context', United Nations Industrial Development Organization, ID/CONF./L/G/63, July 1, 1967.

[84] Triantis, 'Common Market and Economic Development', *op. cit.*, pp. 57–70.

[85] See the discussions in Hoffman, 'The Problems of the Underdeveloped Regions in Southeast Europe', *op. cit.*, p. 664, and 'Thessaloniki, The Impact of a Changing Hinterland', *East European Quarterly*, 2 (March, 1968), pp. 3–9.

[86] Since the publications in Hoffman, 'The Problems of the Underdeveloped Regions in Southeast Europe', *op. cit.*, p. 664, work has further progressed, esp. in making the Meriç navigable along the Greek–Turkish border and protecting large areas from the annual floods.

[87] Resolutions of the Plenum of the Central Committee of the Bulgarian Communist Party, 'Fundamental Trends for the Further Development of the System of Administration of our Society', *Rabotnicheski Delo*, July 27, 1968, translated.

Comments ROGER E. KASPERSON

In my comments on Professor Hoffman's comparative analysis of regional development in Bulgaria and Greece, I shall direct my analysis to Greece for it has been the object of my past research efforts. Three major aspects of economic development arising from Professor Hoffman's analysis are of primary concern: the spatial parameters of development and their implication for national policy priorities, productivity and underemployment in agriculture, and educational and manpower considerations.

Professor Hoffman notes that both Bulgaria and Greece have striven to reduce socio-economic differences among their various regions. In the case of Greece, he notes major regional discrepancies in such variables as *per capita* gross regional product, drinking water installation, *per capita* consumption of electric energy and illiteracy. As even the casual traveller in Greece quickly observes, spatial inequities underlie the extensive rural depopulation prevalent in the Aegean and Ionian islands, Epirus and the Peloponnesus, as well as the strong concentration of population, industry and developmental amenities in the greater Athens metropolitan area. Despite governmental hand-wringing over the oft-dramatic differences indicated so well in the maps in *The Social and Economic Atlas of Greece*, on a variety of socio-economic indices, changes over the last several decades have been in the direction of greater rather than lesser regional inequalities. Given the demographic and political problems intertwined with growing regional discrepancies, the new military regime has, as Professor Hoffman notes, accorded the reduction of such inequities a prominent place in the 1968–72 five-year plan. Yet this decision raises serious issues about the entire developmental process, and the concomitant dilemmas involve basic social and political values.

Perhaps the most basic policy consideration is whether a state should develop programmes designed to stabilize regional population mobility in an area with only limited potential for structural change and markedly increased *per capita* productivity. Would a more efficacious strategy shift populations both occupationally and geographically to achieve a more optimal distribution tailored to greatest projected economic capability and growth? In the event that the population stabilization strategy is chosen, is there any insurance that large-scale regional investment will produce the desired behavioural objectives?

Maximizing overall national development and reducing spatial inequities may well be incompatible goals. To begin with, the geographical pattern of development at any point in time is anything but random; rather, it reflects such variables as the localization of natural resources, taste or living preferences and the accumulated outcomes of previous capital investments. A policy aimed at greater equalization will involve a number of problems. Such a goal, for example, flies in the face of attempts to maximize total benefits accruing from regional specialization and economics of scale and agglomeration. In the long run, the overall effects may be detrimental to the less developed as well as the more developed regions.

Development processes aimed at greater regional equality will unquestionably produce serious short-run occupational and social dislocations for traditionalists who are relatively poorly educated and unskilled. The problem is that those populations which most need change and readjustment are precisely those lacking the qualities required to make change a success. From a national perspective, investment in underdeveloped regions keyed to moving people out of agriculture and into other sectors of the economy may actually produce a time lag before returns are realized; consequently in the short-run increased, rather than reduced, regional inequities may result. Moreover, the time lag involved in investment occurs during a period when capital is particularly scarce and needed for priming the economy. Investment during this time-lag period will inevitably reduce aid to the developed regions, where an immediate return is more possible.

If, on the other hand, the Government chooses to maximize total development at the price of further spatial inequities, other related problems will also arise. Certainly there will be demographic repercussions in the form of continued large-scale rural-to-urban and interregional migration. Few economists would dispute that the heavy concentration of economic activity in the Athens metropolitan area detracts from an optimal national distribution, though they probably would disagree vehemently as to extent and the most satisfactory corrective strategy. Development processes accentuating regional inequalities will doubtlessly increase 'overconcentration' in Athens and farm labour shortages in rural areas. As the 1964–5 campaign platform and programme of George Papandreou indicated, severe inequities in income and other rewards create regional political tensions. Finally, interoccupational as well as interregional population movements will be necessary, yet the skills of the rural population for modernization and the ability of other economy sectors to absorb these people will probably not be commensurate to the task.

As we have noted at other points in this conference, the issues involved in regional development will be settled by a political decision. Having said this, I see no reason for us to adopt the rather undignified posture of the ostrich. There are concepts and tools for us to examine processes in a political economy. Elsewhere I have proposed a general model of environmental stress management which deals with such variables as sets of stresses acting upon the political system, vehicles of articulation, goals, perception and constraints. In reference to the present issue suggested by Professor Hoffman, I would suggest several concrete political variables in Greece. In the public allocation of scarce resources for development, I believe that two decision-making objectives may be recognized: first, all Greek governments strive actively to enlarge their base of electoral support, and the spatial allocation of resources is a potent tool. The reduction of emigration and depopulation and greater equality in benefits are important political issues in Greece. In 1964–5, Papandreou gained heavily among poor farmers in Northern Greece, a traditional base of conservative support. The new regime seeks to broaden its rather limited support in this direction. Second, the present Greek government is involved in a search for legitimacy – here the manipulation of symbols is important.

A final political variable may be thought of as a decision-making constraint. The established set of strategies and tactics arising from past experience and consonance of individual acts with a broader ideology restrict the range of choice for any political body. Certainly the unwillingness of the Government to intervene more actively in the economy and the social structure can be traced in part to the conservative position and beliefs of the men in power.

In both Bulgarian and Greek regional development, extensive underemployment and low labour productivity in agriculture constitute sharp constraints. In 1961, the average output per person employed in agriculture (16,460 drachmès), when compared with that of industry (37,120 drachmès), provides a vivid surrogate of agricultural under-employment. The Ministry of Coordination estimated that with minor changes between 400,000 and 700,000 agricultural workers could be shifted to other economy sectors without any loss in agricultural output (47% of the Greek labour force is still in agriculture – about the same as pre-World War II Greece). Professor Hoffman has indicated the causes of this underemployment and low productivity – the extreme shortage of land (3·5 hectares per family in 1961), land fragmentation (an average of 7·5 separate plots per farm), low mechanization, and shortages of capital for agricultural investment. I should like to add the government intervention into social structure will unquestionably be required to deal with these problems; the role of inheritance and dowry practices, condoned by law, in farm fragmentation is an example. Dowry practices are particularly a problem for their detraction of funds from investment as well as the dispersal of plots.

Yet the shift to higher productivity levels and more efficient employment patterns will not come easily. The migration to the cities has already outstripped the growth of economic opportunities in urban centres and towns. The reasons for this are cultural as well as economic – many Greeks long for a position ensuring καθαρή δουλεια (clean work) and a white shirt and tie. In my field research in the Dodecanese Islands, I found that few young people aspire to be farmers. The rising level of expectations involves amenities concentrated in urban environments. Clearly either the socialization of values must be changed or the system of rewards.

This last point underscores the critical role of education and manpower considerations in the developmental process. The 'overconcentration' of social attributes in Athens is also apparent in education. In 1961, one-third of all gymnasium students and two-thirds of all technical and vocational students were concentrated in Greater Athens. Regional development must also provide an infrastructure of education and appropriate manpower to meet developmental needs. Yet the percentage of the national income devoted to education is still prohibitively low. In 1961, the OECD *Mediterranean Regional Project* indicated that Greece suffered from a shortage of 7,000 teachers. The problem has been especially severe in technical and vocational training; there are still few full-time technical teachers and most lack even basic qualifications. Given the lack of population density in many regions required for efficient gymnasiums and vocational schools and the problems of adequate incentives for teacher and worker alike, if regional development is to involve rapid structural and skill changes, substantial progress must occur in educational and manpower modernization.

Remarks

Dr Thompson in commenting on Professor Hoffman's paper said that if the role of decision-making is to be irrational, it (the decision) will be made in Greece. He pointed out that the Greeks are sentimental and have a rather clear image of the past and this will inevitably colour their thinking because the past is so important in their history and tradition. There is a problem of administrative authority: for example, the *nomar* is like a petty prime minister at a lower level. He brings his family and cousins along with him on his trips. This situation creates quite a problem. If there was a good centralized administration without this sort of thing going on it would be much better.

Professor Baxevanis took issue with the statement that a larger Athens is bad, saying that there is no future in mountain tops; that if he were a Greek Prime Minister he would be forced to evict large numbers of people, and would move them to the cities. It should be pointed out that two-thirds of the total population lives within 20 miles of water in Greece.

If there were no industrial investment model, the cities involved would be called 'catchment' basins. Labour would be encouraged to settle in this area. The particular problem is how to get the political allegiance of those farmers in inaccessible and poor areas.

Professor Sanders said that he had hoped Professor Hoffman would say that in Bulgaria one cannot buy land, whereas in Greece one can. In Bulgaria this money (which in Greece would be used to buy land and/or build apartment houses) is available for consumer goods. The economic effects of this single fact might be demonstrated in the economic prosperity of the farmers.

Professor Kristof then brought up the question of dowry. He also spoke of Romania, and said that there is also the question of regional imbalance. The Romanians have reorganized their whole territory; they have tried to put industry into little towns, putting proportionately more investment into this than into housing, but that certain social problems had not been faced.

Professor Kristof continued that he had noticed in Romania a tendency towards autarky, because, in spite of the national plan, regular local authorities control the locally produced economy. If one wants to buy cabbages and fruits, one has to take into consideration the fact that the prices are the same all over the country, so they have to be produced locally. Thus honey, chickens, etc., have to be produced or else they will not be available for the local market. This makes for a very uneconomical kind of production. Thus, the problem of wages and prices, which have to be equal all over the country, makes this problem of efficient production very acute.

Professor Velikonja posed questions such as: What is the underlying principle which determines an underdeveloped area? What would be the strategy of improvement from which the people would most benefit? He suggested, (1) More water supply; and (2) More frequent services. The consumer could not care less about industrial development. Perhaps with more people in peripheral areas, the amounts of investment could be more balanced.

Professor Hoffman pointed out that, as Dr Kasperson had indicated, education in Greece is a real problem because the Greek system is still geared to the classical type of education. In Athens there are higher technical schools which give degrees, but are not within the scope of their university.

Professor Hoffman said that tremendous amounts of money are being committed to the construction of apartment houses in Greece but many of these will be the slums of the future. The problem for the government is to channel money from the workers in Germany into long-term savings and not into apartment houses.

In commenting on Dr Baxevanis' point, that he would evict masses of people and move them to the cities, Professor Hoffman asked what country would move two to three million people lock, stock and barrel to expedite economic development? This is fine from a theoretical point of view, but politically it is impossible and most unlikely to happen in Greece.

Statistical Summary Tables

STATISTICAL SUMMARY 1: *East-Central and Southeast Europe: Population Distribution Pre-World War II and Post War*

	Latest Census Date	Latest Census Population	1967 Mid-Year Estimate	Annual Rate of Increase 1963-67	Area (km.²)	Density per sq km 1967	Pre-war Urban*	Pre-war Rural	Post-war Urban*	Post-war Rural	Year
							in %				
Albania	2-X-60	1,626,315	1,965,000	2.8	28,748	68	15.4	84.6	33.3	66.7	(1967)
Bulgaria	1-XII-65	8,226,564	8,309,000	0.7	110,912	75	22.3	77.7	45.9	54.1	(1967)
Czechoslovakia	1-III-61	13,745,577	14,305,000	0.6	127,869	112			59.0	41.0	(1966)
East Germany	31-XII-64	15,940,469	16,001,000	−0.2	107,901	148			72.9	27.1	(1966)
East Berlin	31-XII-64	1,071,462	1,081,000	0.4	403	2,682					
Greece	19-III-61	8,388,553	8,716,000	0.7	131,944	66	32.8	67.2	56.2	43.8	(1961)
Hungary	1-I-60	9,961,044	10,212,000	0.3	93,030	110	36.3	63.7	40.0	60.0	(1968)
Poland	6-XII-60	29,775,508	31,944,000	1.0	312,520	102	30.0	70.0	51.2	48.8	(1968)
Romania	15-III-66	19,105,056	19,287,000	0.6	237,500	81	23.6	76.4	40.1	59.9	(1968)
Yugoslavia	31-III-61	18,549,291	19,958,000	1.2	255,804	78	21.0	79.0	40.0	60.0	(1966)

Source: United Nations. Statistical Office of the United Nations. *Demographic Yearbook, 1967*. 19th edition. New York, 1968. pp. 109-112 Various official statistical Yearbooks.

* A variety of definitions is used by the countries of East-Central and Southeast Europe of what constitutes an urban place. See Jerry W. Combs, Jr., 'Urbanization in Eastern Europe', paper presented at meetings of the American Sociological Association, August, 1965, mimeographed, p. 20. Caution is recommended in the use of the urban/rural breakdown.

STATISTICAL SUMMARY 2: Cities of over 100,000 Population in East-Central and Southeast Europe

CZECHOSLOVAKIA

Prague	1,030,000[a]
Brno	333,000[a]
Bratislava	277,000[a]
Ostrava	270,000[a]
Plzeň	143,000[a]
Kosiče	112,000[a]

BULGARIA

Sofia	800,953[b]
Plovdiv	222,737[b]
Varna	180,062[b]
Ruse	128,384[b]
Burgas	106,127[b]

GERMAN DEMOCRATIC REPUBLIC (EAST GERMANY)

Leipzig	594,099[c]
Dresden	505,188[c]
Karl-Marx-Stadt (Chemnitz)	294,897[c]
Halle	276,009[c]
Magdeburg	267,817[c]
Erfurt	191,887[c]
Rostock	186,447[c]
Zwickau	128,184[c]
Potsdam	110,693[c]
Gera	108,990[c]

HUNGARY

Budapest	1,875,000[d]
Miskolc	155,000[d]
Debrecen	137,000[d]
Pécs	125,000[d]

ALBANIA

Tiranë	169,300[e]

ROMANIA

Bucharest	1,365,885
Cluj	185,786
Timişoara	174,388[f]
Braşov	163,348
Iaşi	160,889
Galaţi	151,349
Constanţa	150,436
Craiova	148,821
Ploeişti	146,973

Arad	126,005
Brăila	138,587
Oradea	122,509
Sibiu	109,546

POLAND

Warsaw	1,261,000[g]
Łódź	745,000
Krákow	525,000
Wrocław	477,000
Poznan	441,000
Gdańsk	324,000
Szczecin	315,000
Katowice	287,000
Bydgoszcz	258,000
Lublin	206,000
Zabrze	198,000
Bytom	191,000
Czestochowa	176,000
Gdynia	168,000
Gliwice	164,000
Chorzow	154,000
Radom	145,000
Ruda Slaska	142,000
Białystok	142,000
Sosnowiec	141,000
Toruń	115,000
Kielce	109,000

GREECE

Athens	627,564[h]
Piraeus	183,877
Thessaloniki	250,920

YUGOSLAVIA

Belgrade	598,349[i]
Zagreb	457,499
Sarajevo	175,424
Skopje	171,893
Ljubljana	157,412
Novi Sad	110,877
Rijeka-Susak	100,989

---

[a] 1967 figure, *The Statesman's Yearbook for 1968–69*.
[b] 1965 figure, *The Statesman's Yearbook for 1968–69*.
[c] December 31, 1966, census, *The Statesman's Yearbook for 1968–69*.
[d] 1962 figure, *The Statesman's Yearbook for 1968–69*.
[e] 1967 figure, *Vjetari Statistikor i RP SH 1967 dhe 1968*.
[f] March, 1966, census, *The Statesman's Yearbook for 1968–69*.
[g] December, 1966, census, *The Statesman's Yearbook for 1968–69*.
[h] 1961 census, *Statistical Yearbook for Greece, 1968*.
[i] Official 1965 estimate.

STATISTICAL SUMMARY 3: *Proportion of Active Population by Economic Sector (%), 1960's*

	Manufacture, Mining, Construction	Agriculture, Forestry, Fishing	Services
Albania	22	58	20
Bulgaria	28·5	60·3	11·5
Czechoslovakia	47	20	33
East Germany	47	16	37
East Berlin	—	—	—
Greece	23	47	30
Hungary	34	38	28
Poland	33	42	25
Romania	18	54	28
Yugoslavia	22[b]	53	25

Source: Various national statistical Yearbooks 1966–68; 1967 FAO Yearbook.

STATISTICAL SUMMARY 4: *East-Central and Southeast Europe: Land Use*

	% of Total Area classed as Agricultural	% of Total Land classed as Forests	% of Agricultural land under Cultivation	% of Land classed as Grazing	Fertilizers use of Kg. 1 Hectare of Arable Land
Albania[a]	41·8	43·8	40·6	59·4	7[m]
Bulgaria[b]	52·3	32·6	77·6	22·4	79
Czechoslovakia[c]	56·8	36	74·2	25·8	158[n]
East Germany[d]	58·8	26·7	72	28	255[n]
East Berlin[e]	31·2	20·8	91·3	8·7	n/a
Greece[f]	65·1	20·0	42·3	57·6	65[l]
Hungary[g]	73·8	15·4	82·1	17·9	62[n]
Poland[h]	63·7	27·0	74·1	25·9	71[n]
Romania[i]	62·2	26·8	71·4	28·6	29
Yugoslavia[j]	57·3	34·4	56·2	43·8	56

Source: FAO Yearbook, 1967, National statistics.

[a] 1964
[b] 1966
[c] 1966
[d] 1966
[e] 1964
[f] 1965
[g] 1966
[h] 1966
[i] 1966
[j] 1966
[l] 1965–66
[m] 1965
[n] 1966

Index

NOTE: For topics relating to specific countries see country name.

activity spheres in planning, 195, 199, 204
adaptive location decisions, 204-5
administration, decentralization of, 250-1, 261-2
administrative centres, functions of, 238, 246, 248
 nodality of, 239-40, 243
administrative hierarchy, 229, 235
administrative regions, 230, 244-5, 247
administrative units, divisions of, 232, 244, 245, 251, 253n
 economic role of, 229, 232
 rationale for terms used for, 252n
 size of, 242, 246
adaptive location decisions, 205
agricultural geography, crop or commodity approach to, 142-3, 144, 160
 land use studies approach to, 136-8, 144
 landscape and settlement studies approach to, 129-36, 144
 place of in East-Central and Southeast European geography, 136, 143
 socio-economic studies approach to, 140-2, 144
agricultural typology, 129, 133, 134, 141-3, 144, 145, 163-6 *passim*
 of the I.G.U., 139-40, 164-5
agriculture, collectivization of, 87, 128, 141, 243, 272
 place of in communist planning, 85-6
 inherent characteristics of, 165
 irrigation. *See* Hungary, irrigation

mechanization of, 85, 86, 286-7, 288, 289
 productivity, 296
 types of, in Eastern Europe, 140, 141
Albania, 30, 31, 135, 141, 161
 illiteracy, 4
 industry, 182
 physical landscape, 22, 23, 26
 population, 396
Alföld, 17, 118. Also *see* Great Alföld, Little Alföld
American-Yugoslav Project in Regional and Urban Planning Studies, 312, 314, 355
architectural styles, 61, 100, 105, 108
Athens, 408, 413, 415, 441, 443, 465, 466-7, 468

balak, 2
Balkans, 2, 20, 370
Bar, 29, 367-81 *passim*
behavioural framework for industrial location, 194-7
Belgrade, 53, 97, 365-81 *passim*, 387, 406, 407, 408, 410
Belgrade-Bar Concept, 369-75
Belgrade-Bar Railroad, effects of, 378-81, 385, 389
Bohemia, 11-12, 54, 56, 66, 89, 176, 177, 179, 182
Bosnia and Herzegovina, 21, 30, 64, 182, 190, 366, 369, 378, 386, 402, 406
Bucharest, 68, 80-1, 96, 104, 366
Budapest, 64, 96, 176, 179, 182, 365, 366

497

Bulgaria, administrative divisions, 231–2, 240, 244–5, 419, 453
  agriculture, 141, 161, 441
  dependence on Soviet Union, 465
  foreign trade, 442
  growth poles, 463, 465
  industrial location principles, 185, 187, 191, 198
  industry, 66, 177, 182, 464
  labour force, 161, 440, 450
  land reform, 440
  mineral resources, 30, 31
  physical landscape, 24–8
  planning commissions, 451, 453
  population, 161, 396, 416, 417, 419, 426, 463
  regional planning policies, 434, 437–9, 452, 464
    Economic Plans, 444, 451–5, 464, 471
  structure of economy, 471
  tourism, 460
  urban development, 52–3
bura winds, 28

Carpathian Mountains, 13–16
collective farms, 85–7, 89, 170, 171, 272–3
  purpose of, 292
COMECON, 187, 191–3, 250, 449
command economy, 194, 205
Croatia, 309–20 *passim*, 323, 402, 416
Czechoslovakia, administrative divisions, 227, 230–1, 241, 244–5, 246, 248
  agriculture, 89, 141
  employment, 177, 178
  ethnic minorities, 114
  industrial location principles, 185, 189, 191, 192
  industry, 175, 176
  physical landscape, 13–14
  urban development, 54–6, 59, 61, 64

Dalmatia, 26, 50 79, 407,
Danube River, 16, 18, 102, 276, 277 278,

'dispersed localization' in industry, 183, 215
Dobruja, 19, 104, 118
Dubrovnik (Gruz), 65, 108, 366, 381
Dunaújváros, 46, 70, 99, 190, 191

East-Central Europe, climate, 8, 10, 17–18
  definition of, 2
  industrial location policies in, 190
  industry, 176–9, 182
  physical landscape, 8–19
  population, 396. Also *see* individual countries
  urban development, 54–63, 66–70
  Also *see* Eastern Europe
East Germany. See German Democratic Republic
Eastern Europe, agricultural regions, 161
  behavioural framework for industrial location, 194–7
  definition of, 2
  foreign influence in, 2–4, 101, 123, 125, 432
  industrial concentrations in, 175–176
  'industrial triangle', 176, 179
  industry. *See* industry (major listing)
  emigration, 395, 399, 417
  ethnic landscape (national landscape), 105, 108, 111, 123
  ethnic minorities, communist attitude toward, 113–14, 115, 189–90

'family climate', effect on population of, 422–3
Federal Association of Planners of Yugoslavia, 311
field systems, 87, 131
fields, size of, 87, 105
foreign influences in Eastern Europe. *See* Eastern Europe
fortifications, role of in development of cities, 50–8

Galaţi, 81, 104, 190
Gdańsk, 8, 9, 59, 61, 63, 68
German Democratic Republic (East Germany), agriculture, 86, 87, 141, 142
  cities, medieval, 63
  industrial location principles, 185, 187, 192, 198
  industry, 175, 176, 177, 178, 182
  physical landscape, 8
  population, 426
  rural landscape, 90–3
German settlement in Slavic lands, 58, 60
Gheorghe Gheorghiu-Dej, 84, 99, 104, 111
Great Alföld, 17, 68, 275, 278. Also see Alföld
Greece, agriculture, 439–41, 447
  and European Economic Community, 470
  economic development planning, 446–7
  Economic Development Plans, 434, 447, 455–8, 469
  emigration, 459, 468
  foreign trade, 442
  illiteracy, 4
  immigration, 410, 413, 415
  industrial axes, 470
  Industrial Development Councils, 458
  industry, 441–2
  mineral resources, 30
  physical landscape, 23–4, 25, 28
  population increase, 396, 410
  regional contrasts, 412
  Regional Development Services, 456, 469
  regional planning policies, 434, 437–9
  spatial inequities, 465–9
  tourism, 459
  urban development, 51, 52, 64, 412, 415–16
  U.S. aid to, 444
  vital statistics, 410, 412
grody, 54–5, 56–8

Gruz. See Dubrovnik

House of Culture, 93–4, 111
hrady, 54–6
Hunedoara, 102, 190, 243
Hungarian Autonomous Area, 115
Hungary, administrative divisions, 230–1, 239, 244–5
  agricultural productivity, 283–4, 296
  agricultural techniques, effect of age on, 282, 287, 295n
  agriculture, 86, 141. Also see irrigation
  cities, medieval, 64
  climate, 275–6
  co-operative (collective) farms, 272–3
  evapotranspiration, rates of, 275, 280
  fertilizers, use of, 290
  ground water, supply of for irrigation, 278–9
  industrial location principles, 190, 201
  industry, 176, 178, 182
  irrigation, 276–83
    investment policies for, 281, 296
    limitations to, 277–8, 280–2
  Nadudvar incentive system, 287, 289, 293–4n
  physical landscape, 16–18
  population, 396, 429
  private farms, 283–4
  state farms, 86, 274

I.G.U. Commission for Agricultural Typology, 139–40, 164–5
industrial activity, shifts in, 178–179
industrial location, behavioural framework for, 194–7
  decision-making, 174, 194–205 *passim*
  rationality in, 203–5, 206
  significance of, 175, 189
  socialist principles of, 185–92, 196–7, 203

industrial location—*contd.*
  socialist theory of, 109-11, 174
  trends in, 179, 183
  Western studies of in socialist countries, 173-4
industrial planning. *See* industrial location, decision-making
Industrial Revolution, effects of in Eastern Europe, 66-8
'industrial triangle', 176, 179
industry, concentration of, 179-80, 205
  dispersion of, 182-3, 199, 205
  location of in Eastern Europe, 175-6
  locational planning of, 184-92 *passim*
  regional self-sufficiency of, 187
  regional specialization of, 187
  relocation of factories, 179, 182
infra-red imagery, implications for agricultural geography of, 168-9

karst, 15, 20, 23, 375
Kisalföld. *See* Little Alföld
kosava winds, 28
Košice, 14, 114, 190, 361
Kosovo, 31, 114, 370, 404, 406
Kraków, 56, 57, 60, 63, 131
Kremikovci, 449, 451, 460-1

land use as basis for agricultural typology, 139, 140, 144
land use decisions as factor in regional planning, 303
land use maps, scale of, 137, 138
land use survey, limitations of, 137-8
landscape studies in Eastern Europe, 129-44, *passim*
  trends in, 131, 133
  value of in agricultural geography, 129
Little Alföld, 16-17, 275
Ljubljana, 64, 65, 108, 308, 324-38, 358, 365, 366, 407, 408, 409
location of economic activity, socialist theory of, 109-11, 229
location strategies, 200

Łódź, 10, 46, 66, 70, 176, 183

Macedonia, 26, 48, 51-2, 28, 29, 30, 31, 115, 118, 317, 367, 402, 406, 415, 466
Macedonian-Thracian Massif, 24-5, 27
Machine Tractor Station, position of, 90
mikrorayon. *See* neighbourhood unit
maquis, 23, 27
Moldavia, 18, 68, 81, 104, 191
Montenegro, 64, 115, 312, 316, 317, 369, 371, 373, 402, 406
Morava-Vardar Depression, 22, 25
Moravia, 54, 55, 56, 66

Nadudvar incentive system, 287, 289, 293-4n
Nagyalföld. *See* Great Alföld
NASA, satellite programme of, 167, 170, 171
national landscape. *See* ethnic landscape
neighbourhood unit in socialist countries, 97-8, 117
Niš, 53, 73n, 365, 366, 372
Nowa Huta, 99-100, 183
Nowy Tychy, 46, 70, 99

okrug planning commission, 452-3, 455
'one-axle' tractor, potential for in agricultural production, 286-7

Pannonia, 49, 50, 65, 79, 278-9, 374
Peloponnesus, 27, 413, 415
physical landscape. *See* individual countries
Ploieşti, 81, 84, 101, 104
Poland, administrative divisions, 230-1, 232, 238, 241, 244-5, 248, 267
  agriculture, 86, 141
  Central Industrial Region, 69, 184, 191
  illiteracy, 4
  industrial location principles, 185, 187, 190, 198

industry, 175, 176, 177, 178, 182, 183, 184
Land Use Survey, 137-8, 139, 140, 142, 168-9
physical landscape, 8-13
population, 426
urban development, 54-8, 59, 60-1, 63
Polish Land Use Survey, 137-8, 139, 140, 142, 168-9
population, effect of catastrophes on, 399, 413
effect of economic development on concentrations of, 425-6
effect of emigration on, 399
effect of modernization on, 427
effect of standard-of-living on, 400, 404, 413
political effects, 396
regional contrasts, 400, 412, 419
Poznań, 56, 57, 60, 63, 105, 183
pradoliny, 9, 58. Also see *Urstromtäler*
Prague, 56, 59, 61, 63, 96

railroads in Eastern Europe, 365-87 *passim*
redistricting, administrative divisions, motivation for, 248-9
regional spatial planning, 301-46 *passim*, 355, 358, 359
'regionalized dispersion', 183-4, 215
'rentability', principle of, 117-18, 195
resources, socialist view of, 115, 185, 188
Rijeka, 311, 365, 367, 381, 386, 407
Romania, administrative divisions, 228, 231, 243, 244, 245, 248
agriculture, 87, 89, 141, 161
ethnic minorities, 115
illiteracy, 4
industrial location principles, 185, 189
industry, 102, 104, 177, 182
landscape forms, 101-4, 105, 116
physical landscape, 14-16, 17, 19
population, 396
resources, 30, 69, 115-16
rural settlements, 134

rural landscape, 87-94, 105
studies of, 129-45
rural overpopulation, 99

Salonika. *See* Thessaloniki
Sarajevo, 67, 80, 366, 367, 372, 386, 407
Serbia, 30, 31, 64, 80, 312, 313, 365, 366, 367, 370, 373, 402, 404, 406
settlement geography, 129-36 *passim*, 144, 90-3
'Shatter Belt', 2, 114, 258
Silesia, 12, 66, 69, 130, 176, 179, 183, 185, 187, 198, 242, 361
Skopje, 25, 80, 108, 190, 365, 372, 407, 408, 409
Slovakia, 4, 64, 118, 182, 191, 248
Slovenia, 64, 66, 323-4, 369, 402, 404, 406
regional planning, 307, 311-28 *passim*
tourism, 316-20
'social cost', 110, 118, 201
'socialist' cities, 98-100, 123
socialist landscape, 85-100 *passim*, 111, 118-19, 123, 125, 126
socialist provinces, purposes of, 232, 234, 243
Sofia, 26, 53, 68, 97, 177, 366, 408, 420, 421, 441, 443, 449, 461, 463
Southeast Europe, climate, 28-9
definition of, 2
industry, 177-9, 182
natural resources, 30-1
physical landscape, 13-18
population increase, 396
Roman roads in, 49
soil erosion in, 29-30
spatial contrasts, 432-43
urban development, 48-53, 63-70
Also *see* Eastern Europe
Southeast European Highlands, 20-5, 28-9
space imagery, implications for agricultural geography, 168-9
spatial activity spheres, 195, 199, 204
spatial elasticity of investment planning, 197-8, 200

Index

spatial inelasticity of investment planning, 198, 461
spatial inequities, 431–5
spatial relationships, functional analysis of, 131
spatial structure of administration, 227–8, 240
Stara Planina, 24, 30, 31, 66. Also see Balkans
state farms, 95, 274

territorial planning, 241
Thessaloniki (Salonika), 52, 64, 365, 366, 369, 413, 415
Thessaly, 26–7, 51, 413, 415
Thrace, 25, 48, 52
Tiranë, 97
Tisza River, 18, 274, 276, 277, 278
Titograd, 29, 99, 371, 373, 374
Trans-Danubia, 17, 64, 275
Transylvania, 16, 48, 64, 79, 101, 104, 189, 190
Turkey, influence of in Southeast Europe, 65, 68–9, 80

Ukraine, 68, 232, 234
underdeveloped regions, 182, 183, 185, 190, 197, 205, 223, 375, 388
Upper Silesian Industrial District, 176, 179, 183, 185, 187, 239, 242
urban development, 48–70, 407–8, 412, 415, 416, 426
urban landscape, 94–100, 105, 108
urban planning, 95–6, 97, 100, 308–11, 312, 355
urbanization, socialist theory of, 111, 113
Urstromtäler, 9. Also see *pradoliny*

vardarac winds, 28
Vojvodina, 22, 28–9, 30, 114, 366, 373, 402, 406, 407

Walachia, 68, 80, 81, 104
Warsaw, 10, 70, 96, 100, 108, 176, 182, 183
World Land Use Survey, 139–40, 145
Wrocław, 10, 56, 57, 60, 63, 131

Yugoslavia, administrative divisions, 227, 228, 231, 240, 245, 248, 249, 250–1, 260–1
agriculture, 103, 141, 161
autonomy of Republics, effect of, 307
cities, medieval, 64–8
Council of Nationalities, 114
decentralization, 306–7, 327, 409
emigration, 399
ethnic minorities, 114, 115, 404
Federal banks, role of, 319
Federal Fund for the Developing Areas, 376
illiteracy, 4
immigration, 399
industrial location principles, 188, 189, 191, 199, 204, 206
industry, 177, 182
migration, 399, 406
mineral resources, 30, 31
physical landscape, 17, 22–3, 26
physical planning, 303–5, 307
population, 399, 400, 404
port rivalries, 381
railroads, development of, 365–9
regional contrasts, 400
regional integration, 306, 309, 310
regional planning, 320–46 *passim*, 355
Regional Spatial Planning Act, 321
rentability, principle of, 117–18, 195
self-management of enterprise, 305
soil erosion, 29–30
Soviet influence on planning, 309
tourism, 314–20, 356, 357, 379
UN Development Programme, 378–9
underdeveloped regions, 375–6, 378
Union of Urban Planners, 310
urban development, 407–8
urban landscape, 108
Urban Planning Act, 1967, 321
vital statistics, 323, 395, 396, 400, 417

Zagreb, 64, 182, 188, 240, 308, 310, 311, 312, 355, 365, 366, 367, 372, 407, 408, 409

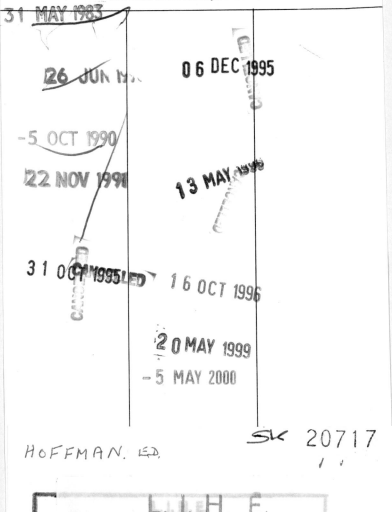

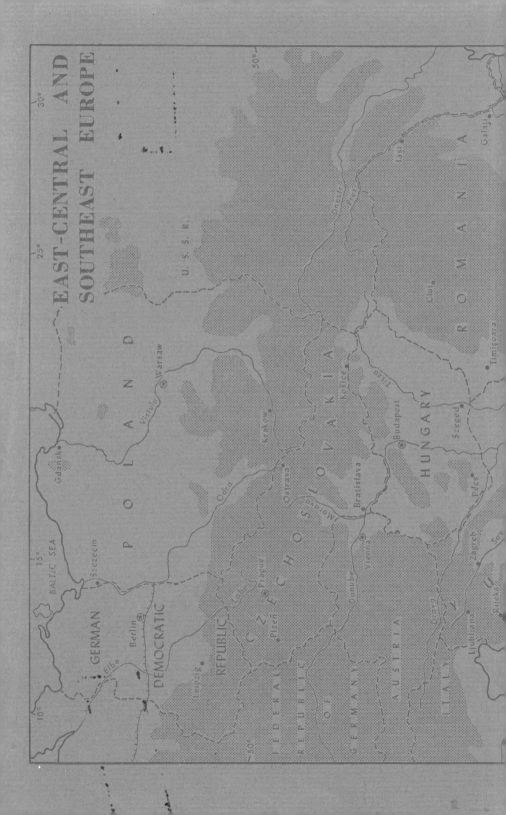